Phoenix Militaria's

AMERICAN MILITARIA SOURCEBOOK & DIRECTORY

First Edition

©1987 Phoenix Militaria, Inc.

All Rights Reserved. No part of this book may be reproduced or utilized in any form or by any means electronic or mechanical, including photocopying, recording or by any information storage or retrieval system without written permission from the Publisher. This sourcebook & directory is for the use of the purchaser only and is not transferable and cannot be rented or sold.

Published by Phoenix Militaria, Incorporated

Postal Box 66/Arcola/Pennsylvania 19420

PRINTED IN THE UNITED STATES OF AMERICA

How to Use This Reference...

There are 43 category sections included in this sourcebook and directory. Each section is listed in the Table of Contents with the appropriate starting page number. Each listing includes all the information available. A typical complete listing which includes an explanation of the "code/abbreviation line" is shown below:

PHOENIX MILITARIA CORPORATION
P.O. Box 66
163 Troutman Road
Arcola, PA 19420, USA
(215)-933-0909
General militaria & publisher of militaria collectors books & magazines.
S6; 19; 23,000; $60.00
Catalogue $3.00. Semi-annually.
CONTACT: Terry Hannon, President
SALES DEPT: Irene Karas
PURCHASING DEPT: Terry Hannon
Send $3.00 or 3 pieces of military insignia for our latest catalogue of U.S. & world-wide militaria. Thousands of items listed including: books, manuals, medals, insignia, equipment, uniform components, badges, etc.

FFL - Federal Firearms License holder
PLEASE NOTE: Each listing is in zipcode order within each section.

Code/Abbreviation Line
EXPLANATION...

Example: S6; 19; 23,000; $60.00

S6; ...Retail and Wholesale Sales Categories. See the codes listed below for a complete explanation. There can be more than one "S" code (Retail & Wholesale Sales Categories) listed.

19; ...Number of Years in Business.

23,000; ...Total number of names on Mailing List

$60.00 ...Cost per Thousand names mailing list rental.

CODES for RETAIL & WHOLESALE SALES CATEGORIES...

S1 ...Retail Sales (over-the-counter only)

S2 ...Wholesale Sales (over-the-counter only)

S3 ...Retail & Wholesale Sales (over-the-counter only)

S4 ...Retail Mail Order Sales

S5 ...Wholesale Mail Order Sales

S6 ...Retail & Wholesale Mail Order Sales

S7 ...Auction Sales

S11 ...Special Order Services

S12 ...Militaria Trade Shows (Gun & Militaria Promoters/Shows)

S13 ...Other

Phoenix Militaria's AMERICAN MILITARIA SOURCEBOOK & DIRECTORY is distributed in the United States and on a limited basis internationally. Products or services advertised may be prohibited, restricted or require licensing by Federal, State, City or International laws.

It is not possible for Phoenix Militaria, Incorporated to guarantee the authenticity of any item or service offered in or through this sourcebook and directory. We cannot accept any liability for incorrect information or false claims made by any company or individual(s) listed.

Readers having a problem should contact our offfice immediately. We will investigate all complaints. Write: Phoenix Militaria, Inc., Postal Box 66, Arcola, Pennsylvania 19420, U.S.A.

TABLE of CONTENTS

Section	Page
Title Page	I
How To Use This Reference	II
Table of Contents	III
Introduction	IV
1. Accoutrements	1
2. Aircraft/Aviation	1
3. Ammunition	2
4. Military Books/Manuals	2
5. U.S. Civil War	7
6. Edged Weaponry	9
7. Military Firearms	11
8. Miltary Film/Video	59
9. Flags/Banners	60
10. Fraternal Items	60
11. General Militaria	60
12. Military Headgear	72
13. Military Insignia	72
14. Military Libraries	74
15. Military Periodicals	76
16. Manufacturers	88
17. Medals/Orders/Decorations	89
18. Military Art & Photographs	92
19. Military Currency	93
20. Military Equipment	93
21. Military Vehicles/Parts	94
22. Military Museums	96
23. Optical Equipment	110
24. Military Paper Items	110
25. Military Book Publishers	111
26. Radio/Electronic Items	113
27. Military Reproductions	113
28. Restoration/Repairs	115
29. Services	116
30. Military Surplus	116
31. Clubs, Societies, Organizations	118
32. Travel/Battlefield Tours	125
33. Uniforms & Componets	125
34. Miscellaneous	126
35. Militaria Show Promoters	127
36. Military Models/Toy Soldiers	134
37. Auction Houses	136
38. Wargamers	137
39. Reenactments/Living History	137
40. Business Card Directory	140
41. Advertising Directory	141
42. Sourcebook & Directory Advertising Information	147
43. Sourcebook & Directory Free Listing Form	148

PHOENIX MILITARIA
MILITARIA & WAR RELICS

- INTRODUCTION -

This sourcebook & directory first began as an accident... in order to expand our business we had to find new companies to do business with. Basically we needed more suppliers of militaria and new advertising media. To help solve these problems we tried to locate militaria sourcebooks and directories.

But we were surprised to find only a directory or two with maybe 400 to 500 business and collector names mixed together. At the time our "business" mailing list included nearly 1,400 militaria businesses (without any collector names). For the next twenty-six months we researched and verified thousands of militaria & military firearms dealers, manufacturers, service companies, military collector publications, museums, clubs, and other militaria organizations throughout the United States.

Originally our plan was to offer a sourcebook & directory with over 3,000 listings. Finding additional companies we changed the number of listings to over 3,500. And now this first edition includes 4,521 listings for militaria companies and organizations.

If your company is not listed in this first edition be sure to fill out the "registration form" on page 148. Your listing is absolutely free. Please note: limited display advertising space is available on a first reserved basis (Page 147). Low cost multiple listings and an additional 25 word description (in addition to your FREE 25 word business description) is available. Details can be found on page 148.

And finally, regardless of whether you're a collector and/or in business, we feel you'll find this comprehensive sourcebook & directory a valuable addition to your reference library. Frankly, it's the only comprehensive militaria "Who's Who" in the marketplace...

Terry Hannon
Terry Hannon
Publisher

P.S. Please feel free to send your comments and suggestions to me personally: Terry Hannon, Publisher/Phoenix Militaria's American Militaria Sourcebook & Directory/Postal Box 66/Arcola/Pennsylvania 19420, U.S.A.

ACCOUTREMENTS

R. ANDREW FULLER COMPANY
P.O. Box 2071-P
Pawtucket, RI 02861, USA
(401)-723-4835
Display cases & mounting devices.
S6; 11;
Catalogue available. Annually.
CONTACT: Richard Fuller
PURCHASING DEPT: Dick
MILITARY AWARD CASES - 5x7 thru 14x20 folded memorial flag cases. Mounting devices for knives, guns, swords, etc. Ribbons, miniature medals. Catalog $1.00 (refundable).

PHOENIX MILITARIA CORPORATION
P.O. Box 66
163 Troutman Road
Arcola, PA 19420, USA
(215)-933-0909
General militaria & publisher of militaria collectors books & magazines.
S6; 19; 23,000; $60.00
Catalogue $3.00. Semi-annually.
CONTACT: Terry Hannon, President
SALES DEPT: Irene Karas
PURCHASING DEPT: Terry Hannon
Send $3.00 or 3 pieces of military insignia for our latest catalogue of U.S. & world-wide militaria. Thousands of items listed including: books, manuals, medals, insignia, equipment, uniform components, badges, etc.

'THE PHOENIX EXCHANGE'
P.O. Box 66
163 Troutman Road
Arcola, PA 19420, USA
(215)-933-0909
Quarterly military collectors magazine.
; 2;
CONTACT: Terry Hannon, Publisher
SALES DEPT: W. R. Bendel
PURCHASING DEPT: Terry Hannon
AD MANAGER: William Bendel
'The Phoenix Exchange' functions as the military collector's & dealer's marketplace. Articles cover a wide spectrum of militaria, but attend to each subject in detail. Send $2.00 for a sample copy.

C & C DISTRIBUTING
P.O. Box 76232
Oklahoma City, OK 73147, USA
New military accouterments.
Catalogue $2.00.

AIRCRAFT/AVIATION

DER SS PANZER BARON
P.O. Box 286
Chicopee, MA 01021, USA
WWII German & WWII aviation collector.
Send catalogues for inspection.

RAY LeKASHMAN
1 Little Fox Lane
Westport, CT 06880, USA
WWII U.S. aviation collector.

ADAM LANGWEILER
250 Gorge Road
Cliffside Park, NJ 07010, USA
WWI aviation collector.

SCHLESINGER
415 East 52nd Street
New York, NY 10022, USA
Lighter-than-air aviation.
Send SASE for list.

KENNETH D. SMITH
42nd Floor
345 Park Avenue
New York, NY 10154, USA
WWI Aeroplane parts.

THE COCKPIT
33-00 47th Avenue
Long Island City, NY 11101, USA
(718)-482-1860
Reproduction aviation gear & insignia.
Catalogue $2.00.
CONTACT: Jeffrey Clyman

WORLD WAR I AEROPLANES
15 Crescent Road
Poughkeepsie, NY 12601, USA
914-473-3679 Aero 818-243-6820 Skyways
Provides service info on WWI aeroplanes.
S6,S10; 25; 2200; 100.00
CONTACT: Leonard E. Opdycke
SALES DEPT: L. E. Opdycke
AD MANAGER: Richard Alden
Provides service (information, names, projects, materials, books) to modellers, builders & restorers of aeroplanes (1900-1919, 1920-1940) through two journals, WW I Aero & Skyways.

DENNIS CLARCQ
RD #1, Jones Road
Cohocton, NY 14826, USA
716-384-5333
Helicopter parts

BOB FORD
2860 Manor Road
Coatesville, PA 19320, USA
215-384-6536
Aviation memorabilia & gun show promoter

DAVID W. OSTROWSKI
5411 Masser Lane
Fairfax, VA 22032, USA
General aviation collector

AJAY ENTERPRISES
P.O. Box 2018, Mosby Branch
Falls Church, VA 22042, USA
703-573-8220
WWI aviation books, films, posters, & prints.
Catalogue $3.00.
CONTACT: A. J. Foster

WALTER G. GREEN, III
105 Lindsay Landing
Crafton, VA 23692, USA
Aviation collector

ROBERT A. ERICKSON
1005 Pebblebrook Drive
Raleigh, NC 27609, USA
WWI aviation collector

J. W. PEKLO
3755 Bonner Road
Pensacola, FL 32503, USA
(904)-432-3757
Current USN flight gear.
S1,S2,S3,S4,S5,S6; 5;
CONTACT: J. W. Peklo
SALES DEPT: J. W. Peklo
PURCHASING DEPT: J. W. Peklo
Current USN flight gear. Largest selection of USN flight helmets in S/E U.S.; probably in U.S. 150 units on hand at all times. Write wants on specific items. Enclose SASE for reply.

EDDIE PFISTER
12032 Boston Road
North Royalton, OH 44133, USA
Military aviation items

AVIATORS CLOTHING SUPPLY
P.O. Box 1216
252 Dakota
Huron, SD 57350, USA
605-352-3245
Aviation clothing

JEFFREY J. SANDS
2136 West 122nd Street
Blue Island, IL 60406, USA
German aviation collector.

AVIATION ANTIQUES, LTD.
P.O. Box 26914
Albuquerque, NM 87101, USA
Antique aviation items

PLANES OF FAME AIR MUSEUM
P.O. Box 17
7000 Merrill Avenue
Chino, CA 91710, USA
(714)-597-3722
Aviation museum.
S1,S4,S8; 30;
SALES DEPT: Karen Hinton
PURCHASING DEPT: Karen Hinton
Planes of Fame Air Museum. You and your entire family will enjoy this special outing that will leave you with memories of a very gallant America.

CRAWFORD-PETERS AERONAUTICA
3702 Nassau Drive
San Diego, CA 92115, USA
(619)-287-3933
Aviation & space books.
S4; 4;
Catalogue available. As needed.
CONTACT: Jim Peters & Mike Crawford
SALES DEPT: Jim Peters

AVIATORS WORLD
Mojave Airport
Tower Building 58
Mojave, CA 93501, USA
805-824-2424
WWI & WWII aviation gear.
6 bi-monthly catalogues $5.00.

PINE MOUNTAIN LAKE AIRPORT
P.O. Box 706
Groveland, CA 95321, USA
209-962-6121
Aviation memorabilia.
40 page catalogue $5.00.
CONTACT: Jon Aldrich

MILITARY AIRCRAFT/AVIATION

ROBERT HANSEN
1053 151st Avenue S.E.
Bellevue, WA 98007, USA
Aviation memorabilia collector

FLYING LEATHER, GIFU H.Q.
256-17 Torikumi
Sakahogi-choh, Kamo-gun
Gifu-ken, 505, Japan
0574-25-8343 0574-26-7644
Aviation items. Repro WW-2 leather jackets. Original WW-2 leather jackets.
S3,S6,S13; 3;
CONTACT: Nariaki Itoh
SALES DEPT: Nariaki Itoh
WANTED!! All aviation items, especially flying jackets, top prices paid. A-2, size 38; J-A-Dubrow, B-3, size 36; Rough Wear. All original new condition, also with SQ-patches, back-paintings, flags.

AMMUNITION

CARTRIDGES UNLIMITED
Rt. 1, Box 50
South Kent, CT 06785, USA
203-927-3053
Ammunition.

RAYMOND GARONE
Hustis Road
Cold Spring, NY 10516, USA
(914)-265-9636
New & used guns, ammo, knives, military surplus, hunting & fishing supplies.
S1; 2;
CONTACT: Raymond Garone
SALES DEPT: Ray Garone
PURCHASING DEPT: Ray Garone

UNITED CARTRIDGES
2 Orchard Hill Drive
Monsey, NY 10952, USA
914-354-7099
Cartridges.
CONTACT: Stephen Svitek

BEIKIRCH AMMUNITION CORP.
930 Linden Avenue
East Rochester, NY 14445, USA
716-248-3434
Ammunition

COLLECTORS CARTRIDGE AND COIN
Court House Office, Box 253
Rockville, MD 20850, USA
Ammo & explosive ordnance consulting.
; 30;
CONTACT: Dr. J. R. C. Schmitt
Buy & trade collectors ammunition, explosive ordnance, artillery, submarine items. Books, manuals, reports - all countries, eras, languages.

THE AMMO DUMP
613 East Gambler Street
Mt. Vernon, OH 43050, USA
614-397-7214
Ammunition

ELLSWORTH GREGOIRE
East Highway 50
Vermilllion, SD 57069, USA

Ammunition and accessories collector.

JOE JELINEK
1201 Cottage Grove Avenue
Chicago Heights, IL 60411, USA
Ammunition

SHORE GALLERIES
SGN 120, 3316 W. Devon
Lincolnwood, IL 60659, USA
312-675-AMMO
Ammunition

MONROE AMMUNITION COMPANY
P.O. Box 142
Monroe City, MO 63456, USA
314-735-4082
Ammunition

MIDWAY ARMS INC.
7450 Old Highway 40 West
Columbia, MO 65201, USA
314-445-9521
Ammunition

BOB'S OFFICE, INC.
505 Jinings Drive
Bellevue, NE 68005, USA
Ammunition

THREE 'R's
7008 Sherwood Lane
El Dorado, AR 71730, USA
(501)-863-4636
Reloading supplies & equipment.
S1; 5;
CONTACT: Ralph W. Rutherford
Reloading supplies for rifles, pistols, and shotguns. Complete line of powder and bullets, primers and wads. Reloading equipment and some reloaded ammunition. Also guns.

R. T. BUTTWEILER
P.O. Box 721793
Houston, TX 77272, USA
Ammunition mail-order auction.
Auction catalogue $10.00.

NEVINS AMMUNITION, INC.
7614 Lemhi, Suite 1
Boise, ID 83709, USA
208-322-8610
Ammunition

BUCK'S WAR SURPLUS
45 N. Lallani
Las Vegas, NV 89110, USA
702-452-8076
Ammunition

REED-LOADS
15743 Via Esmond
San Lorenzo, CA 94580, USA
(415)-278-7039
Custom ammo, gun repair and restocking.
S3,S4,S5,S6,S11,S13; 18;
CONTACT: W. G. Reed
SALES DEPT: W. G. Reed
PURCHASING DEPT: W. G. Reed

CIVIL DEFENSE SURPLUS SALES
North 1501 Fiske
Spokane, WA 99207, USA

509-535-4421
Ammunition & Civil Defense surplus.
List available.

MILITARY BOOKS/MANUALS

RAY D'ADDARIO
195 Central Park Drive
Holyoke, MA 01040, USA
Military bookseller & author.

HERITAGE BOOKS
Bay State West
P.O. Box 15701
Springfield, MA 01115, USA
Military bookseller.

KELLEY'S AUCTION
P.O. Box 125
Woburn, MA 01801, USA
617-935-3389 617-272-9167
Bookdealer, military mail-order auction.
Send $5.00 for current catalogue.

THE BRASS HAT
P.O. Box 97
Foster, RI 02825, USA
401-397-5618
U.S.M.C. books.
List available.
CONTACT: George B. Clark

SNOWBOUND BOOKS
RFD #1, Box 620
Madison, ME 04950, USA
Military books.
List $2.00

KITCH REIS
19 Beach Road
Bloomfield, CT 06002, USA
Boy Scout references.
Write for information.

THE VIETNAM BOOKSTORE
P.O. Box 122
Collinsville, CT 06022, USA
Military books.
Catalogue available.

NOSTALGIA WORLD
P.O. Box 231
North Haven, CT 06473, USA
(203)-269-8502
Military bookseller.
S4,S13; 7;
Catalogue available.
CONTACT: Stan. Lozowski
Publishes a monthly 8-page newspaper listing collectors shows and military shows in the Connecticut area. SASE for Ad rates and listing of military books - videotapes.

MILITARY BOOKS/MANUALS

COLLECTOR'S CASTLE
P.O. Box 51
Wallingford, CT 06492, USA
(203)-269-8502
Military bookseller.
S13; 3;
Catalogue available.
CONTACT: B. Roth
We offer a catalog of military books and videotapes - historic; WWII; Korea; Vietnam; aircraft; firearms; Patton; strategy; naval, etc. Catalog FREE for SASE.

SHUHI-BOOKS
P.O. Box 268
Morris, CT 06763, USA
Used military books.
Free catalogue.

EVANS E. KERRIGAN
P.O. Box 2211
Noroton Heights, CT 06820, USA
Military references.

PUBLISHERS CENTRAL BUREAU
1 Champion Avenue, Dept. 412
Avenel, NJ 07001, USA
Military bookseller.
Catalog available.

PORTRAYAL PRESS
P.O. Box 1913-63
Bloomfield, NJ 07003, USA
(201)-743-1851
Military vehicle books.
S4,S5;
Catalogue $2.00. Bi-annually.
CONTACT: Dennis R. Spence
Gigantic assortment of books, manuals and other publications on military vehicles... jeeps, military Dodges, half-tracks and related references on WW2 uniforms, insignia, history, etc.

C.M.C. MYRIFF, INC.
82 Townsend Drive
Freehold, NJ 07728, USA
Military bookseller.
Catalogue $1.00.
CONTACT: C.M.C. Myriff

PHILLIPS PUBLICATIONS
P.O. Box 168
Williamstown, NJ 08094, USA
609-567-0695
Airborne & elite units books & insignia.
Free catalogue.
CONTACT: Jim Phillips

JERBOA-REDCAP BOOKS
P.O. Box 1058
Hightstown, NJ 08520, USA
(609)-443-3817
British and European naval and military books and collectibles.
S4; 3;
Catalogue available. Quarterly.
CONTACT: Frank Shaw
SALES DEPT: Frank or Maureen Shaw
PURCHASING DEPT: Frank or Maureen Shaw
Specialist British naval and military booksellers, including Royal Naval collectibles and British army cloth and metal badges. Also now dealing in German military books.

EMILIE CALDWELL STEWART
P.O. Box 1792
Toms River, NJ 08754, USA
(201)-286-9575
Author of books detailing WWII German identity documents.

LOKI LTD.
48 Grand Street
New York, NY 10013, USA
Military history bookseller.
Catalogue available.

DA CAPO PRESS
233 Spring Street
New York, NY 10013, USA
800-221-9369
Military book dealer.

WILLIAM MORROW & COMPANY
105 Madison Avenue
New York, NY 10016, USA
Military bookseller.

PRESTON B. VAILS
20 West 115th Street
New York, NY 10026, USA
Military books & periodicals.
List $2.00.

THE MILITARY BOOKMAN
29 East 93rd Street
New York, NY 10128, USA
212-348-1280
Out-of-print & rare military books.
$10 brings 6 catalogues over two years.

CONSERVATIVE BOOK CLUB
15 Oakland Avenue
Harrison, NY 10528, USA
Book club for conservatives.
S4,S10; 22;
Literature available.
CONTACT: Neil McCaffrey
Books of interest to conservatives and anti-communists.

J. H. FABER
P.O. Box 24
Millwood, NY 10546, USA
Used & rare military books.
Free catalogue.

COLUMBIA TRADING COMPANY
2 Rocklyn Drive
Suffern, NY 10901, USA
(914)-357-2368
Out-of-print naval & maritime books.
S4; 4;
Catalogue available. Quarterly.
CONTACT: Bob Glick
SALES DEPT: Bob Glick
PURCHASING DEPT: Bob Glick
Out-of-print & rare naval & maritime books, magazines & ephemera.

EPCO PUBLISHING COMPANY
74-11 Myrtle Avenue
Glendale, NY 11385, USA
(718)-497-1100
Military reference books.
S6; 16;
Catalogue available. Annually.
CONTACT: Edward Siess & Peter Hlinka
SALES DEPT: Edward P. Siess
PURCHASING DEPT: Edward P. Siess

LOUIS KRUH
17 Alfred Road West
Merrick, NY 11566, USA
(516)-378-0263
Cryptography items.
S6; 15;
Catalogue available. Semi-annually.
CONTACT: Louis Kruh
SALES DEPT: Louis Kruh
PURCHASING DEPT: Louis Kruh
Buy and sell items related to cryptography, cryptanalysis, codes, ciphers and secret communications such as cipher machines and devices, books, manuals, documents and letters.

NATIONAL HISTORICAL SOCIETY
P.O. Box 987
Hicksville, NY 11802, USA
Military bookseller.

WILFRED BAUMANN
RD 1, Box 188
Esperance, NY 12066, USA
Military unit histories.

TALBOTHAYS BOOKS
P.O. Box 118
Black Rock Road
Aurora, NY 13026, USA
(315)-364-7550
Military bookseller.
S1,S3,S4,S5; 10;
Catalogue available. Quarterly.
CONTACT: Paul C. Mitchell
SALES DEPT: Paul Mitchell
PURCHASING DEPT: Paul Mitchell
Sell books on all aspects of military history - unit histories, battles, diaries, strategy, manuals, novels. All wars. Catalogs issued. Search service, no obligation to buy.

CARL F. WALSH
P.O. Box 147
Buffalo, NY 14240, USA
716-847-6593
Military bookseller.
Free Catalogue.

THE FAMILY ALBUM
RD 1, Box 42
Glen Rock, PA 17327, USA
(717)-235-2134
Antiquarian bookseller.
S6; 17;
Catalogue available.
CONTACT: Ron Lieberman
SALES DEPT: Ron Lieberman
PURCHASING DEPT: Ron Lieberman
Antiquarian booksellers and library consultants who specialize in developing, appraising, and preserving (binding) subject collections.

MILITARY BOOKS/MANUALS

VALOR PUBLISHING COMPANY
3355 Birch Circle
Allentown, PA 18103, USA
(215)-437-3622
Military books & monograph publisher.
S6; 9;
Catalogue available. Quarterly.
CONTACT: W. Victor Mades
World War Two armies & campaigns. Best (English) German army unit histories. Line includes 24 books. Affiliated with Game-Marketing (200+ books, games, magazines). Free catalogue.

VOLUME CONTROL
955 Sandy Lane
Warminster, PA 18974, USA
(215)-674-0217
Military booksellers.
S4,S12; 2;
Catalogue available. Quarterly.
CONTACT: Jack Hatter
We buy and sell new and used military books specializing in World War 2, with a sub-specialty in European Theatre. Send want-lists.

BOOKSOURCE
P.O. Box 43
7 South Chester Road
Swarthmore, PA 19081, USA
215-328-5083
Old and fine books.
Catalogue available.

PHOENIX MILITARIA CORPORATION
P.O. Box 66
163 Troutman Road
Arcola, PA 19420, USA
(215)-933-0909
General militaria & publisher of militaria collectors books & magazines.
S6; 19; 23,000; $60.00
Catalogue $3.00. Semi-annually.
CONTACT: Terry Hannon, President
SALES DEPT: Irene Karas
PURCHASING DEPT: Terry Hannon
Send $3.00 or 3 pieces of military insignia for our latest catalogue of U.S. & world-wide militaria. Thousands of items listed including: books, manuals, medals, insignia, equipment, uniform components, badges, etc.

'THE PHOENIX EXCHANGE'
P.O. Box 66
163 Troutman Road
Arcola, PA 19420, USA
(215)-933-0909
Quarterly military collectors magazine.
; 2;
CONTACT: Terry Hannon, Publisher
SALES DEPT: W. R. Bendel
PURCHASING DEPT: Terry Hannon
AD MANAGER: William Bendel
'The Phoenix Exchange' functions as the military collector's & dealer's marketplace. Articles cover a wide spectrum of militaria, but attend to each subject in detail. Send $2.00 for a sample copy.

BIEVER
15 Douglass Street
Boyertown, PA 19512, USA
Military manuals.
Send SASE.

ETHICS & PUBLIC POLICY CENTER
1030 Fifteenth Street, N.W.
Washington, DC 20005, USA
202-682-1200
Political & military bookseller

Q. M. DABNEY & COMPANY
P.O. Box 42026
Washington, DC 20015, USA
Out-of-print military books.
Catalogue $1.00.

SMITHSONIAN INSTITUTION PRESS
955 L'Enfant Plaza, #2100
Washington, DC 20560, USA
(202)-287-3765
Military & aviation booksellers.
SS3,S6,S8,S9,S10; 19;
Free catalogue. Annually.
CONTACT: Felix C. Lowe, Director
SALES DEPT: Marian Kester
PURCHASING DEPT: Lisa Kuhn

WILLIAM A. TURNER
6917 Briarcliff Drive
Clinton, MD 20735, USA
Military reference books.
Send for list.

STONE TRAIL PRESS
P.O. Box 34320
Bethesda, MD 20817, USA
Military bookseller

VALLEY BOOK CENTER
720 Pin Oak Road
Hagerstown, MD 21740, USA
Military booksellers

LEONARD J. GARIGLIANO
Rt #6, 633 Robinhood Drive
Salisbury, MD 21801, USA
Reference books.
Free catalogue.

WYVERN PUBLICATIONS
P.O. Box 188
Dumfries, VA 22026, USA
(703)-670-3527
Books on organizations of the U.S. Army.
S6; 10;
CONTACT: James Sawicki
SALES DEPT: James A. Sawicki
PURCHASING DEPT: James A. Sawicki
Mail-order source for books about military organizations (lineages, campaigns, decorations & insignia).

HOWARD R. CROUCH
4812 Village Drive
Fairfax, VA 22030, USA
Military reference books

GUN OWNERS OF AMERICA
8001 Forbes Place, #102
Springfield, VA 22151, USA
Military bookseller

ATHENA PRESS, INC.
P.O. Box 776
Vienna, VA 22180, USA
Military book publisher
S6; 7;
CONTACT: Lee Allen
Athena publishes Rommel's famous book "Infanterie Greiftan". This is the first unabridged edition of the book in English. New title -"Attacks". $16.95 postpaid.

NATIONAL VANGUARD BOOKS
P.O. Box 2264
Arlington, VA 22202, USA
Military bookseller.
Catalogue available.

JOHNSON REFERENCE BOOKS
P.O. Box 7152
Alexandria, VA 22307, USA
Reference books.
List available.
CONTACT: Thomas M. Johnson

TBN ENTERPRISES
P.O. Box 55
Alexandria, VA 22313, USA
703-684-6111
Firearms Books

ROBERT BRUCE
702 Gaulding Road
Mechanicsville, VA 23111, USA
Collector of machine gun manuals.

PARAGON PRESS/DYNAPRESS
Fern Park, FL 32730, USA
Military bookseller.

EUGENE J. BENDER
P.O. Box 636581
Margate, FL 33063, USA
Military reference works.

AMERICANA BOOKSHOP & GALLERY
1719 Ponce de Leon Blvd.
Coral Gables, FL 33134, USA
(305)-442-1776
Old books, maps, prints, toy soldiers, militaria.
S1; 4;
CONTACT: John Detrick
SALES DEPT: John Detrick
Always buying and selling used rare and bargain books, esp. aviation, wars, Napoleonic items, autographs, militaria. Also maps, prints, and toy soldiers, old and new.

DUKE OF BADAM-MAZAR
P.O. Box 370455
Miami, FL 33137, USA
Military bookseller. Islamic & Indian weapons collector.
CONTACT: B. H. White

MICHAEL H. JAKOB
P.O. Box 8685
Fort Lauderdale, FL 33310, USA
Reference books.

MILITARY BOOKS/MANUALS

THE COMPLEAT STRATEGIST
5406 Stirling Road
Davie, FL 33314, USA
(305)-584-8556
Books & magazines on military subjects.
S3,S4; 7;
Catalogue available. Annually.
CONTACT: Danny Kilbert, President
SALES DEPT: Carl Fink
PURCHASING DEPT: Danny Filbert
Dealers in historical and military books and magazines.

HISTORICAL BOOKSHELF,LTD.
4210 S.W. 3rd Street
Plantation, FL 33317, USA
305-583-2378
Military bookseller.

THE BATTERY PRESS
P.O. Box 3107, Uptown Station
Nashville, TN 37219, USA
(615)-298-1401
Airborne & elite unit books.
S3,S4,S5; 10;
Catalogue available. Annually.
CONTACT: Richard S. Gardner
SALES DEPT: Richard Gardner

LARRY JESENSKY
700 Lafayette Drive
Akron, OH 44303, USA
Out-of-print military books.

DARING BOOKS
P.O. Box 526-P
Canton, OH 44701, USA
(216)-454-7519
Books and related items.
S4,S6,S8,S9,S10,S12; 5;
Catalogue available. Quarterly.
CONTACT: Dennis W. Bartow
SALES DEPT: Dennis Bartow
Military books - Vietnam, WWII, Medal of Honor double recipients, Central America, Middle east, Non-fiction and fiction.

HAMMER
P.O. Box 1393
Columbus, IN 47201, USA
German marches on cassettes.
S6,S11,S12; 6;
Catalogue available.
CONTACT: Karl Hammer
Manufacturer/Wholesaler: WW II German march cassettes, flags, pins, and posters. Custom capacity retailer: Videos and books. We buy pre-1945, 78 rpm German march records.

RICHARD'S READERS
Rt. 1, Box 260A
Tennyson, IN 47637, USA
812-362-8462
Vietnam War books.
Catalogue available.

UNIVERSITY OF EVANSVILLE PRESS
P.O. Box 329
Evansville, IN 47702, USA
Military bookseller.

GUNNERMAN BOOKS
P.O. Box 4292
Auburn Heights, MI 48057, USA
313-879-2779
Firearms books

ANTIQUE ORDNANCE PUBLISHERS
3611 Old Farm Lane
Port Huron, MI 48060, USA
Cannon manuals publisher.
CONTACT: Don & Jan Lutz

THE REICH ART
P.O. Box 285
122 1/2 East Main Street
Flushing, MI 48433, USA
(313)-659-8999
Military art prints.
S6; 6;
Illustrated catalogue available. Annually.
CONTACT: James G. Thompson
SALES DEPT: Sybille Palentyn
PURCHASING DEPT: Sybille Palentyn
Presently we are providing military, historical, or ethnic art prints, as well as commissioned art, books and replicas.

DALE ROETHIG
3228 North 84th Street
Milwaukee, WI 53222, USA
Military bookseller

NS PUBLICATIONS
P.O. Box 27486
Milwaukee, WI 53227, USA
703-524-3073
Military booksellers.

MIKE O'CONNOR
1201 N. 3rd Avenue #1
Wausau, WI 54401, USA
Military books

BLAYZE BOOKS
P.O. Box 0913
Oshkosh, WI 54901, USA
414-233-1362
Military bookseller

RAY ROUTHIER
P.O. Box 6654
Great Falls, MT 59406, USA
Military books

REGNERY BOOKS
950 North Shore Drive
Lake Bluff, IL 60044, USA
Military bookseller

ARTICLES OF WAR,LTD.
8806 Bronx Avenue
Skokie, IL 60077, USA
(312)-674-7445
Specializing in all military history books, New & used.
S1,S4,S11; 17;
Catalogue available. 4 or 5 per year.
CONTACT: Robert Ruman & Michael Cobb
SALES DEPT: Bob Ruman
PURCHASING DEPT: Bob Ruman

WARS OF THE WORLD
6010 W. Irving Park Road
Chicago, IL 60634, USA
312-282-0262
Books & general militaria.
Write for details.
CONTACT: A. J. Caliendo

PHIL KRUMHOLZ
22 Lancaster Estates
Mapelton, IL 61547, USA
Military subject author

JOHN R. ANGOLIA
18070 Berryhill Drive
Stilwell, KS 66085, USA
Collector reference author & publisher.

FRANK McGLOTHLIN
405 North Columbia Street
Covington, LA 70433, USA
Author & bookseller

PASEO PRESS
3001 The Paseo
Oklahoma City, OK 73103, USA
(405)-528-1222
Military unit histories publisher
S4,S5,S6,S11,S12;
CONTACT: Claude Hall

DAVID MADIS
2453 W. Five Mile Parkway
Dallas, TX 75233, USA
214-330-7168
Book distributor

AEROEMBLEM PUBLICATIONS,LTD.
#7 March Drive
Wichita Falls, TX 76306, USA
(817)-855-0988
Book sales on Air Force insignia & A.F. insignia for sale/trade.
S6,S9,S12; 15;
CONTACT: Jerome Polder
SALES DEPT: Jerome Polder
PURCHASING DEPT: Jerome Polder
Publisher and distributor of Air Force insignia books. Currently have a motto identification book in print, plus volume I, II, & III of the illustrated USAF insignia guide series.

E & P ENTERPRISE
P.O. Box 2116
San Antonio, TX 78232, USA
Military bookseller

K & S BOOKS COMPANY
P.O. Box 9630
Alpine, TX 79831, USA
915-837-5053
Military bookseller.
Catalogue available.

JIM BAILEY
P.O. Box 745
Castle Rock, CO 80104, USA
Military manuals.
List $1.00.

PALADIN PRESS
P.O. Box 1307
Boulder, CO 80306, USA
Military booksellers.
Catalogue available.

MILITARY BOOKS/MANUALS

THE OLD ARMY PRESS
P.O. Box 2243
Ft. Collins, CO 80522, USA
U.S. military bookseller.
List available.

SIERRA SUPPLY
P.O. Box 1390
Durango, CO 81301, USA
303-259-1822
Manuals

ALLIED (Map Dept.)
P.O. Box 41323-L
Phoenix, AZ 85080, USA
(602)-841-6625
Military maps & books.
S6; 9;
Catalogue available. Bi-annually.
CONTACT: D. Foster
SALES DEPT: Tom Loneir
PURCHASING DEPT: D. Foster
AD MANAGER: Tom

BEACHCOMBER BOOKS
P.O. Box 197
Cortaro, AZ 85652, USA
(602)-744-0487
Military bookseller.
S6;
Free catalogue.
CONTACT: Jim Thorvardson
Book Store. All aviation, naval & military history. Free book search. Best selection of out-of-print, rare books in U.S.A. Specialize in unit histories, imports & hard to find.

PAUL GAUDETTE : BOOKS
1310 N. Alvernon Way
Tucson, AZ 85712, USA
(602)-881-3223
Military bookseller.
S1,S4,S12; 27;
Catalogue available. Monthly.
CONTACT: Paul Gaudette
SALES DEPT: Paul Gaudette
PURCHASING DEPT: Paul Gaudette
Specialist bookseller, Aviation, Military, Naval History, 1900-1986. USMC, Third Reich, WWI-II Aviation, Unit Histories.

TURBO PUBLISHING,INC.
P.O. Box 270
Cornville, AZ 86325, USA
(602)-634-4650 (602)-634-6127
Books - Military & police firearms magazines.
S4,S5,S6; 11; 15,000; $65.00
Catalogue available. Semi-annually.
CONTACT: Everett W. Moore,Jr.
SALES DEPT: Marilyn Moore
PURCHASING DEPT: Everett W. Moore,Jr.
AD MANAGER: Bob Matheny
Publish military & police related books. Also publish two firearms magazines, 'SWAT' magazine & 'FIREPOWER' magazine.

BRUCE W. ORRISDS
8033 Emerson Avenue
Los Angeles, CA 90045, USA
Author

AVIATION BOOK COMPANY
1640 Victory Boulevard
Glendale, CA 91201, USA
(818)-240-1771
Aeronautical books & video-cassettes.
S6; 22;
Catalogue available. Semi-annually.
CONTACT: Walter P. Winner
SALES DEPT: R. W. Harker
PURCHASING DEPT: W. P. Winner
Retail & wholesale distributor of all types of aeronautical books - civil & military aircraft. Publish some. Also sell aviation video-cassettes & aviation paraphenalia.

S. CARWIN & SONS,LTD.
P.O. Box 147
Canoga Park, CA 91304, USA
Military booksellers

BOOK CASTLE,INC.
144 S. Golden Mall
Burbank, CA 91502, USA
(818)-845-6467
Military bookseller.
S4; 7;
Catalogue available. 5 times yearly.
CONTACT: Paul Hunt
SALES DEPT: Chris Nickle
PURCHASING DEPT: Paul Hunt
Books - Mostly out-of-print. Large stock of books on history of Europe, Russia, Far East.

PAUL HUNT
P.O. Box 10907
Burbank, CA 91510, USA
Military unit history collector.

SURVIVAL BOOKS
11106 Magnolia Blvd
N. Hollywood, CA 91601, USA
Survivalist bookseller

B & G FINE BOOKS
P.O. Box 8895
Universal City, CA 91608, USA
Military bookseller

AEOLUS PUBLISHING LIMITED
512-115 West California Avenue
Vista, CA 92083, USA
Military bookseller

MARCELLO PUBLISHING
8421 Beaver Lake Drive
San Diego, CA 92119, USA
(619)-463-7817
Military reference books
S6,S12; 11;
CONTACT: Gerald Marcello
SALES DEPT: G. Marcello
PURCHASING DEPT: G. Marcello
"30-06 We Have Seen", Vol. II $13.00 PP U.S.A., Vol. III $30.00 PP U.S.A.. Both together $38.00 PP U.S.A. Others add $2.00.

CHARGER BOOKS
Box HH
Capistrano Beach, CA 92624, USA
Military bookseller

WORLD WAR II PUBLICATIONS
Postal Drawer 278
Corona Del Mar, CA 92625, USA
Military bookseller

CAPT. L. G. BUSHNELL,(RET)
1825 West Hall
Santa Ana, CA 92704, USA
Military book & postal collector.

VOICES WWII BOOKSHOP
P.O. Box 6397
Los Osos, CA 93402, USA
Military booksellers.
Free catalogue.

PACIFICA PRESS
1149 Grand Teton Drive
Pacifica, CA 94044, USA
Military books

MILITARY BOOKS
125 Bidwell Way
Vallejo, CA 94589, USA
(707)-554-4338 (707)-643-8431
Military books, insignia & T-shirts.
S4,S12; 15;
Catalogue available. 3 times per year.
CONTACT: Leonard P. Rasmussen
SALES DEPT: Leonard P. Rasmussen
PURCHASING DEPT: Leonard P. Rasmussen

MILITARY ARMS RESEARCH
P.O. Box 26772
San Jose, CA 95159, USA
Military reference bookseller.

SHARP & DUNNIGAN PUBLICATIONS
P.O. Box 660
15522 Nopel Avenue
Forest Ranch, CA 95942, USA
(916)-891-6602
Military & firearms references publisher.
S6,S8,S9,S10,S11,S12; 2;
CONTACT: Paul D. Stevens, President
SALES DEPT: Debi Woodbeck
AD MANAGER: Janice Haugh
Book Publisher

HACHIMAN RESEARCH
9160 S.E. Pine Street
Portland, OR 97216, USA
(503)-254-4268
Books relating to Japanese militaria.
S4,S12;
Catalogue available. As needed.
CONTACT: George Gomm
Japan WWII for sale. New, used, out-of-print books, manuals, paper on Japanese military and navy. Hachiman Research, 9160 S.E. Pine Street, Portland, OR 97216.

TOMMY'S BOOKS
647 S.W. 132nd Street
Seattle, WA 98146, USA
Military bookseller

MILITARY BOOKS/MANUALS

LOOMPANICS UNLIMITED
P.O. Box 1197
Port Townsend, WA 98368, USA
Weaponry bookseller & publisher.
S6; 10; 18,000; $80.00
Catalogue available. Annually with quarterly supplimints. CONTACT: Michael Hoy
SALES DEPT: Michael Hoy
PURCHASING DEPT: Michael Hoy
Publishers and sellers of unusual books, many pertaining to weapons.

JAN STILL
P.O. Box 188
Douglas, AK 99824, USA
Military reference books

MILITARY BOOKS
Private Bag No. 7
Burwood, 2134, Australia
Military bookseller.
Catalogue available.

PETER L. JACKSON
23 Castle Green Crescent
Weston, Ontario M9R 1NR, Canada
416-249-4796
Out-of-print military books.
Catalogue $1.00.

THUNDERBIRD BOOKS
P.O. Box 2129
Sidney, B.C. V8L 3S6, Canada
Out-of-print military books.
Quarterly catalogue $1.00.

MILITARY BOOK DISTRIBUTOR
267 Whitegates Crescent
Winnipeg, Manitoba R3K 1L2, Canada
Military bookseller.
Catalogue $1.00.
CONTACT: John Fedorowicz

K. E. SKAFTE
Nykobing
Falster 4, Denmark
Military booksellers.
Catalogue available.

A. A. JOHNSTON -MILITARY BOOKS
Pitney
Langport, Somerset TA10 9AF, Great Britain
0458-72713
Military bookseller.
S3,S6; 24;
List available, Want lists solicited. 8 per year.
CONTACT: A. A. Johnston
SALES DEPT: Arthur Johnson
PURCHASING DEPT: Arthur Johnson
International mail-order military bookseller. Numerous distributor rights in GB and Europe. Lists available upon request. State area of interest.

WOOLCOTT BOOKS
Peacemarsh
Gillingham, Dorset SP8 4EU, Great Britain
07476-2863
Second-hand naval, military & colonial books and manuals.
S4; 8;
Catalogue available. Annually.
CONTACT: H. M. & J. R. St. Aubyn
SALES DEPT: H. M. or J. R. St. Aubyn

PURCHASING DEPT: H. M. or J. R. St. Aubyn
Second-hand and out-of-print books on military and colonial subjects: regimental and formation histories, campaigns, biographies and much on India and Africa.

KEN TROTMAN LTD.
Unit 11, 135 Ditton Walk
Cambridge, CB5 8DQD, Great Britain
0223-211030
Out-of-print military bookseller.
Catalogue available.

WILLEN LTD.
Howard House, Howard Road
London, E11 3PL, Great Britain
01-556-7776
New books on firearms, edged weapons and related subjects.
S3,S6,S12; 40;
Catalogue available. 5 times per year.
CONTACT: W. John Burton
SALES DEPT: Mrs. Jean Boulton
PURCHASING DEPT: W. John Burton
Specialist booksellers dealing in new books on firearms, antique and modern, edged weapons and other related subjects. Catalogue sent on request.

CLARKE'S BOOKSHOP
211 Long Street
Cape Town, 8001, South Africa
021-235739
Military bookseller.
S1,S4,S6; 28;
Catalogue available. Quarterly.
CONTACT: Paul V. Mills, H. M. Dax
SALES DEPT: Paul V. Mills
PURCHASING DEPT: Henrietta M. Dax
Dealers in new and secondhand books, maps, phamphlets and printed ephemera. Specialising in South African military history: Zulu War, Frontier Wars, Anglo-Boer War.

GALAGO PUBLISHING (PTY) LTD.
P.O. Box 404
Alberton, 1450, South Africa
(011) 869-0807
Book publishers & distributors.
S3,S6; 6;
Catalogue available. Bi-annually.
PURCHASING DEPT: Peter Stiff
Military, Special Forces, Hunting books on southern Africa.

U.S. CIVIL WAR

GEORGE A. WILLHAUCK
21 Stevens Road
North Hampton, NH 03862, USA
U.S. Civil War militaria.
$3.00 for 6 Lists.

COOKS ARSENAL WORKS
642 Johnson Avenue
Meriden, CT 06450, USA
(203)-237-7997
Gunsmithing. Manufacturer of cannons, guns; restoration work.
S4; 25;
Catalogue $2.00. Annually.
CONTACT: Lawrence E. Cook
SALES DEPT: Lawrence E. Cook

PURCHASING DEPT: Lawrence E. Cook

AMERICAN COLLECTIBLES
30 Hedgewood Road
Howell, NJ 07731, USA
Civil War collectibles.
Send for catalogue.

MARTIN J. FOWLER
P.O. Box 715
Medford, NJ 08055, USA
U.S. Civil War & Rev. War collector.

MIKE CAVANAUGH
2713 Burgandy Drive
Cinnaminson, NJ 08077, USA
6th Virginia Infantry collector.

ROSA'S CIVIL WAR MILITARIA
Route #1, Box 212 Kipp Road
Goshen, NY 10924, USA
U.S. Civil War militaria.
Catalogue $1.00.

HOWARD SIGLAG
183 Hazelwood Drive
Westbury, NY 11590, USA
G.A.R. & U.C.V. collector.

ARTHUR R. WACK
P.O. Box 194
Manorville, NY 11949, USA
Civil War arms & accoutrements.

RICHARD MATTICE
P.O. Box 310
Pennellville, NY 13132, USA
Civil War militaria.
Write for details.

W. F. SAUNDERS
P.O. Dox 118
Cape Vincent, NY 13618, USA
U.S. Civil War weapon collector

EDWARD H. HAHN
2012 Sampson Street
Pittsburgh, PA 15221, USA
Civil War books & phamphlets.
Send S.A.S.E. for list.

IRWIN RIDER
4420 E. Lake Road
Erie, PA 16511, USA
(814)-898-4517 (814)-725-9372
U.S. Civil War weapons, images & memoriabilia.
S12; 5;
SALES DEPT: Irwin Rider
PURCHASING DEPT: Irwin Rider
Dealer in Civil War weapons, accoutrements, uniforms & images.

CARLISLE MILITARY INSTITUTE
U.S. Army Military History
Carlisle Barracks, PA 17013, USA
Desires GAR & UCV items for donation.

THE REGIMENTAL QUARTERMASTER
P.O. Box 553
Hatboro, PA 19040, USA
Civil War original & repro items.
Catalogue $1.00.
CONTACT: Eugene T. Lomas

U.S. CIVIL WAR

ANTEBELLUM COVERS
P.O. Box 3494
Gaithersburg, MD 20878, USA
(301)-869-2623
Civil War paper items.
S4,S7; 12;
Catalogue available. 5 to 6 per year.
CONTACT: Ron Meininger

MARYLAND LINE TRADER
P.O. Box 190
Linthicum Heights, MD 21090, USA
(301)-859-1852 (301)-465-9669
U.S. Civil War weapons, accoutrements, uniforms, documents & memorabilia.
S1,S4,S11,S12; 14;
Catalogue available. Bi-annually.
CONTACT: Dave Mark
SALES DEPT: Dave Mark
PURCHASING DEPT: Dave Mark
Dealer in general line of Civil War memorabilia, offering items for the beginner to the most advanced collector.

CANNON, LTD.
1316 Lafayette Avenue
Baltimore, MD 21207, USA
Artillery carriages & cannons.
S1,S4; 5;
Catalogue available.
SALES DEPT: P. Miller
PURCHASING DEPT: P. Miller
We build quality and guaranteed black powder cannon and mortar reproductions. We sell cannon hardware, wooden parts and wheels in 1/2, 3/4, and full scale.

FRED SHROYER
Rt. #1, Box 6, Gramlich Rd.
Lavale, MD 21502, USA
Pre-Indian Wars U.S. militaria.

CONFEDERATE MEMORABILIA
P.O. Box 397
Fairfax, VA 22030, USA
703-631-9518
Civil War militaria & collectables.
CONTACT: Lewis Leigh, Jr.

THE CAVALRY SHOP
P.O. Box 12122
Richmond, VA 23241, USA
Civil War leather goods.
Catalogue $1, buckle catalogue $3.

THE CAROLINA TRADER
P.O. Box 26986-PM
Charlotte, NC 28221, USA
(704)-547-1854
Civil War books.
S4,S12; 6;
Catalogue available. 4 to 6 times per year.
CONTACT: Richard E. Shields
SALES DEPT: Richard Shields
PURCHASING DEPT: Richard Shields
Civil War books - catalog of hundreds of books, new and used. Regimentals, battles, leaders, etc. We buy collections or one book, or sell on consignment.

NOR'EAST MILITARIA
3822 Canterbury Road
New Bern, NC 28560-7217, USA
U.S. Civil War militaria.
Catalogue available.

BELLINGER'S MILITARY ANTIQUES
P.O. Box 76371
Atlanta, GA 30328, USA
(404)-252-0267
General militaria 1700-1918.
S4; 6;
Catalogue available. 4 or 5 times per year.
CONTACT: Bill Bellinger
Bellinger's Military Antiques - Fine antique guns, swords, beltplates, leather goods, books, etc. 1918 & earlier. Catalog $2.00. Subscription $6/5 issues. P.O. Box 76371-PM, Atlanta, GA 30328.

BUTTONS MILITARIA
P.O. Box 39
Tallulah Falls, GA 30573, USA
(404)-754-6022 (404)-754-3595
American military buttons.
S5; 15;
Catalogue available. Bi-annually.
CONTACT: R. A. Edmondson
Military buttons of the North and South from the Civil War, before it, and afterwards. A 120-page list for $5.00

THE CIVIL WAR SHOP
P.O. Box 2091
Ft. Oglethorpe, GA 30742, USA
(404)-861-5321
General militaria & Civil War.
S6,S12;
CONTACT: Tim McKeever

STEVE SMITH
P.O. Box 2054
Miami, FL 33143, USA
Civil War paper memorabilia.
List available.

JOSEPH D. ATTWOOD
5415 S.W. 6th Avenue
Cape Coral, FL 33904, USA
G.A.R. memorabilia.

JOHN L. HEFLIN, JR.
5708 Brentwood Trace
Brentwood, TN 37027, USA
(615)-373-2917
Civil War memorabilia.
S4,S8,S9,S10; 25+;
Catalogue available. Monthly.
SALES DEPT: John L. Heflin
PURCHASING DEPT: John L. Heflin
Dealing in Civil War books, letters, photos, documents, and early Tennessee documents. Wish to purchase above items also.

DIXIE GUN WORKS, INC.
P.O. Box 130
Stad Road
Union City, TN 38261, USA
(901)-885-0700 (901)-885-0561
Antique guns, parts, & supplies.
S4,S5,S8,S12; 32; 100,000;
Catalogue available. Three times per year.
CONTACT: J. Lee Fry, President
SALES DEPT: Linda Jackson
PURCHASING DEPT: Linda Jackson
We sell antique guns, antique gun parts and shooting supplies for muzzleloaders. Our parts and supply catalog, that consists of 600 pages, is $3.00.

C & D JARNAGIN COMPANY
Route 3, Box 217
Corinth, MS 38834, USA
800-647-7084
Civil War supplies.
Catalogue $1.00.

INDIAN SHOP
P.O. Box 246
Bannister Road
Independence, KY 41051, USA
Indian artifacts & Civil War relics.
S4,S5,S13; 13;
Catalogue $2.00. 4 to 6 per year.
CONTACT: Vor Hilliard
SALES DEPT: Vor Hilliard
PURCHASING DEPT: Vor Hilliard
Indian artifacts, old & authentic Old World archaeological antiquities & Civil War relics. Huge inventory, send $2.00 for approximately 50 page catalog. Indian Shop, Independence, KY 41051.

LARRY MOESLE
1710 48th Street N.W.
Canton, OH 44709, USA
G.A.R. & U.C.V. collector.

GEORGE FINLAYSON
98 Parkwood Blvd.
Mansfield, OH 44906, USA
G.A.R. & U.C.V. collector.

STEPHEN E. OSMAN
5424 Elliot Avenue South
Minneapolis, MN 55417, USA
(612)-823-4009
Military Furnishings, 1812-1865.
S4;
Catalogue available. 3 times annually.
Books, images, memorabilia, replica items. Civil War to WWI.

JACK KOLLODGE
2812 Sherwood Road
Minneapolis, MN 55432, USA
G.A.R. memorabilia collector.
Write for details.

ROBERT R. ANDERSON, JR.
P.O. Box 213
Arlington Heights, IL 60006, USA
(312)-255-4326 (312)-639-8100
Civil War militaria. Appraisals, museum consultation, and item searches.
S4,S11,S12; 13;
Catalogue available. Quarterly.
SALES DEPT: Bob Anderson
PURCHASING DEPT: Bob Anderson
Buy-Sell-Trade-Research-Appraise Civil War militaria including uniforms, weapons, accoutrements, Horse equipage, personal effects, documents, manuals, unit histories, and musical instruments.

DANIEL J. BINDER
341 California Avenue
Sycamore, IL 60178, USA
Civil War militaria.
List available.

U.S. CIVIL WAR

JAMES MEJDRICH
128 N. Knollwood Drive
Wheaton, IL 60187, USA
G.A.R. memorabilia.
Write for details.

KARL E. SUNDSTOM
2927 N. Lincoln
N. Riverside, IL 60546, USA
Civil War documents & bookseller.

HAWKEYE TRADERS
3238 N. Central Park
Chicago, IL 60618, USA
Civil & Indiam war items.
List available.

KENESAW POST #125
17 Tennessee Street
Danville, IL 61832, USA
G.A.R. & U.C.V. militaria.
Catalogue, $2.00 each. Quarterly.

SABERS EDGE
30 Montauk
St. Louis, MO 63141, USA
U.S. Civil War collector.

JOHN ERTZGAARD
1283 Moncoeur Drive
Creve Coeur, MO 63146, USA
(314)-432-8090 (314)-553-3745
U.S. Civil War photos & tintypes.
S13;
Collector of autographed CDV photos, armed tintypes, Union or Confederate. Buy/trade with copies first. Also want Illinois, MO items, UCV badges,etc.

NOYES HUSTON
Box 449, Elvuelo Road
Rancho Sante Fe, CA 92067, USA
Civil War photo collector.

TIMOTHY A. MILLER
P.O. Box 420882
Sacramento, CA 95842, USA
Imperial Russian & Civil War items.

EDGED WEAPONRY

ROBERT L. KONIOR
P.O. Box 261
Windsor, CT 06095, USA
(203)-525-3777
Edged weapons & German militaria.
S6; 7;
PURCHASING DEPT: Bob Konior
Buy & trade daggers, fighting knives, folding stillettos and Japanese swords. Also, WWII German uniforms, hats, helmets, firearms, ordnance, and other political and military items.

ROBERT BERGER
185 Melba Street, #301
Milford, CT 06460, USA
European dress dagger collector.

GEORGE W. JUNO
535 County Road
Torrington, CT 06790, USA
(203)-489-1575
American military swords 1775-1918.
S6,S12; 10;
Catalogue available. 1 or 2 times per year.
CONTACT: George W. Juno
SALES DEPT: George Weller Juno
The largest and finest selection of American military swords 1775-1918. Enlisted mens, officers, presentations, and Confederate. Buy, sell, trade, layaway, and appraisals.

EMIL R. OLLMANN
P.O. Box 1293
Paramus, NJ 07653, USA
German knives.
Free illustrated list.

THOMAS T. WITTMANN
1253 N. Church Street
Moorestown, NJ 08057, USA
609-235-0622 (WK) 609-866-8733 (HM)
Nazi Daggers.

JIM HART
6554 Irving Avenue
Pennsauken, NJ 08109, USA
609-663-1466
Antique arms & Indian artifacts.

ALBERT N. HARDIN,JR.
5414 Lexington Avenue
Pennsauken, NJ 08109, USA
609-662-2221
U.S. military edged weapon.

GARY HULLFISH
16 Gordon Avenue
Lawrenceville, NJ 08648, USA
Emerson and Silver sword collector.

ROBERT L. BROOKS
235 East 53rd Street
New York, NY 10022, USA
Antique arms & armour.

LUCKY BREAK ENTERPRISES
1500 Main Street
Port Jefferson, NY 11777, USA
Edged weapons.

FRED EDMUNDS
R.D. 2, Box 440
Vorheesville, NY 12186, USA
518-456-3391 518-765-2075
Confederate swords & handgun collector.

ROHER'S
2807 Wilmington Road
New Castle, PA 16105, USA
412-654-5893
Knives

DICK'S GUN ROOM
2621 N. George Street
York, PA 17402, USA
717-846-7746
Antique arms & Indian artifacts
CONTACT: Dick Cary,President

MARK WALBERG
P.O. Box 130
Sunbury, PA 17801, USA
(717)-286-1416 (717)-286-1617
Japanese weaponry & armor.

ALBERT J. STECKER
128 Red Lion Road
Huntingdon Valley, PA 19006, USA
German dagger collector

CHARLES M. PISTOLE
3912 Tedrich Boulevard
Fairfax, VA 22031, USA
Bayonet & combat knife collector.
Send $2.00 for list of militaria.

HOMER BRETT
P.O. Box 111, Old Town Station
Alexandria, VA 22313, USA
Edged weapons & militaria dealer.
Send want lists.

AMERICAN HISTORICAL FOUNDATION
1142 W. Grace Street
Richmond, VA 23220, USA
(804)-353-1812 (800)-368-8080
Commemorative firearms & edged weapons.
S4,S8,S10; 10;
CONTACT: Robert A. Buerlein
The American Historical Foundation produces and offers to its members and the public the worlds finest commemorative firearms and edged weapons.

EK COMMANDO KNIFE COMPANY
601 N. Lombardy Street
Richmond, VA 23220, USA
(804)-257-7272 (800)-468-5575
Manufacturing & sale of EK knives.
S6; 45;
Catalogue available.
CONTACT: Joe Cusumano, Gen. Mgr.
Manufacturing & sales of battle-proven Ek Commando knives since 1941.

FAITH ASSOCIATES,INC.
1139 South Greenville Highway
Hendersonville, NC 28739, USA
704-692-1916
Edged weapons

R P KNIVES
1922 Spartanburg Highway
Hendersonville, NC 28739, USA
(704)-692-3466
Custom knives, gun parts, Colt distributor.
S6; 20;
Catalogue available. Annual updates.
CONTACT: Robert Parrish
SALES DEPT: Robert Parrish
PURCHASING DEPT: Robert Parrish

MATTHEWS CUTLERY
3845-A North Druid Hills Road
Decatur, GA 30033, USA
404-636-3970
Knives.
Catalogue $2.00.

DICKSON KNIVES
2349 Eastway Road
Decatur, GA 30033, USA
404-636-3959
Knives

EDGED WEAPONRY

AZTEC
P.O. Box 1888
Norcross, GA 30091, USA
Exotic weapons

RONNIE BALDWIN, JR.
2556 Cedartown Highway
Rome, GA 30161, USA
British MG & knife collector

ATLANTA CUTLERY
P.O. Box 839
911 Center Street
Conyers, GA 30207, USA
404-922-3700
Edged weapons & knives
CONTACT: Kelly Green

BEN MARTIN
P.O. Box 725062
Atlanta, GA 30339, USA
(404)-565-2598
U.S. military knives & wings.
S6,S11,S12; 2; ; 100.00
Catalogue available. Quarterly.
I collect, buy, sell or trade U.S. military knives and U.S. wings. Catalog costs $2.00 (refundable). Offers, trades and consignments cheerfully considered.

BEAR WHOLESALE SPORTS, INC.
P.O. Box 1698
Thomasville, GA 31799, USA
(800)-34KNIFE
Knife wholesaler.
S5;
Catalogue available.

KNIFECO
P.O. Box 5271
Hialeah Lakes, FL 33014, USA
Knives & edged weapons.

MILITARY REPLICA ARMS, INC.
P.O. Box 36006
Tampa, FL 33673, USA
(813)-237-0764
Replica military edged weapons.
S6,S12; 2;
Catalogue available. Three times per year.
CONTACT: Ron G. Hickox, President
SALES DEPT: Ron G. Hickox
PURCHASING DEPT: Ron G. Hickox
Fine military edged weapons replicas from around the world and from 1800s to W.W.II. For collectors, museums, dealers and reenactors.

ERIC ROSS
111 N. Pine Street
Brewton, AL 36426, USA
Samurai Collector.

PARKER CUTLERY
P.O. Box 22668
6928 Lee Highway
Chattanooga, TN 37422, USA
800-251-7687 615-894-1782
Knives & edged weapons.
Catalogue available.

SMOKY MOUNTAIN KNIFE WORKS
P.O. Box 430
Sevierville, TN 37862, USA
Knives & edged weapons.
Send for catalogue.

THE KING AND SUTLEY COMPANY
11649 South Monticello
Knoxville, TN 37922, USA
Edged weaponry.

ARK CORPORATION
7048 W. Central Avenue
Toledo, OH 43617, USA
419-841-2577
Knives & edged weapons.
CONTACT: Mary Ellen Nelson, AD Director

WILLIAM FAGAN
P.O. Box 425
Fraser, MI 48026, USA
Edged weapons.
Catalogue $5.00 for 6 issues.

G.I.
P.O. Box 282
Caledonia, MI 49316, USA
(616)-891-1980
Edged weapons & manuals.
S6,S12; 3; 15,000; $30.00
Catalogue available. Annually.
CONTACT: Lynn Hillard
SALES DEPT: Lynn Hillard
PURCHASING DEPT: Lynn Hillard
Combat & survival equipment. Hard to get items at a fraction of the cost. Send $5.00 for catalog.
P.O. Box 282, Caledonia, MI 49316.

ROMEX INTERNATIONAL, INC.
P.O. Box 597
Polson, MT 59860, USA
(406)-883-2481
Antique & collectable fighting knives.
S4; 5;
Catalogue available. Three times annually.
CONTACT: Morris Gruenberg
SALES DEPT: M. Gruenberg
PURCHASING DEPT: M. Gruenberg
Dealers in antique and collectable fighting knives, Bowies, daggers, dirks, clandestine blades of the Special Forces, SOE/OSS, etc. Illustrated catalogues $6.00 per year.

BALLARD CUTLERY
1495 Brummel Avenue
Elk Grove Village, IL 60007, USA
Knives & edged weapons.

FREDERICKS MILITARIA
6919 Westview
Oak Forest, IL 60452, USA
Samurai & 3rd Reich edged weapons.
$5.00 for 6 catalogues.

MIDWEST KNIFE COMPANY
9043 S. Western Avenue
Chicago, IL 60620, USA
Knives & edged weapons

JAPANESE SWORD SOCIETY / U.S. INC.
P.O. Box 4387
Grasso Plaza Branch
St. Louis, MO 63123, USA
Japanese sword collectors association.
S10;
The JSS/US is registered with the state of California as a non-profit organization for the appreciation and study of the Japanese art sword and related fields of interest.

HOUSE OF SWORDS & MILITARIA
2804 Hawthorne Avenue
Independence, MO 64052, USA
General militaria & edged weapons.

EXOTIC SPORTS, INC.
Box 129, Lucky Street
Fayette, MO 65248, USA
800-248-5128
Blowguns & exotic weapons

THE DUTCHMAN
9071 Metcalf, Suite 158
Overland Park, KS 66212, USA
800-821-5157
Exotic weapons & accessories.

JIMMY LILE
Route 6, Box 27
Russellville, AR 72801, USA
501-968-2011
Knife maker

REDDICK ENTERPRISES
P.O. Box 314-D38
Denison, TX 75020, USA
(214)-463-1366 (214)-463-1377
Replicas & authenic Nazi edged weapons.
S6,S11,S12,S13; 17; 18,000; $60.00
CONTACT: J. Rex Reddick
SALES DEPT: Telephone order desk
PURCHASING DEPT: J. Rex Reddick
AD MANAGER: Ed Wells
Specializing in original Nazi edged weapons. Manufacturer/Distributor of rare reproduction daggers, display cases, & reprints of rare dagger manufacturers' catalogs. Dealer inquiries invited.

SPYDERCO, INC.
P.O. Box 800
Golden, CO 80402, USA
Knife sharpener

LIFEKNIFE, INC.
P.O. Box 771
Santa Monica, CA 90406, USA
Military & survival knives.

WORLD BALI-SONG SOCIETY
1428 South Gaffrey Street
San Pedro, CA 90731, USA
Knives

DRAGONMASTER
1474 Monterey Pass Road
Monterey Park, CA 91754, USA
Exotic weapons

CROSSBOWS, LTD.
P.O. Box 8177
Fountain Valley, CA 92708, USA
714-548-9443
Crossbows.

EDGED WEAPONRY

KEN NOLAN,INC.,P.M.DIV.
P.O. Box C-19555
16901 P.M. Milliken Avenue
Irvine, CA 92713, USA
(714)-863-1532
Surplus, insignia, medals & uniforms.
S4; 29; 100,000; $60.00
Catalogue $1.00. 3 or 4 times annually.
Serving individuals, military, police for 29 years. 90% of orders shipped within 24 hours. Military field clothing, boots, insignia, nameplates, etc. Send $1.00 for catalog.

COLD STEEL,INC.
2128 Knoll Drive, Unit D
Ventura, CA 93003, USA
Knives

JOHN SKALISKY
P.O. Box 6404
San Jose, CA 95150, USA
Edged weapons & armour.
Write for details.

DARRYL KINNISON
P.O. Box 521
Westwood, CA 96137, USA
Swords.
List available.

SELECT LINE
98148 Kauhihau Plaza
Pearl City, HI 96782, USA
Military knives.
Catalogue available.

GERBER LEGENDARY BLADES
14200 S.W. 72nd Avenue
Portland, OR 97223, USA
Knives

WELLS CREEK KNIFE & GUN WORKS
32956 Street, Highway 38
Scottsburg, OR 97473, USA
Knives

EDRED A. F. GWILLAIM
Candletree House
Crickdale
Nr. Swindon, Wiltshire SN6 6AX, Great Britain
0793-750241
Arms & Armour.

ARBOUR ANTIQUES LTD.
Poet's Arbour, Sheep Street
Stratford-On-Avon, Warwickshire CV37 6EF, Great Britain 0789 3453
Arms & Armour.
CONTACT: R. J. Wigington

PETER DALE LTD.
11 & 12 Royal Opera Arcade
Pall Mall
London, SW1, Great Britain
01-930-2695
Arms & Armour.

ARMADA ANTIQUES
Gray's Antique Market
58 Davies Street, Stand 122
London, W1, Great Britain
01-499-1087
Arms & Armour.

EDGED MILITARIA
11 Dacres Road
London, SE23 2NW, Great Britain
01-699-2527 01-291-1318
Edged weaponry.
Catalogue $1.00.

MICHAEL C. GERMAN
38B Kensington Church Street
London, W8, Great Britain
01-937-2771
Arms & armour.

MILITARY FIREARMS

ROBERT P. HANAFEE
29 Bedford Court
Amherst, MA 01002, USA
413-253-2583
Military firearms & accessories.

RICKY'S GUN ROOM
45 Deroy Drive
Chicopee, MA 01020, USA
FFL Firearms Dealer

BRUCE D. KEYES
45 Scott Street
Ludlow, MA 01056, USA
FFL Firearms Dealer.

BRAZAS SPORTING ARMS,INC.
P.O. Box 22
Rt. 32
Monson, MA 01057, USA
413-267-4006
Firearms Dealer.

DOAKES GUN SHOP
555 Holyoke Road
Westfield, MA 01085, USA
FFL Firearms Dealer.

FRAN'S PLACE
433 East Main Street
Westfield, MA 01085, USA
FFL Firearms Dealer.
CONTACT: Westfield Shops

PETES GUN SHOP
31 Columbia Street
Adams, ME 01220, USA
FFL Firearms Dealer

MARINE ARMS COMPANY
P.O. Box 7311
Fitchburg, MA 01420, USA
Firearms dealer
CONTACT: Mr. Goguen

MID-STATE SECURITY SERVICES
14 Summer Street
Fitchburg, MA 01420, USA
FFL Firearms Dealer

DOUGLAS E. SAGE
40 Boutelle Street
Leominster, MA 01453, USA
FFL Firearms Dealer

PODUNK GUN SHOPPE
P.O. Box 135
East Brookfield, ME 01515, USA
FFL Firearms Dealer

MAIN STREET FIREARMS SALES
125 Main Street
S. Lancaster, MA 01561, USA
FFL Firearms Dealer

JOHN T. MURPHY
415 Arlington Street
Acton, ME 01720, USA
FFL Firearms Dealer

STUART JACOBS CO.
10 Edith Road
Hudson, MA 01749, USA
FFL Firearms Dealer

J. DeDOMINICIS
3 Diane Lane
Chelmsford, MA 01824, USA
FFL Firearms Dealer

ELKO POLICE & FIRE EQUIPMENT
89-B N. Lowell Street
Methuen, MA 01844, USA
FFL Firearms Dealer

JADFYE SPORTING & GUN SHOP
50 Viola Street
Lowell, MA 01851, USA
FFL Firearms Dealer

THE SHOOTERS DEN
15 High Street
Reading, MA 01867, USA
FFL Firearms Dealer

FERRO
P.O. Box 73
West Boxford, MA 01885, USA
FFL Firearms Dealer

NOAH'S MOTORS
622 Broadway
Saugus, MA 01906, USA
(617)-233-1616
U.S. service rifles, sales & service.
S1,S4,S12; 10;
CONTACT: Harry G. Cakounes
SALES DEPT: Harry G. Cakounes
PURCHASING DEPT: Harry G. Cakounes
U.S. service rifle sales and service. Custom built national match rifles. Special purpose sniper rifles. M.O.A. accuracy. M-14 parts and kits in stock.

JOE N. MURRAY & SOHN
10 Winthrop Street
Essex, MA 01929, USA
Firearms dealer.

THE GUNRUNNER
6 Moulton Street
Newburyport, MA 01950, USA
FFL Firearms Dealer

LARRY BROWN
2 Marston Road
Walpole, MA 02032, USA
FFL Firearms Dealer

MILITARY FIREARMS

FIREARMS UNLIMITED
100 Green Street
Foxboro, MA 02035, USA
FFL Firearms Dealer

STEVE'S GUNS & SUPPLIES
31 North Avenue
Norwood, MA 02062, USA
FFL Firearms Dealer

MATT RARA
150 Crescent Avenue
Revere, MA 02151, USA
FFL Firearms Dealer

WILLIAM E. TRUDEAU
37 Old Brook Circle
Melrose, MA 02176, USA
FFL Firearms Dealer

CHARLES H. HAMPE
18 Maugua Avenue
Wellesley Hills, MA 02181, USA
FFL Firearms Dealer

INTER-STATE ARMS
6 Brook Road
Needham, MA 02194, USA
617-449-1713
Firearms

NORTHEAST SHOOTERS
64 Pilgrim Road
Pembroke, MA 02359, USA
FFL Firearms Dealer

G. J. FIREARMS
P.O. Box 63
27 Overy Drive
North Falmouth, MA 02556, USA
FFL Firearms Dealer
CONTACT: Gil Johnson

BISSON ARMS COMPANY
93B County Road
East Freetown, MA 02717, USA
FFL Firearms Dealer

RON RESDEN
100 Plain Street
Norton, MA 02766, USA
FFL Firearms Dealer

INDEPENDENCE ENTERPRISES
RR 2, Box 503
Foster, RI 02825, USA
FFL Firearms Dealer

RHODE ISLAND RADIATOR SERVICE
3688 Quaker Lane
North Kingston, RI 02852, USA
FFL Firearms Dealer

EARL W. DURAND
RFD 1, 15 Calef Road
Auburn, NH 03032, USA
FFL Firearms Dealer

FIREARMS FOR LAW ENFORCEMENT
36 Bockes Road
Hudson, NH 03051, USA
FFL Firearms Dealer
CONTACT: D. R. McCrady

NORM'S GUN ROOM
29 Simpson Road
Pelham, NH 03076, USA
FFL Firearms Dealer
CONTACT: Norman G. Clermont

ANTHONY'S GUNS & AMMO
RFD 1, Box 483, Forest St.
Bradford, NH 03221, USA
FFL Firearms Dealer
CONTACT: Tony Salers

VILLAGE ARMS
RFD #2, Box 394
Contoocook, NH 03229, USA
603-746-3551
Firearms
CONTACT: Ray Mock

THE GREAT OUTDOORS, INC.
Pumpkin Hill Road
RD 1, Box 226
Warner, NH 03278, USA
FFL Firearms Dealer

PAN GUNS
P.O. Box 235
Keene, NH 03431, USA
FFL Firearms Dealer

DENNIS BACHAND
RR #1, Hill Road
Dummer, NH 03588, USA
FFL Firearms Dealer

NORTH COUNTRY SPECIALTIES
287 Elm St., RFD 3
Claremont, NH 03743, USA
FFL Firearms Dealer

HAROLD HURD
4 Upland Road
Atkinson, NH 03811, USA
FFL Firearms Dealer

CHOCORUS GUN SHOP
Rt. 16, Box 127
Chocorus, NH 03817, USA
FFL Firearms Dealer

BOX LAKE TRADING POST
P.O. Box 651
Hampstead, NH 03841, USA
FFL Firearms Dealer

CLARK D. TOWLE
6 Dorre Road
Kingston, NH 03848, USA
FFL Firearms Dealer

ROCHESTER HUNTING & FISHING
10 Cove Court
Rochester, NH 03867, USA
FFL Firearms Dealer

JERRY'S SPORT SHOP
East Baldwin, ME 04024, USA
FFL Firearms Dealer
CONTACT: Douglas H. Sebago

THE FOX SHOP
RFD 3, Box 61-A
Freeport, ME 04032, USA
207-865-3388
Gunsmith.
CONTACT: Roland A. Fraser

JOE'S GUN SHOP
RR 5, Box 383, Elliott Road
Gorham, ME 04038, USA
FFL Firearms Dealer

GERALD J. J. GENEST
3 Harding Street
Sanford, ME 04073, USA
FFL Firearms Dealer

QUARTERMASTER SALES
51 Chase Street
South Portland, ME 04106, USA
FFL Firearms Dealer

GERALD S. HUSSAR
53 Pinewood Drive
Auburn, ME 04210, USA
FFL Firearms Dealer

RAY J. CHRISTMAN
RFD 2, Box 270
Greene, ME 04236, USA
FFL Firearms Dealer

MINOT GUNSHOP
RFD 2, Box 4964
Mechanic Falls, ME 04256, USA
FFL Firearms Dealer

ARTHUR PRUE
Route 100
New Gloucester, ME 04260, USA
FFL Firearms Dealer

VINCENT S. GEORGE
RD 1, Box 504
South Paris, ME 04281, USA
FFL Firearms Dealer

BREWER GUN EXCHANGE
1087 North Main Street
Brewer, ME 04412, USA
207-989-1429
Firearms
CONTACT: R. E. McLeod

RIDEOUTS LODGE
East Grand Lake
Danforth, ME 04424, USA
FFL Firearms Dealer

PINE TREE GUNS & AMMO
RFD #1, Box 849
E. Holden, ME 04429, USA
FFL Firearms Dealer
CONTACT: Wayne Gragg

KEN'S GUN SHOP
33 Shackford Street
Eastport, ME 04631, USA
FFL Firearms Dealer
CONTACT: Kenneth Yankauer

SCHMIDT GUNSHOP
P.O. Box 288
Island Falls, ME 04747, USA
FFL Firearms Dealer

MILITARY FIREARMS

ROLAND T. PELLETIER
53 Mt. View
Madawaska, ME 04756, USA
FFL Firearms Dealer

ROY L. STRAHAN
45 Joe Avenue
Winslow, ME 04901, USA
FFL Firearms Dealer

GEORGE DAVIS, JR.
RD. 2, Box 665, Elm Street
Mercer, ME 04957, USA
FFL Firearms Dealer

STEVE'S SPORT SHOP
142 Union Street
Springfield, VT 05156, USA
FFL Firearms Dealer

JAMES F. LEIBFRIED
RD 2, Box 178, Miller Rd.
Putney, VT 05346, USA
FFL Firearms Dealer

PAUL J. MAYER
64 Colchester Avenue
Burlington, VT 05401, USA
FFL Firearms Dealer

ARMS & AMMO
15 Warner Avenue
Essex Junction, VT 05452, USA
FFL Firearms Dealer

CHASE GUNSHOP
Box 191
Jeffersonville, VT 05464, USA
FFL Firearms Dealer

MIDSTATE SHOOTING SPORTS
RD 1, Colonial Drive
Middlebury, VT 05753, USA
FFL Firearms Dealer

KARENS GUN SHOP
RFD 1, Box 343A
Poultney, VT 05764, USA
FFL Firearms Dealer

GARY H. SMITH
Star Rt. 14
Hardwick, VT 05843, USA
FFL Firearms Dealer

AVON GUN ROOM
5 Haynes Road
Avon, CT 06001, USA
FFL Firearms Dealer

THE GUN COMPANY
P.O. Box 232
Collinsville, CT 06022, USA
FFL Firearms Dealer

DONALD EMERSON
28 Garden Street
New Britain, CT 06050, USA
FFL Firearms Dealer

NICHOLSON ENTERPRISES
P.O. Box 372
Ledge Road
Plainville, CT 06062, USA
FFL Firearms Dealer

THE LOWER DECK
79 Hublard Drive
Vernon, CT 06066, USA
FFL Firearms Dealer

GARY'S GUNSMITHING & SUPPLIES
P.O. Box 882
Enfield, CT 06082, USA
FFL Firearms Dealer

KEN'S GUNSMITH
P.O. Box 3295
Enfield, CT 06082, USA
FFL Firearms Dealer

THOMAS L. DOHERTY
47 Notch Road
West Simsbury, CT 06092, USA
FFL Firearms Dealer

TANGUAY GUNS & SUPPLIES
133 Brookside Circle
Wethersfield, CT 06109, USA
FFL Firearms Dealer

COUNTRY ARMS
173 South Windham Road
Willimantic, CT 06226, USA
FFL Firearms Dealer

LAND & SEA SPORTS
Canada Lane
Chaplin, CT 06235, USA
FFL Firearms Dealer
CONTACT: Steve Winick

SHIPMAN'S
4 Oakridge Drive
Gales Ferry, CT 06335, USA
FFL Firearms Dealer

J & J ARMS
6 Caribou Drive
Norwich, CT 06360, USA
FFL Firearms Dealer
CONTACT: John W. Monarski, Sr.

BILL CAMPELL
23 Strosberg Road
Waterford, CT 06385, USA
FFL Firearms Dealer

LEISURE TIME SPORTS
11 Lawrence Avenue
Milford, CT 06460, USA
FFL Firearms Dealer
CONTACT: Neil Tanis

ANDREW B. FLORENCE
21 Round Hill Road
Monroe, CT 06468, USA
FFL Firearms Dealer

STANLEY J. KOS
200 Parsonage Hill
Northford, CT 06472, USA
FFL Firearms Dealer

MILLERIDGE ARMS COMPANY
238 Main Street
Portland, CT 06480, USA
FFL Firearms Dealer

JOHNSON'S GUNS & AMMO
63 Fairlane Drive
Sheldon, CT 06484, USA
FFL Firearms Dealer

KNOTT ASSOCIATES
503 Main Street
Southington, CT 06489, USA
FFL Firearms Dealer

STURM, RUGER & COMPANY, INC.
Lacey Place
Southport, CT 06490, USA
(203)-259-7843
Firearms manufacturer.
S13; 37;
Catalogue available. Bi-annually.
CONTACT: William B. Ruger, Sr.
SALES DEPT: J-T. Ruger
Complete line of Police, Government and Service training rifles and revolvers.

EASTERN FIREARMS
89 Edson Avenue
Waterbury, CT 06705, USA
FFL Firearms Dealer
CONTACT: Tom Boiano

THE GUN RACK
30 Joy Road
Waterbury, CT 06708, USA
(203)-729-6938 (203)-756-1312
Firearms, ammo & surplus.
S1, S6; 5;
Catalogue available. Annually.
CONTACT: Bob Beaudry
SALES DEPT: Bob
PURCHASING DEPT: Bob
Class I and III firearms, ammo, military surplus items, out of state Class III transfers.

STEVE'S GUNS & KNIVES
179 Elm Street
Thomaston, CT 06787, USA
FFL Firearms Dealer

GREGG'S GUN SHOP
122 Winsted Road
Torrington, CT 06790, USA
FFL Firearms Dealer

HOLLOWELL & COMPANY
340 West Putnam Avenue
Greenwich, CT 06830, USA
FFL Firearms Dealer

SLOAN'S SPORTING GOODS COMPANY
10 South Street
Ridgefield, CT 06877, USA
203-438-7341
Firearms

RIVERSIDE GUN SALES & SERVICE
25 Carrona Drive
Riverside, CT 06878, USA
FFL Firearms Dealer
CONTACT: Peter W. Pintek

MILITARY FIREARMS

BILL PETRI GUNS
43 Bald Hill Road
Wilton, CT 06897, USA
Firearms

DOUGLAS L. OEFINGER
19 Vassar Avenue
Stamford, CT 06902, USA
203-324-2952
Machine guns, all services.
Write for prices & details.

DUNCAN & FIELD
233 Broadway
Bayonne, NJ 07002, USA
FFL Firearms Dealer

RAVEN ROCK ARMS CO.
93 Laurel Avenue
Kearny, NJ 07032, USA
FFL Firearms Dealer

R. J. MARONI & SON, INC.
P.O. Box 848
Upper Montclair, NJ 07043, USA
FFL Firearms Dealer

JOE LA PLACA
39 Tintle Road
Kinnelon, NJ 07405, USA
FFL Firearms Dealer

ARMOUR PRODUCTS
Godwin Ave., Box 56
Midland Park, NJ 07432, USA
Firearms accessories.

WILDCAT FIREARMS, INC.
1160 Hamburg Turnpike
Wayne, NJ 07470, USA
FFL Firearms Dealer

S. J. ELMI
43 Hawthorne Avenue
Hawthorne, NJ 07506, USA
FFL Firearms Dealer

DYNAMIT NOBEL OF AMERICA, INC.
105 Stonehurst Court
Northvale, NJ 07647, USA
Airgun manufacturer.

MORTON E. KARP
554 Shaler Blvd.
Ridgefield, NJ 07657, USA
FFL Firearms Dealer

MICHAEL DeSTEFANO
RR 1, Box 409, Foxhill Road
Branchville, NJ 07826, USA
FFL Firearms Dealer

FOX HARDWARE
12 Mine Brook Road
Bernardsville, NJ 07924, USA
FFL Firearms Dealer

SARCO, INC.
323 Union Street
Stirling, NJ 07980, USA
201-647-3800
Firearms

R & R GUNS & SUPPLIES
8 Duchess Ave
Marlton, NJ 08053, USA
FFL Firearms Dealer

LOG CABIN ARMS
7 Over Hill Road
Rancocas Woods
Mount Holly, NJ 08060, USA
FFL Firearms Dealer

CROSSROADS GUN SHOP
4201 Maple Avenue
Pennsauken, NJ 08109, USA
609-488-0811 609-488-0760
Firearms

SOUTH JERSEY ORDNANCE
Tilton Shopping Center
Northfield, NJ 08225, USA
FFL Firearms Dealer

CARIBOU GUN CLUB
P.O. Box 213
Rio Grande, NJ 08242, USA
FFL Firearms Dealer

R & R GUNSMITH
35 W. Groveland Avenue
Somers Point, NJ 08244, USA
FFL Firearms Dealer

JIM STOCKTON
Box 1, 716 Maxim Avenue
Forked River, NJ 08731, USA
FFL Firearms Dealer

GUNWORKS, INC.
75 N. Main Street
Milltown, NJ 08850, USA
FFL Firearms Dealer

SOSTA GUN SHOP
P.O. Box 621
Main Street
Neshanic Station, NJ 08853, USA
FFL Firearms Dealer

RUTGERS BOOK CENTER
127 Raritan Avenue
Highland Park, NJ 08904, USA
201-545-4344
Antique firearms.

MIDTOWN SPORTS CLUB
481 8th Avenue
New York, NY 10001, USA
FFL Firearms Dealer

BASTION ARMS
U.S. Coast Guard Base
Bldg. 877/308
Governors Island, NY 10004, USA
FFL Firearms Dealer
CONTACT: Michael J. & Mary McGuire

JOHN JOVINO COMPANY, INC.
5 Centre Market Place
New York, NY 10013, USA
800-221-3350
Firearms

ALLSPORTS
3472 Webster Avenue
Bronx, NY 10462, USA
FFL Firearms Dealer

NO GA PLACE
Putham Drive Noga Place
Lake Carmel, NY 10512, USA
FFL Firearms Dealer

NORTH WESTCHESTER POLICE & MILITARY SUPPLY CENTER
RD #10, Ellen Ave., Box 444A
Mahopac, NY 10541, USA
FFL Firearms Dealer

EAST COAST ANTIQUES
655 North Barry Avenue
Mamaroneck, NY 10543, USA
(914)-834-9565 Eve. (212)-295-4966 Day
Fine modern and antique weaponry.
CONTACT: Joe or Don

EL RANCHERO
195 N. Bedford Road
Mt. Kisco, NY 10549, USA
FFL Firearms Dealer

WALTER G. BAHR
133 Hillside Avenue
Mt. Vernon, NY 10553, USA
FFL Firearms Dealer

HARRY S. MORELLI
Oscawanna Heights Road
Putnam Valley, NY 10579, USA
FFL Firearms Dealer

F. RISSE
Rt. 4, Box 73
Putnam Valley, NY 10579, USA
FFL Firearms Dealer

JOHN R. DELORENZO
17 E. Taylor Square
White Plains, NY 10604, USA
FFL Firearms Dealer

J. RADZINSKI
Burhans Avenue
Yonkers, NY 10701, USA
FFL Firearms Dealer

ROBERT G. ALFF
46 Amherst Drive
Yonkers, NY 10710, USA
FFL Firearms Dealer

JOSEPH P. RACANELLO
38 Lincoln Street
New Rochelle, NY 10801, USA
FFL Firearms Dealer

LEATHERNECK VENTURES
32 Smith Clove Road
Central Valley, NY 10917, USA
914-928-9559
Firearms

CUSTOM ARMS
RD 3, Margaret Road
Monroe, NY 10950, USA
FFL Firearms Dealer
CONTACT: T. A. Schellberg

MILITARY FIREARMS

GLODE M. REQUA INC.
98 Saddle River Road
Monsey, NY 10952, USA
Firearms from 1600-1890.

NICK'S GUN SHOP
P.O. Box 92
Stony Point, NY 10980, USA
FFL Firearms Dealer

JOTCO
P.O. Box 51
Tomkins Cove, NY 10986, USA
FFL Firearms Dealer

RICK ZINO
175 Court House Road
Franklin Square, NY 11010, USA
FFL Firearms Dealer

DE SANTIS GUNHIDE
149 Denton Avenue
New Hyde Park, NY 11040, USA
516-354-8000
Gun leather & holsters.

B & B
24-70 49th Street
Astoria, NY 11103, USA
FFL Firearms Dealer

ALPINE ARMS CORP.
6716 Ft. Hamilton Parkway
Brooklyn, NY 11219, USA
FFL Firearms Dealer

BOB VADALA SALES COMPANY, INC.
1100 Franklin Avenue
Garden City, NY 11530, USA
FFL Firearms Dealer

DISCOUNT GUNS, INC.
541 W. Merrick Road
Valley Stream, NY 11580, USA
FFL Firearms Dealer

STEVE HAYES
65 Braham Avenue
Amityville, NY 11701, USA
FFL Firearms Dealer

PATRICK CARPENTER
70 Midway Street
Babylon, NY 11702, USA
FFL Firearms Dealer

JOSEF J. PRILLER
55 Thomas Avenue
Bethpage, NY 11714, USA
FFL Firearms Dealer

GEORGE W. FERRIS
224 Claywood Drive
Brentwood, NY 11717, USA
FFL Firearms Dealer

ELWOOD SPORTS
166 Warner Road
Elwood, NY 11743, USA
FFL Firearms Dealer
CONTACT: Thomas W. Burns

DEPOT FIREARMS, INC.
The 211 Bldg. - Depot Road
Huntington Station, NY 11746, USA
FFL Firearms Dealer

ANTHONY H. BATSON
797 Centre Avenue
Lindenhurst, NY 11757, USA
FFL Firearms Dealer

LAD'S FIREARMS
2145 Rt. 112
Medford, NY 11763, USA
FFL Firearms Dealer

STEVEN G. MAULDIN
5 Middle Court
Miller Place, NY 11764, USA
FFL Firearms Dealer

NEAL SEAMAN
220 Adirondeck Drive
Selden, NY 11784, USA
FFL Firearms Dealer

GEORGE R. RANIOLO
1212 Rt. 25 A
Stony Brook, NY 11790, USA
FFL Firearms Dealer

PRESTON L. REMSEN
15 Beverly Lane
E. Moriches, L.I., NY 11940, USA
FFL Firearms Dealer

MAIN STREET MAGAZINE
74 Main Street
Westhampton Beach, NY 11978, USA
FFL Firearms Dealer
CONTACT: Dean Speir

THOMAS ENGLE
172 Guy Park Avenue
Amsterdam, NY 12010, USA
FFL Firearms Dealer

C K GUN SHOP
RD 2, Lake Road (Zani Lane)
Ballston Lake, NY 12019, USA
518-399-5536
Firearms
CONTACT: Chet Krosky

JOHN MULLEN
P.O. Box 146-A
Bernie, NY 12023, USA
518-872-0750
Percussion Colt firearms collector.

STEVEN BROWER
Wendell Avenue
Broadalin, NY 12025, USA
FFL Firearms Dealer

LARRY'S SPORTING GOODS
Box 356, Walnut Creek, Mall
Cobleskill, NY 12043, USA
FFL Firearms Dealer

THE QUEST
Rt. 9, Box 1247
Clifton Park, NY 12065, USA
(518)-383-2414 Shop (518)-383-0071 Home
Antique firearms and accessories.

CONTACT: Paul Weisberg, Prop.

C. & D. PETRONIS, INC.
16 River Road
Mechanicville, NY 12118, USA
518-664-7610
Firearms
CONTACT: David Petronis, President

GEORGE W. PORTER
228 Pinehurst Avenue
Rensselaer, NY 12144, USA
518-449-2898
Firearms

AL ROSTOHAR
E. Main Street
Scheneuus, NY 12155, USA
FFL Firearms Dealer

GRUSS COUNTRY BAIT & TACKLE
Box 288, Miller Road
Selkirk, NY 12158, USA
FFL Firearms Dealer

KARL JUUL
199 Blessing Road
Slingerlands, NY 12159, USA
FFL Firearms Dealer

GENE'S GUNROOM
3059 Lindenwald Avenue
Stuyvesant Falls, NY 12174, USA
FFL Firearms Dealer

BELANGER ARMS
459 Brunswick Road
Troy, NY 12180, USA
FFL Firearms Dealer

CATSKILL GUN SHOP
Rte. 9W, Box 227-A
West Coxsackie, NY 12192, USA
FFL Firearms Dealer

WILLIAM COUSER
3 Kasper Drive
Loudonville, NY 12211, USA
FFL Firearms Dealer

DAVE'S GUNS
34 Hadel Road
Scotia, NY 12302, USA
518-399-9438
Antique firearms.

PETE'S GUN SHOP
1221 Vinewood Avenue
Schenectady, NY 12306, USA
FFL Firearms Dealer

NUMRICH ARMS CORPORATION
P.O. Box 6
West Hurley, NY 12491, USA
914-679-2417
Firearms

MILITARY FIREARMS

AUTO ORDNANCE CORPORATION
P.O. Box PMD
West Hurley, NY 12491, USA
(914)-679-7225
Firearms & related accessories.
S6; 12;
CONTACT: Ira Trast
SALES DEPT: Steven Trast
PURCHASING DEPT: Steven Trast
Manufacturer of authentic semi-auto models of Tommy Gun (Thompson Sub-machine gun), 1911A-1 Government Model Pistol, plus related parts and accessories.

DON'S SPORTING GOODS
318 Union Street
Hudson, NY 12534, USA
(518)-828-0955
Ammo, guns, and hunting supplies.
S3,S5,S12; 3;
CONTACT: Donald T. LaValley
Wholesale sporting goods, guns, ammo, clothing, hunting and fishing supplies, archery, reloading, decoys and game calls. Quality goods at a fair price. Ask for Don.

PETER LICIS
10 Gardiner Park
New Paltz, NY 12561, USA
FFL Firearms Dealer

THE TOY SOLDIER GALLERY, INC.
P.O. Box 1054
Wappingers Falls, NY 12590, USA
Toy soldiers.
SASE for listing of available sets.

F. D. CARLSEN
P.O. Box 32-A
Grahamsville, NY 12740, USA
FFL Firearms Dealer

ED'S REPAIR SERVICE
RR 1, Box 166, Neversink Road
Liberty, NY 12754, USA
FFL Firearms Dealer

DONALD J. LETIZIA, SR.
RD 2, Box 1, South Road
Cambridge, NY 12816, USA
FFL Firearms Dealer

LARRY R. LaRUE
R.D. #1, Box 164
Greenfield Center, NY 12833, USA
FFL Firearms Dealer

JOHN GUNDERSON
64 Phillip
Lake George, NY 12845, USA
518-668-2221
Antique longarms.

W. J. LA SARSO
P.O. Box 1442
Lake George, NY 12845, USA
FFL Firearms Dealer

POWDERHORN RIFLE COMPANY
4100 Stony Creek Road
Stony Creek, NY 12878, USA
(518)-696-3697
Antique & old military firearms & ammo.

S1,S4,S12,S13; 20;
Catalogue available. Quarterly.
CONTACT: Bruce Strohm
SALES DEPT: Neal Mattson
PURCHASING DEPT: Bruce Strohm
Powderhorn specializes in finding antique and old military firearms for the collector. We have many calibers of military ammo in stock. $10.00 membership fee.

AL'S GUN SHOP
164 N. Williams Street
Whitehall, NY 12887, USA
FFL Firearms Dealer

M & M GUNS & SPORTS, INC.
P.O. Box 70
20 Palmer Street
Ausable Forks, NY 12912, USA
518-647-8132
Firearms
CONTACT: Mark Cross

J & S GUN SUPPLY
RR 1 Box 129, 7729 Meaney Road
Cleveland, NY 13042, USA
FFL Firearms Dealer

BILL BUIS
P.O. Box 433
Liverpool, NY 13088, USA
315-622-2386
Firearms

RENE HECKLER
Rt. 1, Box 128, Furnace Road
Red Creek, NY 13143, USA
FFL Firearms Dealer

MATHEW J. ZABINSKI
P.O. Box 773
Skaneateles, NY 13152, USA
(315)-685-3576
Antique & collectable weaponry.

ANDY'S GUN SHOP
P.O. Box 88
1899 Route 414 N
Waterloo, NY 13165, USA
FFL Firearms Dealer

LUCKY'S
P.O. Box 549
Weedsport, NY 13166, USA
FFL Firearms Dealer

THE RELOADER
141 Gertrude Street
North Syracuse, NY 13212, USA
FFL Firearms Dealer

C. J. ENTERPRISE
100 Woodland Drive
N. Syracuse, NY 13212, USA
FFL Firearms Dealer
CONTACT: C. Nuss

GILL'S PLACE
108 Wells Avenue
N. Syracuse, NY 13212, USA
Firearms, ammo, special equipment for firearms, night sights, special sights.
S3; 4;

CONTACT: Bill Gillmore
SALES DEPT: Bill Gillmore
PURCHASING DEPT: Bill Gillmore

CHARLES S. CRESSWELL
1054 South St., RD 2
Clinton, NY 13323, USA
FFL Firearms Dealer

SILLIMAN ENTERPRISES
Gridley Road
Deansboro, NY 13328, USA
FFL Firearms Dealer

THOMAS H. WELSH
6239 Rt. 291, Box 390-A
Marcy, NY 13403, USA
FFL Firearms Dealer

SJP SHOOTERS SUPPLY
744 Blandina Street
Utica, NY 13501, USA
FFL Firearms Dealer
CONTACT: S. J. Perrifano, Jr.

FRANK'S GUNS
2107 Whitesboro Street
Utica, NY 13502, USA
FFL Firearms Dealer

SALVAGE I
RD 2, Box 1117
Alton, NY 13730, USA
FFL Firearms Dealer

WEST MANUFACTURING CO.
301 Oak Hill Ave., Suite 2
Edicott, NY 13760, USA
FFL Firearms Dealer
CONTACT: Marty Ziskind

V. Z. ENTERPRISES
Rt. 1, Box 198
Richford, NY 13835, USA
FFL Firearms Dealer

VILLAGE GUNS
81 Liberty Street
Walton, NY 13856, USA
FFL Firearms Dealer
CONTACT: Charles Turner

ROBERT J. KEIBEL
P.O. Box 166
367 Walter Road
Whitney Point, NY 13862, USA
FFL Firearms Dealer

A.P.J. UNLIMITED
60 Grant Street
Depew, NY 14043, USA
FFL Firearms Dealer

ROBERT KORALEWSKI
226 Bryant
N. Tonawanda, NY 14120, USA
FFL Firearms Dealer

KUSTOM KRAFT KREATIONS, INC.
P.O. Box 307
Buffalo, NY 14223, USA
FFL Firearms Dealer

MILITARY FIREARMS

JAMES A. ANZALONE
136 Fairfield Avenue
Tonawanda, NY 14223, USA
FFL Firearms Dealer

TROPHY ROOM
47 Edlewood Drive
West Seneca, NY 14224, USA
FFL Firearms Dealer
CONTACT: Len Bogumil

JAMES WILKINS
6306 Redman Road
Brockport, NY 14420, USA
FFL Firearms Dealer

JOHN W. REICHERT
39 Black Watch Trail
Fairport, NY 14450, USA
FFL Firearms Dealer

BUB'S GUN SHOP
P.O. Box 149
745 Alderman Road
Macedon, NY 14502, USA
FFL Firearms Dealer

JOHN H. SCHMITT
227 Longview Drive
Webster, NY 14580, USA
FFL Firearms Dealer

RSR WHOLESALE GUNS,INC.
P.O. Box 15395
552 Avis Street
Rochester, NY 14615, USA
716-458-8820
Firearms

ROBERT CARL
24 Sandhurst Drive
Rochester, NY 14617, USA
FFL Firearms Dealer

MOHAWK SPORTING ARMS
2435 Chili Avenue
Rochester, NY 14624, USA
FFL Firearms Dealer

R. T. PONESS
155 Brooklea Drive
Rochester, NY 14624, USA
FFL Firearms Dealer

AJ WHOLESALE GUN DISTRIBUTORS
P.O. Box 25469
Rochester, NY 14625, USA
(716)-227-7280
Firearms

D & B GUNS & SUPPLIES
P.O. Box 163
5271 Wolf Run Road
Campbell, NY 14821, USA
FFL Firearms Dealer

JOHN HARLEY,JR.
RD 2, Box 69
Campbell, NY 14821, USA
FFL Firearms Dealer

AL ROB'S GUN SHOP
RD 2, Box 190, Searsburg Road
Trumansburg, NY 14886, USA
FFL Firearms Dealer

THE SHOOTER'S ARK
219 Franklin Street
Watkins Glen, NY 14891, USA
FFL Firearms Dealer

CRAIG'S GUN SHOP
RD 1, Box 174
Woodhill, NY 14898, USA
FFL Firearms Dealer

VACCARO INDUSTRIES
RD 5, Box 187C
Belle Vernon, PA 15012, USA
(412)-929-2190
Firearms & accessories
S1,S11; 8;
CONTACT: Dennis Vaccaro
SALES DEPT: Dennis Vaccaro

J. D. FIREARMS
737 William Drive
Trafford, PA 15085, USA
FFL Firearms Dealer
CONTACT: John Dengler

BILL TUCKER
P.O. Box 351
W. Mifflin, PA 15122, USA
FFL Firearms Dealer

FRED HRABAK
249 Poplar Avenue
Monroeville, PA 15146, USA
FFL Firearms Dealer

SPORTSMEN'S SUPPLY & SERVICE
2321 Haymaker Road
Monroeville, PA 15146, USA
FFL Firearms Dealer
CONTACT: James M. Aakenase

CLARKE D. EDENHART
603 Hemlock Street
Pittsburgh, PA 15202, USA
FFL Firearms Dealer

ISLAND FIREARMS,INC.
7500 Grand Avenue
Neville Island, PA 15225, USA
FFL Firearms Dealer

RONALD P. LINHART
5054 McAnnulty Road
Pittsburgh, PA 15236, USA
FFL Firearms Dealer

SETTER ARMS
7856 Saltsburg Road
Pittsburgh, PA 15239, USA
FFL Firearms Dealer

WILLIE'S WEAPONS
P.O. Box 16283
Pittsburgh, PA 15242, USA
FFL Firearms Dealer
CONTACT: W. F. Stilts,Jr.

D & D GUNS
Thomas-Eighty Four Road
RD 3, Box 194
Eighty Four, PA 15330, USA
FFL Firearms Dealer

DAVID PERSHING,JR.
578 Shuey Avenue
Greensburg, PA 15601, USA
FFL Firearms Dealer

JAMES E. FERACE
503 Arthur Place
Greensburg, PA 15601, USA
FFL Firearms Dealer

FRONTIERLAND
RD 1, Box 207
Hunker, PA 15639, USA
FFL Firearms Dealer

ARM BLACK POWDER PRODUCTS
156 Lincoln Way West
Jeannette, PA 15644, USA
(412)-527-5590
Black powder guns, kits, & accessories.
CONTACT: Bob Davis, Owner

SALANDRE SPORTS SHOP
RD 2, Box 419
Latrobe, PA 15650, USA
FFL Firearms Dealer

A & M GUNSMITHS
RD 3, Box 360
Leechburg, PA 15656, USA
FFL Firearms Dealer

RON CALIGUIRE & ASSOCIATES
P.O. Box 964
207 Oak Street
Mt. Pleasant, PA 15666, USA
FFL Firearms Dealer

JOHN S. EBERT
RD 3, Box 214
New Alexandria, PA 15670, USA
FFL Firearms Dealer

GUN SHOP
Star Route, Box 64
Spring Church, PA 15686, USA
FFL Firearms Dealer
CONTACT: Tom P. Rupert

SPRINGFIELD SPORTERS,INC.
RD #1
Penn Run, PA 15765, USA
412-254-2626
Firearms

J. DAVID TRUBY
RD #3, Box 15-B
Shelocta, PA 15774, USA
(412)-726-5140
Automatic weapons & sound suppressors.
S3,S6,S11,S13; 3;
SALES DEPT: J. David Truby
Full auto weapons and suppressors at fair prices
from a known name you can trust to be honest.

SPORTING GOODS DISCOUNTERS
770 Goucher Street
Johnstown, PA 15905, USA
FFL Firearms Dealer

MILITARY FIREARMS

JOHN'S FIREARMS
RD 4, Box 15
Ebensburg, PA 15931, USA
FFL Firearms Dealer

DENNIS BLOOM
RD 2, Box 301
Hollsopple, PA 15935, USA
FFL Firearms Dealer

VALENCIA POLICE EQUIPMENT & SUPPLY
Box 1, RD 2
Valencia, PA 16059, USA
FFL Firearms Dealer
CONTACT: Ray C. Soergel

WEST SIDE GUN SHOP
84 North 3rd Street
Greenville, PA 16125, USA
FFL Firearms Dealer

ACME FIREARMS
285 Sherma Avenue
Sharow, PA 16146, USA
FFL Firearms Dealer
CONTACT: Arthur Widmyer

LOWVILLE ARMS
11170 Firethorn Road
Wattsburg, PA 16442, USA
FFL Firearms Dealer
CONTACT: Gary Kanis

ODROSKY'S ARMS
935 Brisbin Street
Houtsdale, PA 16651, USA
FFL Firearms Dealer
CONTACT: D. L. Odrosky

THE TIME MACHINE
Star Rt. 1, Box 53
Lewis Run, PA 16738, USA
(814)-362-2642
Automatic weapons & accessories.
S6; 9;
Catalogue available. Bi-annually.
CONTACT: Greg Souchik
We carry a complete line of automatic weapons and accessories for police agencies and qualified individuals and collectors. Inventory changes weekly.

GRICE WHOLESALE
P.O. Box 1028
Clearfield, PA 16830, USA
(814)-765-9334 (814)-765-2601
Firearms and accessories.

JEFFREY L. KNOPP
P.O. Box 187, Main Street
Kylertown, PA 16847, USA
FFL Firearms Dealer

TERRY'S GUN SHOP
RD #2, Box 172
Wellsboro, PA 16901, USA
717-724-7918
Firearms dealer & gun show promoter.

HITCHCRAFT
235 Green Lane Drive
Camp Hill, PA 17011, USA
FFL Firearms Dealer

ROSEMONT ARMS
3804 Rosemont Avenue
Camp Hill, PA 17011, USA
FFL Firearms Dealer

LOVELL'S LONGRIFLE SHOP
827 S. Humer Street
Enola, PA 17025, USA
( )-732-0283
Muzzleloader firearms, supplies, parts, kits, and accessories.
CONTACT: Gerald L. Lovell, Outfitter

RON SHIRK'S SHOOTERS SUPPLIES
R. D. #2, Box 1775
Lebanon, PA 17042, USA
Firearms

THE SECOND HAND HUNTER
121 East Weidman Street
Lebanon, PA 17042, USA
FFL Firearms Dealer

GEORGES OUTDOOR STORE
100 S. Front Street
Wormleysburg, PA 17043, USA
FFL Firearms Dealer

BURROUGHS BUSINESS SYSTEMS
P.O. Box 543
Mt. Gretna, PA 17064, USA
FFL Firearms Dealer

FORT CHAMBERS GUN SHOP
3292 Black Gap Road
Chambersburg, PA 17201, USA
(717)-263-2223
Muzzleloaders, kits, parts, & accessories.

CONOCOCHEAGUE TRADE COMPANY
1482 Black Gap Road
Fayetteville, PA 17222, USA
(717)-352-7212
Black powder, modern guns, gun kits and accessories.

JONH K. BENCHOFF
P.O. Box 156
Shady Grove, PA 17256, USA
717-762-3036
Antique military firearms

CULLISON'S SPORTING GOODS
2275 Mummasburg Road
Gettysburg, PA 17325, USA
FFL Firearms Dealer
CONTACT: Rick C. Cullison

GEM GUNS
8 Hanover Street
Glen Rock, PA 17327, USA
FFL Firearms Dealer

M. F. & C.
RD 2, Box 45
Glen Rock, PA 17327, USA
FFL Firearms Dealer
CONTACT: Edward Lamparski

F. A. SHIPLEY & SONS GUN SHOP
RD 2, Box 7
5426 W. Market Street
Thomasville, PA 17364, USA
FFL Firearms Dealer

CHRIS L. ALTLAND
151 Penwood Road
York, PA 17402, USA
FFL Firearms Dealer

FEDERAL GUN SHOPPE
Mt. Vernon Road
Gap, PA 17527, USA
(717)-442-8578
Custom built Black Powder Guns and accessories. Restoration work.
CONTACT: Paul N. Allison, Jr., Prop.

R-P SALES
P.O. BOX 387
Leola, PA 17540, USA
FFL Firearms Dealer

JOHN E. KUHNS
737 E. Marion Street
Lancaster, PA 17602, USA
FFL Firearms Dealer

G. R. 'BUD' STAUFFER
23 Glendale Drive
Lancaster, PA 17602, USA
(717)-392-7767
New & used firearms bought, sold & traded; Gunsmithing; Accessories.

VICTOR J. VARE
RD 2, Box A 283
Montoursville, PA 17754, USA
FFL Firearms Dealer

THOMAS A. STREINER
RD 2, Box 447
Muncy, PA 17756, USA
FFL Firearms Dealer

HENRY A. TRUSLOW
P.O. Box 40
Sunbury, PA 17801, USA
Percussion revolver collector

D. MOTTO
R. D. 3, Box 324
Catawissa, PA 17820, USA
Custom muzzleloading rifles.

R. A. POLLOCK, JR.
R. D. 1
Danville, PA 17821, USA
(717)-275-1125
Custom muzzleloading rifles.

CLARK'S GUNS
514 S. Market Street
Selinsgrove, PA 17870, USA
FFL Firearms Dealer
CONTACT: c/o Arbogast & Son

ANTHRACITE ORDNANCE
200 Mahantongo Street, Ste. 23
Pottsville, PA 17901, USA
(717)-622-6880
Class 1 and Class 3 firearms.
CONTACT: James J. Davidson

MILITARY FIREARMS

BIXLER'S GUNS
615 Seminole Street
Bethlehem, PA 18015, USA
FFL Firearms Dealer

RANDY KOPPLK
RD 1, Box 582
Breinigsville, PA 18031, USA
FFL Firearms Dealer

LINDA S. WARFEL
12 Little Creek Circle
Breinigsville, PA 18031, USA
FFL Firearms Dealer

ARDLE'S WEAPONS SPECIALIST
Rt. 248
Cherryville, PA 18035, USA
(215)-767-8430
Class I & III firearms and supplies.
CONTACT: Bob Ardle

BILL'S ARCO
4755 Main Street
Whitehall, PA 18052, USA
FFL Firearms Dealer

NORMAN P. BACHMAN,JR.
RD 1, Box 12
New Tripoli, PA 18066, USA
FFL Firearms Dealer

NEMETH'S ANTIQUES
Rt. 309, Box 172
Orefield, PA 18069, USA
FFL Firearms Dealer

HERRING'S SPORTING GOODS
RD 1, Rt. 940, Box 1760
Freeland, PA 18224, USA
717-636-2424 717-636-2929
Firearms

D & M GUNS & SUPPLY
RD 1, Box 1735
Freeland, PA 18224, USA
FFL Firearms Dealer
CONTACT: David J. Stutzman

TRIAD SECURITY SYSTEMS
RD 1, Box 351
Moscow, PA 18444, USA
FFL Firearms Dealer

TIMS SPORT CENTER
Box 1416
Scranton, PA 18501, USA
FFL Firearms Dealer

ROBERT W. HART & SON,INC.
401 Montgomery Street
Nescopeck, PA 18635, USA
(717)-752-3481
Muzzleloader supplies and equipment.

MICHAEL A. KALINOSKY
238 Academy Street
Luzerne, PA 18709, USA
FFL Firearms Dealer

WRIGHT ASSOCIATES
P.O. Box 56, Rt. 220
Milan, PA 18831, USA

FFL Firearms Dealer

TARGET WORLD,INC.
Rt. 202 & County Line Road
Chalfont, PA 18914, USA
FFL Firearms Dealer

NATIONWIDE SPORTS DISTRIBUTORS
70 James Way
Southampton, PA 18966, USA
215-322-2054
Firearms

TRADE-WELL DISTRIBUTERS,INC.
PO Box 228
Tylersport, PA 18971, USA
FFL Firearms Dealer

DONALD F. EISMANN
Box 168
Abington, PA 19001, USA
215-659-1030
Remington firearms collector

AMERICAN GUNS INTERNATIONAL
3338 Richlieu Road, Apt F-92
Bensalem, PA 19020, USA
(215)-638-1591
Buy, sell & trade. Guns, militaria, reloads, gunsmithing.
CONTACT: Bob Sperber

R. V. WEAPONS COMPANY
2311 Broadway
Hatboro, PA 19040, USA
FFL Firearms Dealer

BEAL'S BULLETS
170 W. Marshall Road
Lansdowne, PA 19050, USA
215-259-1220
Firearms
CONTACT: R. S. Beal,JR.

JOHN F. DUSSLING
P.O. Box 63
Media, PA 19063, USA
-566-0902
Colt collector

DENNIS A. TODD
10 S. Harwood Avenue
Upper Darby, PA 19082, USA
215-449-0575
Automatic weapons

DEAN MIOLI
218 N. Lynn Blvd.
Upper Darby, PA 19082, USA
FFL Firearms Dealer

PEPPERBOX
P.O. Box 768
Havertown, PA 19083, USA
Pepperboxes.
No list, write your needs.

ACTION ARMS LTD.
P.O. Box 9573
Philadelphia, PA 19124, USA
Firearms

RICHARD T. FOX

6274 Large Street
Philadelphia, PA 19149, USA
FFL Firearms Dealer

TRADIN' TOM'S
3535 Cottman Avenue
Philadelphia, PA 19149, USA
FFL Firearms Dealer

HARRY V. PAGNOTTI
920 Vienna Avenue
Fairview Village, PA 19403, USA
(215)-631-5592
Military weapons, modern guns, new & used. Military related items.

JEAN CROSS GUNSMITH
275 Elbow Lane
Chester Springs, PA 19425, USA
(215)-827-9244
Blackpowder arms & parts.

TRUE GRIT GUN SHOP
1245 Clymer Road
Hatfield, PA 19440, USA
FFL Firearms Dealer

GOULD'S GUNSMITHING SERVICE
935 Parkview Drive
Phoenixville, PA 19460, USA
FFL Firearms Dealer

DIXON MUZZLELOADING SHOP,INC.
R.D. 1, Box 175
Kempton, PA 19529, USA
(215)-756-6271
Muzzleloaders, parts, kits, & supplies.

SCHEARER'S GUNS & AMMO
2117 Frush Valley Road
Temple, PA 19560, USA
FFL Firearms Dealer

GEORGE'S GUNS
738 Ritter Street
Reading, PA 19601, USA
FFL Firearms Dealer
CONTACT: George Ehrgood

FISHERS STEREO
2807 Perkiomen Avenue
Mt. Penn, PA 19606, USA
FFL Firearms Dealer

THE SMOKEPOLE SHOP INC.
3706 Old Capitol Trail
Marshallton, DE 19808, USA
(302)-998-3554
Black powder firearms, accessories & supplies.
CONTACT: Gary W. Kimball, President

V5 ENTERPRISES
RR 2, Box 160-I
Indian Head, MD 20640, USA
FFL Firearms Dealer
CONTACT: Arthur A. Van Hecke

RICHARD L. GETZ
Route 2, Box 134
Pomfret, MD 20675, USA
FFL Firearms Dealer

MILITARY FIREARMS

AMHERST
9747 Washington Blvd.
Laurel, MD 20707, USA
301-490-5394
Firearms

CHARLIE'S GUN SHOPPE
5525 Volta Avenue
Bladensburg, MD 20710, USA
FFL Firearms Dealer

EAST COAST HUNTING SUPPLIES
P.O. Box 1044
Bowie, MD 20715, USA
FFL Firearms Dealer
CONTACT: Jim Corbett

MILLER'S GUNWORKS
708 Kelly Road
Fort Washington, MD 20744, USA
FFL Firearms Dealer

DISCOUNT GUNS
7720 Wisconsin Avenue, B-103
Bethesda, MD 20814, USA
FFL Firearms Dealer

J & B ENTERPRISES
P.O. Box 805
Gaithersburg, MD 20877, USA
FFL Firearms Dealer

B. C. GUNWORKS
11613 Collegeview Drive
Wheaton, MD 20902, USA
FFL Firearms Dealer
CONTACT: Robert Keese

BERNARD BOX
241 Carroll Road
Glen Burnie, MD 21061, USA
FFL Firearms Dealer

ARMAS, LTD.
5235 Hydes Road
Hydes, MD 21082, USA
FFL Firearms Dealer
CONTACT: Napoleon Fernando

PIONEER POLICE PRODUCTS
1648 Millersville Road
Anne Arundel, MD 21108, USA
FFL Firearms Dealer

M & T SPORTING GOODS, INC.
1707 West Bancrift Lane
Crofton, MD 21114, USA
FFL Firearms Dealer
CONTACT: R. H. Myers

SHOOTER'S DISCOUNT
417 London Court
Westminster, MD 21157, USA
FFL Firearms Dealer
CONTACT: Louis B. Nichols,Jr.

CANNON,LTD.
1316 Lafayette Avenue
Baltimore, MD 21207, USA
Artillery carriages & cannons.
S1,S4; 5;
Catalogue available.
SALES DEPT: P. Miller
PURCHASING DEPT: P. Miller
We build quality and guaranteed black powder cannon and mortar reproductions. We sell cannon hardware, wooden parts and wheels in 1/2, 3/4, and full scale.

CHESAPEAKE SPORTING GOODS
2120 Maryland Avenue
Baltimore, MD 21218, USA
FFL Firearms Dealer

JERRY'S LOCK STOCK & BARREL
1907 Old Eastern Avenue
Baltimore, MD 21221, USA
FFL Firearms Dealer

H. F. S.
1302 Seabright Drive
Cape St. Claire, MD 21401, USA
1903 Springfield rifle parts dealer. Restoration work in $15.00 per hour.
; 4;
Catalogue available.
CONTACT: Larry R. Hartman
SALES DEPT: Larry R. Hartman
PURCHASING DEPT: Larry R. Hartman

JOHN DIVICO SPORTS
Route 8, Box 232
Cumberland, MD 21502, USA
FFL Firearms Dealer

PLUTSCHAK'S
P.O. Box 1516
Easton, MD 21601, USA
FFL Firearms Dealer

DEAN ENTERPRISES
4976 Lingamore Woods Drive
Monrovia, MD 21770, USA
FFL Firearms Dealer
CONTACT: Phillip B. Dean

PROTEUS PRODUCTS
7561 Morgan Road
Woodbine, MD 21797, USA
FFL Firearms Dealer
CONTACT: William A. McKnight,Jr.

HENDERSON FUR COMPANY
P.O. Box 202
Marion Station, MD 21838, USA
FFL Firearms Dealer

MIKE BURNETT
Rt. 1, Box 845
Amissville, VA 22002, USA
FFL Firearms Dealer

MICHAEL CINNAMON
6108 O'Day Drive
Centreville, VA 22020, USA
FFL Firearms Dealer

HECKLER & KOCH,INC.
14601 Lee Road
Chantilly, VA 22021, USA
703-631-2800
Firearms manufacturer & retailer

RODNEY R. SANDEL
4803 Jennichelle Court
Fairfax, VA 22032, USA
FFL Firearms Dealer

ROBERT M. & MICHAEL P. O'BOYLE
6609 Quinten Court
Falls Church, VA 22043, USA
FFL Firearms Dealer

JON N. LECHEVET
9445 Lapstrake Lane
Burke, VA 22105, USA
FFL Firearms Dealer

EAGLEHEAD ARSENAL
5711 Sudley Road
Manassas, VA 22110, USA
Artillery carriages
CONTACT: Dick Katter

PAUL SALENZ
P.O. Box 2635
Manassas, VA 22110, USA
FFL Firearms Dealer

DAVID CONDON,INC.
109 E. Washington Street
Middleburg, VA 22117, USA
Antique & sporting arms.
6 issues of catalogue $15.00.

BILLY'S GUNS & SUPPLIES
6460 Franconia Road
Springfield, VA 22150, USA
FFL Firearms Dealer

JOHN MORRIS
7204 Danford Lane
Springfield, VA 22152, USA
Cannons & artillery material collector.
Write for details.

FIBICH'S FIREARMS
7126 Hamor Lane
Springfield, VA 22153, USA
FFL Firearms Dealer
CONTACT: A. Lawrence Fibich

ELMAR P. BORKOWSKI
225 Laura Anne Court
Sterling, VA 22170, USA
FFL Firearms Dealer

THE POWER COMPANY
120 Penny Lane
Sterling, VA 22170, USA
FFL Firearms Dealer

CHARLES L. GRANT
P.O. Box 647
71 S. Fifth Street
Warrenton, VA 22186, USA
(703)-347-3702
Class III weapons, military ammunition.
S4;
CONTACT: Charles L. Grant
SALES DEPT: Charles L. Grant
PURCHASING DEPT: Charles L. Grant
Original WWII machineguns, parts and ammunition, restoration of military vehicles.

MILITARY FIREARMS

TOIGOS DISCOUNTS
1203 Rope Court
Woodbridge, VA 22191, USA
FFL Firearms Dealer
CONTACT: Mark M. Toigo

THE HOUSE OF GRIER
13201 Conrad Court
Woodbridge, VA 22191, USA
FFL Firearms Dealer

THOMAS RINEHART
6323 - 22nd Road North
Arlington, VA 22205, USA
FFL Firearms Dealer

DONALD ELDER
3321-B South Wakefield St.
Arlington, VA 22206, USA
FFL Firearms Dealer

JIM TUCKER
4201 31st St., Apt 806
Arlington, VA 22206, USA
FFL Firearms Dealer

GAC,LTD
1510 North 12th St. PH3
Arlington, VA 22209, USA
FFL Firearms Dealer

P & L FIREARMS COMPANY
3243 Gunston Road
Alexandria, VA 22302, USA
FFL Firearms Dealer

E. J. WIELICZKIEWICZ
8839 Camfield Drive
Alexandria, VA 22308, USA
FFL Firearms Dealer

SPORTSMAN'S TRADE & CONSULTING SERVICE
6482 Silver Ridge Circle
Alexandria, VA 22310, USA
FFL Firearms Dealer

RAYMAR PRODUCTIONS
P.O. Box 10252
Alexandria, VA 22310, USA
FFL Firearms Dealer

RALPH H. LOVETT,SR.
10807 Leavells Road
Fredericksburg, VA 22401, USA
FFL Firearms Dealer

J. A. JACKSON
Rt. 7, Box 130
Spotsylvania, VA 22553, USA
FFL Firearms Dealer

PROFESSIONAL ARMAMENT SERVICES
244 Raintree Blvd.
Stafford, VA 22554, USA
FFL Firearms Dealer

W. J. HELM
311 Fairmont Avenue
Winchester, VA 22601, USA
WWI machine gun & ordnance collector.

CARLOS VILLAR-GOSALGEZ

P.O. Box 192
Ivy, VA 22945, USA
FFL Firearms Dealer

WHITAKER OIL COMPANY,INC.
N. Bayard Avenue
Waynesboro, VA 22980, USA
FFL Firearms Dealer

JOHN'S GUNS
P.O. Box 802
Columbia, VA 23038, USA
Search for special weapons, Class III firearms.
S1,S11,S12; 2;
CONTACT: John S. Fornaro
SALES DEPT: John Fornaro
PURCHASING DEPT: John Fornaro
We are a new business specializing in military and NFA weapons. We sell modern arms on a special order basis.

GLOUCESTER SHOOTERS SUPPLY
Rt. 1246, Box 38
Hayes, VA 23072, USA
FFL Firearms Dealer
CONTACT: Ken Lewis

F. THOMAS CROSS
3906 Cloyd Avenue
Richmond, VA 23221, USA
FFL Firearms Dealer

KENSINGTON ARMS
3019 Kensington Avenue
Richmond, VA 23221, USA
FFL Firearms Dealer

ROBERT D. STINNETTE
1101 North Hamilton Street, #8
Richmond, VA 23221, USA
FFL Firearms Dealer

HOWARD L. ARNOLD
2507 Lincoln Avenue
Richmond, VA 23228, USA
FFL Firearms Dealer

GENTRY L. JOHNSON
1117 Kings Mill Court
Chesapeake, VA 23320, USA
FFL Firearms Dealer

DAVE'S GUN SHOP
4605 Chickadee Street
Chesapeake, VA 23322, USA
FFL Firearms Dealer

F. E. WOOD,JR.
Box 6015, Crittenden Sta.
Suffolk, VA 23433, USA
FFL Firearms Dealer

STEVE'S GUNS
1317 Whitemarsh Road
Suffolk, VA 23434, USA
FFL Firearms Dealer

FIREARMS FOR LESS
921 Wildflower Court
Virginia Beach, VA 23452, USA
FFL Firearms Dealer

JAMES M. C. BONAVITA

3924 Old Farm Lane
Virginia Beach, VA 23452, USA
FFL Firearms Dealer

BLANDFORD'S GUN SHOP
5636 Sedgemoor Road
Virginia Beach, VA 23455, USA
FFL Firearms Dealer

BILL BURGESS
4790 Christopher Arch
Virginia Beach, VA 23464, USA
FFL Firearms Dealer

STUART C. MAGUIRE
1468 Kempsville Road
Norfolk, VA 23502, USA
FFL Firearms Dealer

LARRY'S GUN SHOP
1215 - 74th Street
Newport News, VA 23605, USA
FFL Firearms Dealer

PAUL W. CROCKETT,JR.
1121 Wormley Creek Drive
Yorktown, VA 23690, USA
FFL Firearms Dealer

JAMES K. POWELL
817 Burgess Street
Martinsville, VA 24112, USA
FFL Firearms Dealer

G.W.'s GUN SHOP
Rt. 4, Box 239
Ridgeway, VA 24148, USA
FFL Firearms Dealer

SAM'S GUN SHOP,INC.
415 Commonwealth Avenue
Bristol, VA 24201, USA
(703)-466-8903
Guns, sporting goods, & military items.
S1,S3,S11,S12; 10;
Catalogue available. Annually.
CONTACT: Sam Sorah
SALES DEPT: Sam Sorah & George Warren,III
PURCHASING DEPT: Sam Sorah
We stock a full line of handguns, shotguns & rifles, plus our 700 page catalog for those hard to find items. We're Virginia's Uzi headquarters.

T. MARK FLEENOR
Rt. 3, Box 146
Abingdon, VA 24210, USA
FFL Firearms Dealer

JAMES W. HITE
808 Monroe Street
Staunton, VA 24401, USA
FFL Firearms Dealer

J. D. YOUMANS
142 Riverview Drive
Verona, VA 24482, USA
FFL Firearms Dealer

GILD'S GUN SHOP
7110 Richland Drive
Lynchburg, VA 24502, USA
FFL Firearms Dealer

MILITARY FIREARMS

VAUGHN ENTERPRISES
Rt. 3, Box 585-D
Rustburg, VA 24588, USA
FFL Firearms Dealer

LAWRENCE A. FRENCH, II
Rt. 1, Box 259-H
Bluefield, VA 24605, USA
FFL Firearms Dealer

STEVE'S TRADING POST
2118 Jefferson Street
Bluefield, WV 24701, USA
FFL Firearms Dealer

JOHN D. PROFFITT
P.O. Box 246
Leslie Br. Rd.
Gary, WV 24836, USA
FFL Firearms Dealer

WILLIAM D. LOWRY, JR.
Rt. 1, Box 174
Forest Hill, WV 24935, USA
FFL Firearms Dealer

CHET'S SHOOTING SUPPLIES
P.O. Box 182
Jone's Branch
Jeffrey, WV 25114, USA
FFL Firearms Dealer

LOCK 'N' LOAD
Rt. 1, Box 678
St. Albans, WV 25177, USA
FFL Firearms Dealer

AJ's MILITARY SURPLUS
520 D Street
South Charleston, WV 25303, USA
FFL Firearms Dealer

CHEMCITY FIREARMS
2108 Geary Road, South
Charleston, WV 25303, USA
FFL Firearms Dealer

RUSSELL SAYEGH
5243 Dalewood Drive #95
Charleston, WV 25313, USA
FFL Firearms Dealer

THE 45 SHOP
340 Goff Mt. Road
Charleston, WV 25313, USA
FFL Firearms Dealer
CONTACT: Jim Ingram

ADKINS GUNS & RELOADING SUPPLIES
Rt. 2, Box 202, Trace Creek Rd
Hamlin, WV 25523, USA
FFL Firearms Dealer

BECKLEY HONDA
Plaza S/C
Beckley, WV 25801, USA
FFL Firearms Dealer
CONTACT: Bill R. Via

SNUFFER ENTERPRISES
116 Queensbury Street
Beckley, WV 25801, USA
FFL Firearms Dealer

DOVE'S
P.O. Box 138-A
105 N. Blue Jay Drive
Beaver, WV 25813, USA
FFL Firearms Dealer

FIL PISHNER
Rt. 9, Box 104-L
Beaver, WV 25813, USA
FFL Firearms Dealer

FRENCH A. MAYNOR
Box 15
Pax, WV 25904, USA
FFL Firearms Dealer

COOK ENTERPRISES
P.O. Box 28
Quinwood, WV 25981, USA
FFL Firearms Dealer

ROGER E. SPARKS
Rt. 1, Box 195A, Lot #4
Wheeling, WV 26003, USA
FFL Firearms Dealer

NORM'S PLACE
1917 Green Hills
Glendale, WV 26038, USA
FFL Firearms Dealer

JOHN W. KOVACS, JR.
Box 139
Valley Grove, WV 26060, USA
FFL Firearms Dealer

STEVE'S
128 Belmont Street
Weirton, WV 26062, USA
FFL Firearms Dealer
CONTACT: Steve Pawlock

M. G. BOGGESS
203 Gibbs Street
Ravenwood, WV 26164, USA
FFL Firearms Dealer

ROBERT D. GARRISON
Rt. 1, Box 120
Buckhannon, WV 26201, USA
FFL Firearms Dealer

GAULEY RIVER TRADING COMPANY
Box 177
Bolair, WV 26288, USA
FFL Firearms Dealer
CONTACT: John B. Bell

PAUL'S GUN & REPAIR
1621 Hamill Avenue
Clarksburg, WV 26301, USA
FFL Firearms Dealer

ENERGY SPECIALTIES
Rt. 2, Box 39
Shinnston, WV 26431, USA
FFL Firearms Dealer

NEELEY'S GUNS & AMMO
Rt. 1, Box 265
West Union, WV 26456, USA
FFL Firearms Dealer

DAVID S. HYLTON
Rt. 7, Box 645
Morgantown, WV 26505, USA
FFL Firearms Dealer

BELLOTTE HUNTING & FISHING
Rt. 13, Box 62
Morgantown, WV 26505, USA
FFL Firearms Dealer

WILLIAM E. DEVINE
Rt. 4, Box 11
Mannington, WV 26582, USA
FFL Firearms Dealer

RICHARD LEMAY
P.O. Box 430
Ft. Ashby, WV 26719, USA
FFL Firearms Dealer

JERRY HOBSON
8436 Stanwick Drive
Tobaccoville, NC 27050, USA
FFL Firearms Dealer

J. S. HINSHAW GUN SALES/REPAIR
Rt. 1, Box 444
Franklinville, NC 27248, USA
FFL Firearms Dealer

CAM'S GUN & KNIFE SHOP
538 Linville Road
Kernersville, NC 27284, USA
FFL Firearms Dealer

CHIGGER RIDGE RELOADING
Rt. 10, Box 147
Reidsville, NC 27320, USA
FFL Firearms Dealer

TIM TALLEY
Rt. 5, Box 93
Reidsville, NC 27320, USA
FFL Firearms Dealer

RAY'S GUN & AMMO SUPPLY
2823 John Rosser Road
Sanford, NC 27330, USA
FFL Firearms Dealer

G. F. 'CHIP' MEADOWS
1106 Parkwood Drive
Siler City, NC 27344, USA
FFL Firearms Dealer

BUD ELVIN
P.O. Box 172
Bahama, NC 27503, USA
FFL Firearms Dealer

OWEN R. EDWARDS
Rt. 3, Box 290
Louisburg, NC 27549, USA
FFL Firearms Dealer

JOHN R. DAVIS
10512 Tree Bark Court
Raleigh, NC 27612, USA
FFL Firearms Dealer

MILITARY FIREARMS

SPORTSMANS SUPPLY
Rt. 2, Box 567
Durham, NC 27705, USA
FFL Firearms Dealer
CONTACT: Kent Goodwin

TOM PEARSE
154 Montrose
Durham, NC 27707, USA
FFL Firearms Dealer

COCKRELL'S FIREARMS & ACCESSORIES
Rt. 1, Box 326 A
Rocky Mount, NC 27801, USA
FFL Firearms Dealer

THOMAS L. ARVIN
P.O. Box 754
115-A S. Railroad Street
Bethel, NC 27812, USA
FFL Firearms Dealer

SWANKS SPORTING ARMS
21 Westhills
Greenville, NC 27834, USA
FFL Firearms Dealer

POWERS AUTO REPAIR
Rt. 2, Box 196
Grimesland, NC 27837, USA
FFL Firearms Dealer
CONTACT: Hughie C. Powers

BROWN'S GUN SHOP
P.O. Box 328
Littleton, NC 27850, USA
FFL Firearms Dealer

THE ARSENAL
Rt. 3, Box 418-F
Nashville, NC 27856, USA
FFL Firearms Dealer

EUDIS HARPER
P.O. Box 112
Pikeville, NC 27863, USA
FFL Firearms Dealer

JAMES E. PATRICK, JR.
Rt. 2, Box 440-C
Roanoke Rapids, NC 27870, USA
FFL Firearms Dealer

J. A. SHARPE
1114 Sunset Road
Cherryville, NC 28021, USA
FFL Firearms Dealer

ADVANCED WEAPONS SYSTEMS
3116 Olde Creek Trail
Matthews, NC 28105, USA
FFL Firearms Dealer

DAVID L. DIXON
7121 Edgefield Court
Matthews, NC 28105, USA
FFL Firearms Dealer

BUCK RORIE
503 Arborlea Court
Matthews, NC 28105, USA
FFL Firearms Dealer

D & B CRAFTS
402 Rutledge Road
Mount Holly, NC 28120, USA
FFL Firearms Dealer

JERRY W. WOOD
1713 Seifert Cr.
Charlotte, NC 28205, USA
FFL Firearms Dealer

MARTIN'S GUN WORKS
525 Mallard Creek Church Road
Charlotte, NC 28213, USA
FFL Firearms Dealer

EASTOVER GUNS
1728 Eastover Street
Fayetteville, NC 28301, USA
FFL Firearms Dealer
CONTACT: Ronald Matthews

ACE PAWN SHOP
5671 Bragg Blvd.
Fayetteville, NC 28303, USA
FFL Firearms Dealer

F. W. GROSS
P.O. Box 35213
Fayetteville, NC 28303, USA
FFL Firearms Dealer

DAN HENSON
5230 Pala Verde Circle
Fayetteville, NC 28304, USA
FFL Firearms Dealer

HAL IVERSON
39 Congaree Road, Route 3
Carthwage, NC 28327, USA
FFL Firearms Dealer

PATRICKS GUNS & ARCHERY
Rt. 3, Box 234-A
Leland, NC 28451, USA
FFL Firearms Dealer

TEN MILE GUN SHOPPE
Rt. 1, Box 163-B
Maple Hill, NC 28454, USA
FFL Firearms Dealer

MARINE GUN & LOCK SHOP
309 Marine Blvd.
Jacksonville, NC 28540, USA
FFL Firearms Dealer

CHUCK'S REPAIR SERVICE
312 Neuse Forest Avenue
New Bern, NC 28560, USA
FFL Firearms Dealer

COST-PLUS GUNS
Rt. 1, Box 146 H
New Bern, NC 28560, USA
FFL Firearms Dealer
CONTACT: Tim Gibbons

STAN MARSNICK
542 10th St. Dr. N.W.
Hickory, NC 28601, USA
FFL Firearms Dealer

G. R. SHELL
P.O. Box 3004
840 First Avenue, S.W.
Hickory, NC 28601, USA
FFL Firearms Dealer

RICHARD HAYES
P.O. Box 51
Hays, NC 28635, USA
FFL Firearms Dealer

JERRY'S GUNSHOP
197 East Street, C12-7
Hudson, NC 28638, USA
FFL Firearms Dealer

PIEDMONT HARDWARE CO.
10 E. Main Street
Maiden, NC 28650, USA
FFL Firearms Dealer

JELLISON'S FIREARMS
133 Sunrise Terrace
Black Mountain, NC 28711, USA
FFL Firearms Dealer
CONTACT: Gregg Jellison

JEROME R. FINKE
Route 6 Box 706, #C1
Marion, NC 28752, USA
FFL Firearms Dealer

DANIELS REPAIR SERVICE
Rt. 2, Box 475-B
Spruce Pine, NC 28777, USA
FFL Firearms Dealer
CONTACT: Marshall E. Daniels

RON'S GUNS & AMMO SUPPLY
Bee Ridge Road
Rt. 5, Box 1061-A
Asheville, NC 28803, USA
FFL Firearms Dealer
CONTACT: Ronald T. Blaine

FRANCES JOHNSON
Rt. 3, Box 307
Batesburg, SC 29006, USA
FFL Firearms Dealer

ANTHONY E. ROMYAK
119 Dickert Drive
Lexington, SC 29072, USA
FFL Firearms Dealer.

WOODY'S PAWN SHOP
303 Russell Street, S.W.
Orangeburg, SC 29115, USA
FFL Firearms Dealer

RICHARD H. WELL
613 Baldwin Avenue
Sumter, SC 29150, USA
FFL Firearms Dealer

DONALD T. MARTIN
2017 Dew Avenue
West Columbia, SC 29169, USA
FFL Firearms Dealer

DANIEL HOOK
1805 Devine St., Suite 913
Columbia, SC 29201, USA
FFL Firearms Dealer

MILITARY FIREARMS

WILDERNESS SPORTSMAN SUPPLY
163 Bristown Lane
Spartanburg, SC 29301, USA
FFL Firearms Dealer
CONTACT: Dennis Crocker

WILLIAM A. GUNTER
Rt. 1, Box 374, River Rd.
Cowpens, SC 29330, USA
FFL Firearms Dealer

T.U.F. ENTERPRISES
793 Dills Bluff Road
Charleston, SC 29412, USA
FFL Firearms Dealer
CONTACT: Tony G. Foessel

PAINTERS SPORTS
4508 Garwood Street
Ladson, SC 29456, USA
FFL Firearms Dealer
CONTACT: Danny L. Painter

STEPHEN C. PAGE
Rt. 4, Patriot Park #5
Florence, SC 29501, USA
FFL Firearms Dealer

RICK'S GUN BARN
22 Bent Twig Drive
Greenville, SC 29605, USA
FFL Firearms Dealer
CONTACT: Rick Greer

MIKE'S GUN SHOP
PO Box 3460
Greenwood, SC 29646, USA
FFL Firearms Dealer

JOHNNY'S GUNS
Rt. 6, Box 610
Greer, SC 29651, USA
FFL Firearms Dealer

MARION A. TAYLOR
60 N. Greenwood Ave. Extn.
Ware Shoals, SC 29692, USA
FFL Firearms Dealer

BOBBY L. ANGLIN
P.O. Box 3543 CRS
1168 Ebibport Road
Rock Hill, SC 29731, USA
FFL Firearms Dealer

PALMETTO ANTIQUES
Hwy. 301 & 321
Rt. 1, Box 75-E
Ulmer, SC 29849, USA
FFL Firearms Dealer

WILLEO CREEK ARMORY
5269 West Bank Drive N.E.
Marietta, GA 30067, USA
FFL Firearms Dealer
CONTACT: Charles H. Cahan

JOSEPH A. KRISS
350 Briarwood Court
Marietta, GA 30067, USA
FFL Firearms Dealer

DAVID F. GREENE

P.O. Box 504
Redan, GA 30074, USA
FFL Firearms Dealer

THE GUN ROOM, INC.
2468 Spring Road
Smyrna, GA 30080, USA
(404)-432-2324
Antique & moden firearms & accessories.
S1; 25;
CONTACT: R. N. Kennedy, Jr.
SALES DEPT: R. N. Kennedy, Jr.
PURCHASING DEPT: R. N. Kennedy, Jr.
Antique & modern firearms, new & used, accessories, swords, bayonets, repair service, taxidermy, ammo, knives, consignments, appraisals, and reloading supplies.

DELTA ARMS & AMMUNITION
6355 Memorial Drive L2-203
Stone Mountain, GA 30083, USA
FFL Firearms Dealer
CONTACT: Thomas L. Harrison

AZTEC INTERNATIONAL LTD.
1256-E Oakbrook Drive
Norcross, GA 30093, USA
(404)-446-2304 (404)-446-2350
Pyrotechnics.
S4,S5,S6; 2;
Catalogue available. Annually.
CONTACT: M. Arent, President
Signal alarms, flares, smoke grenades, simulators, booby traps, firearms accessories, pengun flares, etc.

DEMARCO'S
6528 N.E. Expressway
Norcross, GA 30093, USA
FFL Firearms Dealer

J. D. ENTERPRISES
Rt. 4, Box 267
Cumming, GA 30130, USA
FFL Firearms Dealer

LAZY H ENTERPRISES
Route 9, Box 419
Cumming, GA 30130, USA
FFL Firearms Dealer
CONTACT: Larry Hill

GEORGE W. MOORE
129 13th Street
Cumming, GA 30130, USA
FFL Firearms Dealer

DOUG BALDEN
Rt. 5, Box 295, Mt. Tabor Rd.
Dallas, GA 30132, USA
FFL Firearms Dealer

MIKE'S GUNS & AMMO
Route 6, Box 497-B, Hwy. 92
Dallas, GA 30132, USA
FFL Firearms Dealer

WILDMAN'S CIVIL WAR SURPLUS
2879 North Main Street
Kennesaw, GA 30144, USA
-422-1785
Guns, Books, & Herb Dealer.
List $1.00.

CONTACT: Dent Myers

GUN REPAIR SERVICE
1203-B Calhoun Avenue N.E.
Rome, GA 30161, USA
FFL Firearms Dealer
CONTACT: Jerry W. Branton

SFL
P.O. Box 1111
Conyers, GA 30207, USA
Automatic weapons

SOUTHERN GUNRUNNERS
114 Shannon Chase Lane
Fairburn, GA 30213, USA
FFL Firearms Dealer
CONTACT: Lee Wood

CALDWELL SHOOTERS SUPPLY
1314 Monroe Drive N.E.
Atlanta, GA 30306, USA
FFL Firearms Dealer

RAYMOND K. PADEN
1397 Blashfield Street S.E.
Atlanta, GA 30315, USA
FFL Firearms Dealer

WILLIAM C. SPEIDEL
1866 Independence Square
Atlanta, GA 30338, USA
FFL Firearms Dealer

DANIEL H. WILLARD
1829 Chancery Drive
Chamblee, GA 30341, USA
FFL Firearms Dealer

WILLIAM S. GLISSON
GA. 107 Highway, Route 1
Ailey, GA 30410, USA
FFL Firearms Dealer

CENTURION ARMS, INC.
P.O. Box 1329
Stateboro, GA 30458, USA
FFL Firearms Dealer

JAMES MILNE
814 Laurel Drive
Vidalia, GA 30474, USA
FFL Firearms Dealer

GARY MEEKS
Route 2
Dahlonega, GA 30533, USA
FFL Firearms Dealer

CHRISTY'S MINI SERVICE
P.O. Box 186
Lexington, GA 30648, USA
FFL Firearms Dealer
CONTACT: Lamal Cofer

JR'S GUNS
Rt. 2, Box 103
Chatsworth, GA 30705, USA
FFL Firearms Dealer

MILITARY FIREARMS

DANIEL J. JONES
102 Clark Street
Chickamauga, GA 30707, USA
FFL Firearms Dealer

MARK'S GUN SHACK
Route 3, Box 283-A
Summerville, GA 30747, USA
FFL Firearms Dealer

T & M FIREARMS
2358 Old Alabama Road
Thomaston, GA 30747, USA
FFL Firearms Dealer

K. L. M. GUNWORKS
P.O. Box 92
East Railroad Street
Byron, GA 31008, USA
FFL Firearms Dealer

DAVE'S GUN SHOP
102 Westminister Drive
Eatonton, GA 31024, USA
FFL Firearms Dealer

McDONALD FIREARMS SALES
103 Brookwood Drive
Warner Robins, GA 31093, USA
FFL Firearms Dealer

CHARLES H. SHURTLEFF
1107 Kelly Drive
Hinesville, GA 31313, USA
FFL Firearms Dealer

RYKARD'S FIREARMS
P.O. Box 279
704 Holly Drive
Valdosta, GA 31601, USA
FFL Firearms Dealer

ED'S PAWN SHOP
318 East Water Street
Bainbridge, GA 31717, USA
FFL Firearms Dealer

JOHN R. CLARKE
2417 N. Patterson Street
Thomasville, GA 31792, USA
FFL Firearms Dealer

HUNTER ARMS COMPANY
419 Glenwood Drive
Thomasville, GA 31792, USA
(912)-226-3492 (904)-648-5197
Gunsmithing.
S1,S13; 40;
CONTACT: G. R. Hunter
SALES DEPT: G. R. Hunter
PURCHASING DEPT: G. R. Hunter
Gunsmithing; retail pertinent items.

POWER PLUS ENTERPRISES,INC.
P.O. Box 6070
Columbus, GA 31907, USA
FFL Firearms Dealer
CONTACT: Robert Lewis

HARLEY J. NOLDEN
1235 Brandy Wine Drive
Columbus, GA 31907, USA
FFL Firearms Dealer

LIGHT ENTERPRISES
507 1/2 Seabreeze Blvd.
Daytona Beach, FL 32018, USA
Firearms accessories.

III SALES
P.O. Box 1270
203 East Howard Street
Live Oak, FL 32060, USA
(904)-362-7101
Firearms, Class 3; investments.
S3,S4; 5;
CONTACT: Herman Gunter,III
SALES DEPT: Herman Gunter,III
PURCHASING DEPT: Herman Gunter,III

WILLIAMSON'S TAXIDERMY
159 Ponce Deleon Drive
Ormond Beach, FL 32074, USA
FFL Firearms Dealer

INSTANT WHEELS USED CARS
5242 Normandy Blvd.
Jacksonville, FL 32205, USA
FFL Firearms Dealer

LANE AVENUE PAWN SHOP
6505 Commonwealth Avenue
Jacksonville, FL 32205, USA
FFL Firearms Dealer

T & T GUN REPAIR
4810 Highway Avenue
Jacksonville, FL 32205, USA
FFL Firearms Dealer

J's DISCOUNT WEAPONS
Suite 456 South Shop
9570 Regency Square Blvd.
Jacksonville, FL 32211, USA
FFL Firearms Dealer

MARSHALL CRACKER'S GUNS
6020 Bowden Dale Road
Jacksonville, FL 32216, USA
FFL Firearms Dealer

MICHAEL DRAYOVITCH,JR.
1230 Oakwood Lane
Jacksonville, FL 32223, USA
FFL Firearms Dealer

GUNS & AMMO
1902 Kathryn, Speed Court
Tallahassee, FL 32303, USA
FFL Firearms Dealer

DAN'S GUN SALES
1221 Piedmont Drive
Tallahassee, FL 32312, USA
FFL Firearms Dealer

G & A SALES COMPANY
2233 East 15th St., Suite A
Panama City, FL 32405, USA
FFL Firearms Dealer

HOT SHOT GUN WORKS
3102 Wood Vally Road
Panama City, FL 32405, USA
FFL Firearms Dealer

INTERSTELLAR FRONTIER CORP.
P.O. Box 728
Lynn Haven, FL 32444, USA
(904)-265-4480
Firearms, edged weapons & accessories.
S5; 4;
Catalogue available. Annually.
CONTACT: Neal F. Anderson
SALES DEPT: Vivian Godwin
PURCHASING DEPT: Amy S. Anderson
Distributor of: Desert Eagle, Detonics, Iver Johnson, Alessi holsters, Cold Steel knives, Browning & Ft. Knox safes, Golden Rod dehumidifiers, Harris Bipods - and accessories.

CHARLES H. FOSSLER
Rt. 2, Box 551-E
Seagrove Beach, FL 32454, USA
FFL Firearms Dealer

DAVE'S GUN SHOP
7847 Coronet Way
Pensacola, FL 32514, USA
FFL Firearms Dealer

HAL SHUTE
P.O. Box 3116
Pensacola, FL 32516, USA
FFL Firearms Dealer

POP'S AMMO DUMP
243 North Chipper Road
Cantonment, FL 32533, USA
FFL Firearms Dealer

GREAT GUNS
17 Sherwood Road
Ft. Walton Beach, FL 32548, USA
(904)-862-5499 (904)-862-0442
Firearms & Civil War items.
S1,S4,S12; 4;
Catalogue available. Bi-annually.
CONTACT: John T. Lindsay
SALES DEPT: John Lindsay
PURCHASING DEPT: Chris Lindsay
Antique and modern firearms, buy and sell. Specializing in Colt and Harpers Ferry firearms and accoutrements.

DANA ANDREW TEASLEY
330 Fort Pickens Road, 9-C
Pensacola Beach, FL 32561, USA
FFL Firearms Dealer

CARDENAS & SON
SR 21, Box 606
Keystone Heights, FL 32565, USA
FFL Firearms Dealer

WILLIAM F. GREEN PLANS COMPANY
P.O. Box 892
Shalimar, FL 32579, USA
Cannon plans.
Catalogue $1.00.

J. WALLACE GUNS & AMMO
Rt. 3, Box 487, SR 241
Alachua, FL 32615, USA
FFL Firearms Dealer

HAMILTON GUNS
Rt 4 Box 172-m
Chipley, FL 32628, USA
FFL Firearms Dealer

MILITARY FIREARMS

JAMES GILBERT
11021 North Citrus Avenue
Crystal River, FL 32629, USA
FFL Firearms Dealer

CHARLES H. SNYDER
4053 N.E. 20th Avenue
Ocala, FL 32670, USA
FFL Firearms Dealer

COUNTRY GUN SHOP
12186 N.W. Highway 27
Ocala, FL 32675, USA
FFL Firearms Dealer

INTERNATIONAL ARMS SERVICE
66 Lake Wood Circle
Ocala, FL 32675, USA
FFL Firearms Dealer

ROBERT L. MATHIS
915 W. New York Avenue
DeLand, FL 32720, USA
FFL Firearms Dealer

GUN'S WITH ATEN
1777 Providence Blvd.
Deltona, FL 32725, USA
FFL Firearms Dealer
CONTACT: Gregory W. Aten

LARC INTERNATIONAL
736 Industry Road
Longwood, FL 32750, USA
305-339-6699
BB guns.

SANDY SCOTT
221 Silver Leaf Court
LOngwood, FL 32779, USA
FFL Firearms Dealer

JONATHAN ARTHUR CIENER, INC.
6850 Riveredge Drive
RD 2, Box 66Y6
Titusville, FL 32780, USA
(305)-268-1921
Firearms accessories.
S3,S6; 10;
Catalogue available.
CONTACT: Jonathan A. Ciener
Manufacturer of the finest in suppressed firearms and dealer in automatic weapons. Manufacturer of Hohrein .22LR Conversion Kit for the Ruger Mini-14 Rifle.

J. PHILIP OWENS GUN SHOP
Rt. 2, Box 680
1441 West Ocala Street
Umatilla, FL 32784, USA
FFL Firearms Dealer

BLAKEMORE & SONS
P.O. Box 28028
Orlando, FL 32867, USA
FFL Firearms Dealer
CONTACT: H. Noel Blakemore

DON'S RELOADING
2875 Palm Bay Road N.E.
Palm Bay, FL 32905, USA
FFL Firearms Dealer

WILDERNESS GUN SHOP
391 Crestview Street, N.E.
Palm Bay, FL 32907, USA
FFL Firearms Dealer

WAGNER'S WEAPONS
1011D South A1A
Patrick AFB, IL 32925, USA
FFL Firearms Dealer
CONTACT: Mrs. Peggy S. Wagner

D. E. WARRENSFORD
40 Uranus Avenue
Merritt Island, FL 32953, USA
FFL Firearms Dealer

OLD DIXIE GUN SHOP
1715 Dixie Avenue
Vero Beach, FL 32960, USA
FFL Firearms Dealer
CONTACT: Grady W. Tyner

RECONDO EQUIPMENT INC.
527 Unit 1 East 9th Street
Hialeah, FL 33010, USA
FFL Firearms Dealer

JOHN A. SUBIC
P.O. Box 103
Islamorada, FL 33036, USA
FFL Firearms Dealer

CHARLES L. McALLISTER
P.O. Box 10534
Pompano Beach, FL 33061, USA
FFL Firearms Dealer

JOHN PRIBIL
1545 Saragossa Avenue
Coral Gables, FL 33134, USA
FFL Firearms Dealer

BRIAN L. BALKANY
1553 San Ignacio Avenue
Coral Gables, FL 33146, USA
FFL Firearms Dealer

A & M GUN SHOP
10740 Northeast Court
Miami, FL 33161, USA
FFL Firearms Dealer

TAMIAMI DISTRIBUTORS, INC.
11315 S.W. 40th Street
Miami, FL 33165, USA
305-552-5358
Firearms

SPRINGS IMPORT/EXPORT COMPANY
P.O. Box 660431
Miami Springs, FL 33166, USA
305-635-8194
Firearms

SER DISTRIBUTORS, INC.
P.O. Box 55851
Miami, FL 33255-8951, USA
305-264-9321
Firearms

R. CUNNINGHAM
4509 'F' Treehouse Lane
Tamarac, FL 33319, USA
FFL Firearms Dealer

GUNS OF MERRITT
375 N. University Drive
Plantation, FL 33324, USA
305-473-2166
Firearms

DENNIS A. JECK
12450 S.W. 10 Court
Davie, FL 33325, USA
FFL Firearms Dealer

HAROLD L. WIENER
1110 F3 Green Pine Boulevard
West Palm Beach, FL 33409, USA
FFL Firearms Dealer

NORMAN R. McCLELLAND, III
2325 Sunset Drive
West Palm Beach, FL 33415, USA
FFL Firearms Dealer

ANTIQUE ARMS
3191 East Road
Lexahatchee, FL 33470, USA
FFL Firearms Dealer
CONTACT: Bill Kline

W.C.R.N. MARKETING
1745 Kanner Hwy. S.W.
Stuart, FL 33494, USA
FFL Firearms Dealer
CONTACT: Wayne M. Gerhardt

THE MENAGERIE
2201 S.E. Indian St., F-4
Stuart, FL 33497, USA
FFL Firearms Dealer
CONTACT: Frank C. Farris

DAVE'S GUNSMITHING
301 43rd Street West
Bradenton, FL 33505, USA
FFL Firearms Dealer

RICHARD J. BOUDREAU
5120-14th Street West #26
Bradenton, FL 33507, USA
FFL Firearms Dealer

GEORGE J. PARKS
2501 Florida Blvd.
Bradenton, FL 33507, USA
FFL Firearms Dealer

SURVIVE ALIVE FIREARMS
P.O. Box 5487
Clearwater, FL 33518, USA
FFL Firearms Dealer
CONTACT: John Hamilton

PAUL'S GUNSMITHING SERVICE
296 Lake Joyce Drive
Land O'Lakes, FL 33539, USA
FFL Firearms Dealer
CONTACT: Paul W. Rowe

CROUNSE FIRE ARMS
15597 Newport Road
Clearwater, FL 33546, USA
FFL Firearms Dealer

MILITARY FIREARMS

JACKIE L. JENKINS
Route 3, Box 3230
Plant City, FL 33566, USA
FFL Firearms Dealer

PASCO HARDWARE, INC.
16034 US Highway 19
Hudson, FL 33567, USA
FFL Firearms Dealer

BUD NORWOOD
220 N. Tuttle Avenue
Sarasota, FL 33577, USA
FFL Firearms Dealer

TY BRIGNER
2121 Dodge Avenue
Sarasota, FL 33580, USA
FFL Firearms Dealer

ALAN C. PLUSH
4109 Palau Drive
Sarasota, FL 33583, USA
FFL Firearms Dealer

DIAMOND 'J' ARMS
2101 W. Hillsborough Avenue
Tampa, FL 33603, USA
FFL Firearms Dealer

GREGORY W. BROWN
1019 E. Hamilton Avenue
Tampa, FL 33604, USA
FFL Firearms Dealer

C & F FIREARMS SALES
1813 Marvy Avenue
Tampa, FL 33612, USA
FFL Firearms Dealer

STEVE'S PAWN SHOP
12726 N. Florida Avenue
Tampa, FL 33612, USA
FFL Firearms Dealer

VILLAGE STAMP & COIN
1747 West Fletcher Avenue
Tampa, FL 33612, USA
FFL Firearms Dealer
CONTACT: Ed Howard

ABBOTT'S GUNS
6900-17 Way North
St. Petersburg, FL 33702, USA
FFL Firearms Dealer
CONTACT: Walter R. Abbott

SAWYER'S GUNSMITH SHOP
774 - 32nd Avenue North
St. Petersburg, FL 33704, USA
FFL Firearms Dealer

TAMPA BAY ARMS COLLECTORS
2461 - 67th Ave., S.
St. Petersburg, FL 33712, USA
813-867-1019
FFL Firearms Dealer
CONTACT: John Turvell

C & D GUN SHOP
7640 Green Road
Lakeland, FL 33805, USA
FFL Firearms Dealer

RONALD D. DICKS
P.O. Box 1809
Dundee, FL 33838, USA
FFL Firearms Dealer

JOSEF'S SALES & SERVICE
5358 Malibu Court
Cape Coral, FL 33904, USA
FFL Firearms Dealer

JAMES W. HAVERFIELD
Route 25
31 Sophmore Lane
Ft. Myers, FL 33912, USA
FFL Firearms Dealer

BOYDS PAWN SHOP
408 2nd Avenue East
Oneonta, AL 35121, USA
FFL Firearms Dealer

TALLADEGA ORDNANCE
Rt. 1, Box 524
Sulacauga, AL 35150, USA
FFL Firearms Dealer
CONTACT: Tommy Bunn

GUN SOUTH, INC.
P.O. Box 129
108 Morrow Avenue
Trussville, AL 35173, USA
(205)-655-8299
Firearms dealer, Steyr & FN products.
S5; 5;
Catalogue available. Annually.
CONTACT: Donald F. Wood
SALES DEPT: Tony J. Rosetti, Jr.
PURCHASING DEPT: Tony J. Rosetti, Jr.
U.S. agent for Steyr, Daimler, Puch AG and Fabrique National, SA weapons.

AEROMARINE, INC.
7605 Eastwood
Birmingham, AL 35210, USA
Firearms

THOMAS M. WEST, JR.
4165 Old Leeds Lane
Mount Brook, AL 35213, USA
Pre-1898 British Webley collector.

BIRMINGHAM PISTOL PARLOR
1833 Pinson Valley Parkway
Tarrant, AL 35217, USA
FFL Firearms Dealer
CONTACT: John G. Walker

GUNS & AMMO
Rt. 6, Box 375
Jasper, AL 35501, USA
FFL Firearms Dealer
CONTACT: H. W. Robertson

GLENN E. BOMAN
P.O. Box 593
Vernon, AL 35592, USA
FFL Firearms Dealer

COTTON HILL WORKS
Rt. 3, Box 477
Athens, AL 35611, USA
205-232-7194
Artillery carriages.

CONTACT: Bill Summerfelt

RICHARD C. CROY, JR.
Rt. 4, Box 202-0
Eufaula, AL 36027, USA
FFL Firearms Dealer

GEORGE N. BOYD, JR.
1361 Federal Drive
Montgomery, AL 36107, USA
205-262-6936
FFL Firearms Dealer

G & N GUNS & PAWN
Federal Drive
1361 Coliseum Shopping Center
Montgomery, AL 36110, USA
FFL Firearms Dealer

M & M GUNS
Rt. 1, Box 898A
Alexandria, AL 36250, USA
FFL Firearms Dealer

LINDY'S GUN SHOP
Rt. 3, Box 103
Enterprise, AL 36330, USA
FFL Firearms Dealer

GREG EASOM
Rt. 1, Box 265-A
Midland City, AL 36350, USA
FFL Firearms Dealer

GUYS SEWING MACHINE SHOP
Rt. c, Box 27, Hwy. 77
Evergreen, AL 36401, USA
FFL Firearms Dealer
CONTACT: Guy N. Blackburn

LEATH GUN SALES
Rt. 1, Box 375
Chunchula, AL 36521, USA
FFL Firearms Dealer
CONTACT: Barry Leath

ROBERT L. WINHAUER
Rt. 2, Box 307
Coden, AL 36523, USA
FFL Firearms Dealer

GOLD MINE
Daphne Shopping Center
614 Whispering Pines Road
Daphne, AL 36526, USA
FFL Firearms Dealer

CUSTOM PURCHASING SERVICE
P.O. Box 188
Semmes, AL 36575, USA
FFL Firearms Dealer
CONTACT: D. D. Brewer

BUFORD E. HOOPER
1361 Odette Avenue
Mobile, AL 36605, USA
FFL Firearms Dealer

HUGH S. KITE, JR.
P.O. Box 57
Cottonton, AL 36851, USA
FFL Firearms Dealer

MILITARY FIREARMS

SURVIVAL SUPPLIES
6212 Bridlewood Lane
Brentwood, TN 37027, USA
FFL Firearms Dealer
CONTACT: Frederick S. Arnold

TIMOTHY R. THOMAS
5120 Paddock Village Court B-1
Brentwood, TN 37027, USA
FFL Firearms Dealer

AUTOMATIC WEAPONRY
P.O. Box 1124
Brentwood, TN 37027, USA
615-331-9655
Firearms

JIM'S GUNS
216 Kraft Street
Clarksville, TN 37040, USA
FFL Firearms Dealer

MILITARY SPECIAL ORDER DEPT.
1818 Memeorial Drive, #17
Clarksville, TN 37043, USA
FFL Firearms Dealer

SCOTT PILKINGTON
Box 125
U.S. Hwy. 127 & River
Dunlap, TN 37043, USA
FFL Firearms Dealer

RIDGERUNNER GUNS
Box 318, Wilson Park
Franklin, TN 37064, USA
FFL Firearms Dealer

FULL AUTO
P.O. Box 1881
Murfreesboro, TN 37133, USA
Firearms

JERRY'S RELOADING & SHOOTING
SUPPLIES
Rt. 1, Box 178-A
Springfield, TN 37172, USA
FFL Firearms Dealer

THE GUN & TACKLE SHOP
Rt. 1, Box 422 B
Dunlap, TN 37327, USA
FFL Firearms Dealer
CONTACT: Lewis P. Roberson,Jr.

DARNELL'S GUN SHOP
Rt. 1, Box 178, Good Branch Rd
Lynchburg, TN 37352, USA
FFL Firearms Dealer

ASKEW'S FIREARMS
P.O. Box 266
Chattanooga, TN 37401, USA
FFL Firearms Dealer
CONTACT: Gregory L. Askew

NATCHEZ SHOOTERS SUPPLIES
P.O. Box 22247
Chattanooga, TN 37422, USA
615-899-0499
Firearms & accessories.

WIDENER'S RELOADING SUPPLY, INC.
2309 Nave Drive
Johnson City, TN 37601, USA
615-282-6786
Firearms & accessories.

TC's FIREARMS
101 24th Street
Bristol, TN 37620, USA
FFL Firearms Dealer

CHARLES W. BURLING
306 Range Street
Elizabethton, TN 37643, USA
FFL Firearms Dealer

BOB STITT,JR.
P.O. Box 384
160 Rine Street
Mountain City, TN 37683, USA
FFL Firearms Dealer

MICHAEL PHILLIPS
P.O. Box 345
Lenoir City, TN 37771, USA
FFL Firearms Dealer

WALTER J. TEEPFER
Rt. 1, Box 247H
New Market, TN 37820, USA
FFL Firearms Dealer

ALAN C. McCOY
Rt. 3, Box 325A
Sneedville, TN 37869, USA
FFL Firearms Dealer

DON SMITH
2704 Spencer Street
Knoxville, TN 37917, USA
FFL Firearms Dealer

DONELSON FIREARMS
828 Wood Harbor Road
Knoxville, TN 37922, USA
FFL Firearms Dealer
CONTACT: P. M. Clariday

MID-CONTINENT ARMAMENT
1798 Madison Avenue
Memphis, TN 38104, USA
FFL Firearms Dealer

WILSON G. POPE
1760 Borden Road
Memphis, TN 38116, USA
FFL Firearms Dealer

RON AYERS
Rt. 4, Box 28
Martin, TN 38237, USA
FFL Firearms Dealer

BROWN'S DISCOUNT GUNS
Rt. 2, Box 26
Troy, TN 38260, USA
FFL Firearms Dealer

DIXIE GUN WORKS
Union City, TN 38261, USA
Firearms & accessories.
CONTACT: Attn: Mr. T. Kirkland

DONALD N. SMITH
402 Gale Lane
Lawrenceburg, TN 38464, USA
FFL Firearms Dealer

JOHN F. BRUMFIELD
Rt. 1, Box 47
Lunnville, TN 38472, USA
FFL Firearms Dealer

FRANKIE WILLIAMSON
Box 5675
Greenville, MS 38704, USA
601-335-3924
FFL Firearms Dealer

AG MAINTENANCE, INC.
Indianola Municipal Airport
P.O. Box 311
Indianola, MS 38751, USA
FFL Firearms Dealer
CONTACT: George W. Gallaspy

BRADFORD GUNSMITH SALES
Route 1, Box 362
Calhoun City, MS 38916, USA
FFL Firearms Dealer
CONTACT: Billy Bradford

BRANDON GUN WORKS
317 Martin Road
Brandon, MS 39042, USA
FFL Firearms Dealer
CONTACT: Larry D. Quillen

SHENANDOAH VALLEY ARMS
227 Shenandoah Road
Florence, MS 39073, USA
FFL Firearms Dealer

J. HODGES
C/O Stuckeys
Rt. 1, Box 122
Madison, MS 39110, USA
FFL Firearms Dealer

STAN'S GUNS & SUPPLIES
2814 Itasca Drive
Natchez, MS 39120, USA
FFL Firearms Dealer
CONTACT: Stan Owens

McGEE INDUSTRIES
P.O. Box 56
Edwards Road
Raymond, MS 39154, USA
FFL Firearms Dealer

GEORGE L. HOLDEN
106 Oak Park Drive
Pearl, MS 39208, USA
FFL Firearms Dealer

WILLIAM LEAMOND
2014 Carolyn Lane
Pearl, MS 39208, USA
FFL Firearms Dealer

ROCK HOUSE TRADING POST
Highway 80 West
Meridian, MS 39301, USA
FFL Firearms Dealer

MILITARY FIREARMS

SILVER DOLLAR GUN EXCHANGE
721 Acme Plaza
Meridian, MS 39301, USA
FFL Firearms Dealer

HILLTOP FIREARMS
P.O. Box 967
Hattiesburg, MS 39401, USA
FFL Firearms Dealer

RAY RAYBURN
200 Weathersby Road
Hattiesburg, MS 39401, USA
FFL Firearms Dealer

RILEY, LTD.
123 West Front Street
Hattiesburg, MS 39401, USA
FFL Firearms Dealer

J. CLIFF REDDOCH
P.O. Box 1586
Laurel, MS 39440, USA
FFL Firearms Dealer

RICKY HOLDER
Route 4, Box 390
Laurel, MS 39442, USA
FFL Firearms Dealer

SALES, INC.
Route 8, Box 297
Lucedale, MS 39452, USA
FFL Firearms Dealer

LEE'S GUNWORKS
Route 2, Box 227
Picayune, MS 39466, USA
FFL Firearms Dealer

THE PROFESSIONAL GUNSMITHS
1010 Jackson Avenue
Pascagoula, MS 39567, USA
FFL Firearms Dealer

SOUTHWIND CUSTOM GUNS
344 Menge Avenue
Pass Christian, MS 39571, USA
FFL Firearms Dealer
CONTACT: Michael W. Bryant

WILSON GUNSMITHING
Route 1, Box 473
Brookhaven, MS 39601, USA
FFL Firearms Dealer

MIKE ENNIS
218 Wilson Drive
McComb, MS 39648, USA
FFL Firearms Dealer

J. M. FISCHER
2348 Saratoga Drive
Louisville, KY 40205, USA
FFL Firearms Dealer

THE GUN CAGE
2525 Gardiner Lane
Louisville, KY 40205, USA
FFL Firearms Dealer

KAREN'S CUTLERY SHOPPE
1701 Mellwood Avenue
Louisville, KY 40206, USA
FFL Firearms Dealer

'THE AMERICAN WAY' GUN SHOP
5207 Baywood Drive
Louisville, KY 40222, USA
FFL Firearms Dealer
CONTACT: William Gehring,Sr.

MORGAN VANCE
143 Tanglewood Trail
Louisville, KY 40223, USA
FFL Firearms Dealer

JOE N. BARROW
7208 Sky Blue Avenue
Louisville, KY 40258, USA
FFL Firearms Dealer

SEIBERT'S
59 Hiawatha Trail
Winchester, KY 40391, USA
FFL Firearms Dealer

LES H. SIMMONS,JR.
624 Eden Road
Lexington, KY 40505, USA
FFL Firearms Dealer

MONTY'S GUN SHOP
P.O. Box 23428
Lexington, KY 40523, USA
FFL Firearms Dealer

KY-VA HUNT CLUB
HC66 Box 2215
Wurtland, KY 41144, USA
FFL Firearms Dealer
CONTACT: George Bennett

MICHAEL T. McCANN
1160 Jefferson Street
Paducah, KY 42001, USA
FFL Firearms Dealer

FULTON TRUE VALUE HARDWARE
208 Lake Street
Fulton, KY 42041, USA
FFL Firearms Dealer

D & K ENTERPRISES
P.O. Box 327
113 W. South Street
Mayfield, KY 42066, USA
FFL Firearms Dealer

JERRY'S SPORTING GOODS
6th & Walnut Streets
Mayfield, KY 42066, USA
FFL Firearms Dealer

MAR-CAL STORES
202 East Locust Street
Scottsville, KY 42164, USA
FFL Firearms Dealer

WOODROW F. HOWELL,II
2933 Seminole Avenue
Hopkilnsville, KY 42240, USA
FFL Firearms Dealer

ACME PLUMBING & HEATING,INC.
2103 Wink Court
Owensboro, KY 42301, USA
FFL Firearms Dealer

RONALD G. SWEAT
2311 North York
Owensboro, KY 42301, USA
FFL Firearms Dealer

THOMAS BAIT & TACKLE
Rt. 5, Box 460
Morganfield, KY 42437, USA
FFL Firearms Dealer

J. M. CLONTZ
141 Beauchamp Drive
Somerset, KY 42501, USA
FFL Firearms Dealer

PHILIP A. SELLERS
307 Rosewood Avenue
Somerset, KY 42501, USA
FFL Firearms Dealer

E.S.K. AUTOMATIC WEAPONS
P.O. Box 2
Bonnieville, KY 42713, USA
FFL Firearms Dealer

D & R ORDNANCE & SUPPLY
157 N. Main Street
Fredericktown, OH 43019, USA
FFL Firearms Dealer

DAVID G. CAMP
P.O. Box 94
Ostrander, OH 43061, USA
614-663-3404
Firearms

J. C. GUN REPAIR
3928 Three Rivers Lane
Groveport, OH 43125, USA
FFL Firearms Dealer

LONG RIFLE GUN SHOP
1243 Grandview Avenue
Columbus, OH 43212, USA
Firearms, edged weapons, & field equipment.

ZANES GUN RACK
4167 N. High Street
Columbus, OH 43214, USA
Firearms

PRECISION,LTD.
P.O. Box 21530
Columbus, OH 43221, USA
FFL Firearms Dealer

COLUMBUS GUN HEADQUARTERS
5339 W. Broad Street
Columbus, OH 43228, USA
FFL Firearms Dealer

CRAIG E. PEMBERTON
P.O. Box 28285
Columbus, OH 43228, USA
FFL Firearms Dealer

R & M ENTERPRISES
401 Knob Hill East
Columbus, OH 43228, USA
FFL Firearms Dealer

MILITARY FIREARMS

GUYAN FUR COMPANY
1020 1/2 S. Main Street
Bellefontaine, OH 43311, USA
FFL Firearms Dealer

PARSONS SPORT SHOP
806 E. Gypsy Lane Road
Bowling Green, OH 43402, USA
FFL Firearms Dealer

ED MATYAS
1306 Buckeye Street
Genoa, OH 43430, USA
FFL Firearms Dealer

DAVID O. ANDERSON
127 Hardan Drive
Wallbridge, OH 43465, USA
FFL Firearms Dealer

EASLEY'S GUN SHOP,INC.
209 Main Street
Defiance, OH 43512, USA
(419)-784-4751
Firearms & accessories.
S1,S11; 20;
CONTACT: Richard T. Easley,Sr.
SALES DEPT: Richard T. Easley,Sr.
PURCHASING DEPT: Richard T. Easley,Sr.

OLE SARGE'S GUN SHOP
411 W. Main Street
Napoleon, OH 43545, USA
FFL Firearms Dealer
CONTACT: Lester J. Joseph

JIM'S SPORT SUPPLY
606 Main Street
Cochocton, OH 43812, USA
FFL Firearms Dealer

KIRK M. BROWN
108 S. Main Street
Chagrin Falls, OH 44022, USA
FFL Firearms Dealer

ROBERT PASCHAL
10024 Sunset Drive
Chagrin Falls, OH 44022, USA
FFL Firearms Dealer

THE MUZZLELOADER SHOP
15009 S. Island Road
Columbia Station, OH 44028, USA
FFL Firearms Dealer

DAF COMPANY
P.O. Box 566
Mentor, OH 44060, USA
FFL Firearms Dealer

QUALITY FIREARMS
5839 South Shandle
Mentor, OH 44060, USA
FFL Firearms Dealer

MICKEYS GUN CARE
27208 Lorain Road
North Olmsted, OH 44070, USA
FFL Firearms Dealer
CONTACT: Harry Cook

JOHN A. FALKENSTEIN,III
27504 Blossom Blvd.
North Olmsted, OH 44070, USA
FFL Firearms Dealer

WESTON SALES COMPANY
14251 Gifford Road
Oberlin, OH 44074, USA
FFL Firearms Dealer

CLEVELAND AREA SURVIVAL,INC.
2228 Northland Avenue
Lakeland, OH 44107, USA
FFL Firearms Dealer

NORTH COAST SHOOTING SUPPLY
10010 Lorain Avenue
Cleveland, OH 44111, USA
FFL Firearms Dealer

WORLD OF LUGERS
5838 Mayfield Road
Mayfield Heights, OH 44124, USA
Firearms.
List $2.00, 5 issues for $5.00.
CONTACT: Sam Costanzo

NORTH COAST SHOOTING SUPPLY
7314 Grant Blvd.
Cleveland, OH 44130, USA
FFL Firearms Dealer

SMALL ARMS SPECIALITIES
9531 Yorkview Drive
North Royalton, OH 44133, USA
FFL Firearms Dealer
CONTACT: Gary Szechy

RR & MM COMPANY
24827 Rockledge Lane
Richmond Heights, OH 44143, USA
(216)-486-8886
Firearms & ammunition
S6; 10;
CONTACT: Robert F. McSween
SALES DEPT: Robert F. McSween
PURCHASING DEPT: Robert F. McSween

KURTZ'S SUPPLY
1277 Overlook Drive
Norton, OH 44203, USA
FFL Firearms Dealer
CONTACT: Bradley S. Kurtz

THOMAS L. FALKENSTEIN
3010 Grill Road
Clinton, OH 44216, USA
FFL Firearms Dealer

DALTON ENTERPRISE
2432 Shaw Avenue
Cuyahoga Falls, OH 44223, USA
FFL Firearms Dealer

TERRY LAPCHYNSKI
9975 N. Delmonte Blvd.
Streetsboro, OH 44240, USA
FFL Firearms Dealer

LOG CABIN SHOP
P.O. Box 275
Lodi, OH 44254, USA
Black powder guns & accesssories.

MATHENY'S HUNTING & FISHING
4965 E. Garfield Road
Petersburg, OH 44454, USA
FFL Firearms Dealer

RANDY M. BAUER
4894-A Warren-Sharon Road
Vienna, OH 44473, USA
FFL Firearms Dealer

SPECIAL WEAPONS
766 West Summit Street
Alliance, OH 44601, USA
FFL Firearms Dealer

HARTER'S GUNS & AMMO
5257 Dover Road
Apple Creek, OH 44606, USA
FFL Firearms Dealer

TERRITORY ARMAMENT COMPANY
1230 Battlesburg St., S.E.
East Sparta, OH 44626, USA
FFL Firearms Dealer

ARMENI'S FINE GUNS & AMMO
1969 Twp. Rd. 247
Arcadia, OH 44804, USA
FFL Firearms Dealer

WINTERHAWK PISTOLSMITHING AND
548 Star Route 58
Ashland, OH 44805, USA
FFL Firearms Dealer
CONTACT: CUSTOM RELOADING

ASHLAND SHOOTING SUPPLIES,INC.
209 Orange Street
Ashland, OH 44805-0391, USA
Firearms

FISHER'S SPORTING GOODS
505 N. Yhoman Street
Crestline, OH 44827, USA
FFL Firearms Dealer

KEN'S STOCK & BARREL GUN SHOP
8503 E. Co. Road 38
Republic, OH 44867, USA
FFL Firearms Dealer

LARRY'S SPORTS EQUIPMENT
P.O. Box 755
1000 Maple Avenue
Sandusky, OH 44870, USA
FFL Firearms Dealer

JEFF PENWELL
232 Motson Street
Willard, OH 44890, USA
FFL Firearms Dealer

TIMOTHY EVANS
219 South D Street
Hamilton, OH 45013, USA
FFL Firearms Dealer

DOUG'S GUN SALES
5677 Auberger Drive
Fairfield, OH 45014, USA
FFL Firearms Dealer

MILITARY FIREARMS

CLIFF'S GUN ROOM
4411 Hope Drive
Middletown, OH 45042, USA
FFL Firearms Dealer

RODNEY L. WAGNER
1955 Bethel - Maple Road
Hamersville, OH 45130, USA
FFL Firearms Dealer

John E. Carrigan
48 W. McMicken Avenue
Cincinnati, OH 45210, USA
FFL Firearms Dealer

NORTH SIDE ARMS
4219 Kirby Road
Cincinnati, OH 45223, USA
FFL Firearms Dealer

DEL'S GUNS
6749 Parkview Drive
Cincinnati, OH 45224, USA
FFL Firearms Dealer

BILL ERNST
7034 Noble Court
Cincinnati, OH 45239, USA
FFL Firearms Dealer

JACK JEFFERS
10080 Sturgeon Lane
Cincinnati, OH 45239, USA
FFL Firearms Dealer

STEVE WITHROW
1019 Heathwood Drive
Englewood, OH 45322, USA
FFL Firearms Dealer

EDWIN A. WINGFIELD
8138 Wescott Avenue
Fairborn, OH 45324, USA
FFL Firearms Dealer

SMITTY'S GUN SHOP
1143 Fisher Drive
Piqua, OH 45356, USA
FFL Firearms Dealer
CONTACT: Scott J. Smith

LOUDAS GUN SHOP
1144 Richfield Center
Dayton, OH 45400, USA
Firearms

SPECIALTY GUN SERVICE
2730 Linden Avenue
Dayton, OH 45410, USA
FFL Firearms Dealer

MIKE'S GUN REPAIR & SALES
4777 Taylorsville Road
Huber Heights, OH 45424, USA
FFL Firearms Dealer

DAN BANASIAK
4188 Grouse Court
Dayton, OH 45430, USA
FFL Firearms Dealer

JIM'S GUNS & WHATEVER
1148 Richfield Center
Dayton, OH 45430, USA
FFL Firearms Dealer

KEMP SPORTING GOODS
2159 Pine Knott Drive
Beavercreek, OH 45431, USA
FFL Firearms Dealer

RLK GUN REPAIR
7985 Washington Park Drive
Centerville, OH 45459, USA
FFL Firearms Dealer

PATRIOT GUN & COIN SHOP
152 E. Main Street
Chillicothe, OH 45601, USA
FFL Firearms Dealer

LORETTA'S
2310 Eastern Avenue
Gallipolis, OH 45631, USA
FFL Firearms Dealer

ROBERT C. KRAMER
1133 Coles Blvd.
Portsmouth, OH 45662, USA
FFL Firearms Dealer

FRANKS FRONTIER SPORTS
102 Mann Avenue
South Point, OH 45680, USA
FFL Firearms Dealer

HERBERT H. DeVOL
1339 Eastman Street
Zanesville, OH 45701, USA
FFL Firearms Dealer

HEFNER'S HUNTING & SHOOTING SUPPLIES
1133 W. Bluelick Road
Lima, OH 45801, USA
FFL Firearms Dealer

WILLIAM A. KLAUSING
102 W. Main Street
Beaverdam, OH 45808, USA
FFL Firearms Dealer

LEFTHANDED GUN SHOP
424 W. 10th Street
Anderson, IN 46016, USA
FFL Firearms Dealer
CONTACT: Cecil C. Carmichael,III

CLAUDE A. BLACK
2100 W. Green
Frankfort, IN 46041, USA
FFL Firearms Dealer

BOB'S GUN SHOP
Rt. 1, Box 319
Kirklin, IN 46050, USA
FFL Firearms Dealer
CONTACT: Robert L. Earl

BULLSHOOTER SUPPLY
105 Jordan Court
Martinsville, IN 46151, USA
FFL Firearms Dealer

STEVEN W WILSON
7135 State Road 39
Martinsville, IN 46151, USA
FFL Firearms Dealer

FLINTLOCKS,INC.
RR #1, Box 263-E
Nineveh, IN 46164, USA
317-933-3441
Firearms

MARK HIGHSMITH
5222 Judith Ann Drive
Indianapolis, IN 46236, USA
FFL Firearms Dealer

RICHARD EMBRY
6510 Marble Lane
Indianapolas, IN 46237, USA
FFL Firearms Dealer

FIREBALL FIREARMS
P.O. Box 2289
East Chicago, IN 46312, USA
(219)-398-1858 (219)-322-3879
Flareguns, firearms, and military items.
S13; 20;
Catalogue available. Quarterly.
CONTACT: Howard Resnick
SALES DEPT: Howard (Signalman) Resnick
Buy and sell flareguns - world's largest selection; odd, early, unusual, pistols - semi-auto's and revolvers; military relics such as cigarette cases, flasks, bayonets, etc.

THE SURVIVAL SHOP
1601 LWE
South Bend, IN 46613, USA
FFL Firearms Dealer

SOUTH BEND REPLICAS,INC.
61650 Oak Road
South Bend, IN 46614, USA
(219)-289-4500
Antique artillery reproductions, miniature to full scale.
S4,S11; 14;
Catalogue available. Every two years.
CONTACT: J. P. Barnett
SALES DEPT: J. P. Barnett, President
PURCHASING DEPT: J. P. Barnett
Manufacturing of antique artillery reproductions, miniature to full scale, for ship and fort restorations, museums, etc.

ST. JOE RIVER TRADING CO.
556 River Avenue
South Bend, IN 46618, USA
FFL Firearms Dealer

EADER'S SPORT SHOP
Rt. 3, Box 37-A
Angola, IN 46703, USA
FFL Firearms Dealer

SPORTLAND SPORTING GOODS
1408 W. Adams Street
Decatur, IN 46733, USA
FFL Firearms Dealer

K. E. HIRSCHY,JR.
4129 Thorton Drive
Ft. Wayne, IN 46815, USA
FFL Firearms Dealer

MILITARY FIREARMS

ROTHMAN'S
7101 Silverthorn Run
Ft. Wayne, IN 46815, USA
FFL Firearms Dealer

SAMOAN TRADERS
1507 Cadillac Drive East
Kokomo, IN 46902, USA
FFL Firearms Dealer
CONTACT: Ronald P. Gavala

GARY E. RALSTIN
P.O. Box 35
Oakford, IN 46965, USA
FFL Firearms Dealer

REICHARD FIREARMS
415 A Main Street
Rochester, IN 46975, USA
FFL Firearms Dealer

RIPLEY COUNTY GUNS
Rt. 1, Box 423
Milan, IN 47031, USA
FFL Firearms Dealer
CONTACT: Joseph P. Thompson

JAMES LOONEY
234 Pearl Street
New Albany, IN 47150, USA
FFL Firearms Dealer

BILL ROBERTSON
137 East Spring Street
New Albany, IN 47150, USA
FFL Firearms Dealer

RON'S TRADING POST
Rt. 1, Box 844
Palmyra, IN 47164, USA
FFL Firearms Dealer

B & J GUN REPAIR
P.O. Box 1697
Columbus, IN 47202, USA
FFL Firearms Dealer
CONTACT: Gerald Tekulve

MILES PER GALLON CO.
PO Box 87
Freetown, IN 47235, USA
FFL Firearms Dealer
CONTACT: Ed Wolter

JOHN A. CARPANINI
400 Elmhurst Lane
Madison, IN 47250, USA
FFL Firearms Dealer

GENE B. CRUM
509 Harvey Drive
Bloomington, IN 47401, USA
FFL Firearms Dealer

MELVIN L. WENDELL
1339 Busseron
Vincennes, IN 47591, USA
FFL Firearms Dealer

MARCHAND'S GUNS & SURPLUS
P.O. Box 37
409 W. Cherry Street
Tennyson, IN 47637, USA
FFL Firearms Dealer

D M GUNS & AMMO
RR 2, Box 291
Wadesville, IN 47638, USA
FFL Firearms Dealer
CONTACT: David W. Motz

LESTER H. HIGHMAN
3036 Cottage Drive
Evansville, IN 47711, USA
FFL Firearms Dealer

BENNY'S CUSTOM GUNS
11801E Carol Heights Drive
Evansville, IN 47712, USA
FFL Firearms Dealer

KEITH'S ACCESSORIES
RR 53, Box 144
Terre Haute, IN 47805, USA
FFL Firearms Dealer

SCOT DEL MASTRO
P.O. Box 562
Clinton, IN 47842, USA
FFL Firearms Dealer

DOUG ORMAN
Rt. 1, Box 53-B
Dana, IN 47847, USA
FFL Firearms Dealer

LYFORD RIFLE SHOP
RR 1
Rosedale, IN 47878, USA
FFL Firearms Dealer

RAY'S SPORTING GOODS
RR 1, Box 233
Fowler, IN 47944, USA
FFL Firearms Dealer

R. H. COLTER
3019 Chewton Cross
Birmingham, MI 48010, USA
FFL Firearms Dealer

THE BLOOMFIELD ARMORY
3905 Oakland Drive
Birmingham, MI 48010, USA
FFL Firearms Dealer
CONTACT: Jerome Widmer

WILLIAM C. MELTON
29316 Lancaster Dr., Ste. 108
Southfield, MI 48034, USA
(313)-350-1024
FFL transfers for modern weapons.
S12; 5;
SALES DEPT: William C. Melton
PURCHASING DEPT: William C. Melton
Third Reich small arms, holsters, magazines, optics, militaria, modern and curio-relic firearms, Class III firearms.

ANDREW A. ANSON
33 Hudson
Pontiac, MI 48058, USA
FFL Firearms Dealer

CLARE WEINAND
19018 Rockport
Roseville, MI 48066, USA
FFL Firearms Dealer

KLEINS FIREARMS
23888 Merrill
Southfield, MI 48075, USA
FFL Firearms Dealer
CONTACT: Randy Klein

RONALD ARBUCKLE
34895 Carbon Drive
Sterling Heights, MI 48077, USA
FFL Firearms Dealer

CHARLES STOVER
23337 Playview
St. Clair Shores, MI 48082, USA
FFL Firearms Dealer

NEWKS GUN SHOP
1343 Esther Road
Scottville, MI 48082, USA
FFL Firearms Dealer
CONTACT: O'Neil J. Newkirk

D & D GUN SHOP
363 Elwood Avenue
Troy, MI 48084, USA
313-435-0160
General gunsmithing & firearms.
CONTACT: Dan Kurkowski

R & S GUNS
9560 Lowe Pine
Union Lake, MI 48085, USA
FFL Firearms Dealer

F. M. G.
22828 Sharrow
Warren, MI 48089, USA
FFL Firearms Dealer

WILLIAM R. HANNA
5119 Fedora
Troy, MI 48098, USA
FFL Firearms Dealer

JAMES M. PETERS
1410 Broadway
Ann Arbor, MI 48105, USA
FFL Firearms Dealer

ANTIQUE ARMS
P.O. Box 672
Dearborn, MI 48120, USA
(313)-425-0578
Antique arms
S4; 6;
CONTACT: William R. Johnson
SALES DEPT: William R. Johnson
American and European arms, armourment, and edged weapons. Antique to pre-1945. Liquidator of sporting and military accoutrements.

HERB JAMES
24427 Currier
Dearborn Heights, MI 48125, USA
FFL Firearms Dealer

NICHOLAS CARSTEA
P.O. Box 1633
Dearborn, MI 48126, USA
Firearms

MILITARY FIREARMS

ROBERT E. DRABICKI
6735 Whitby
Garden City, MI 48135, USA
FFL Firearms Dealer

D & S GUN SHOP
2168 Hartwick
Lincoln Park, MI 48146, USA
FFL Firearms Dealer

DAVID MYERS
305 Hack Street
Milan, MI 48160, USA
FFL Firearms Dealer

WILLIAM FRANK
357 Cambridge
South Lyon, MI 48178, USA
FFL Firearms Dealer

BUCKAROO SPORTING GOODS
9722 Dudley
Taylor, MI 48180, USA
FFL Firearms Dealer

P & J GUNS
28108 Rose
Trenton, MI 48183, USA
FFL Firearms Dealer
CONTACT: Paul Fuscoe

OTTO'S REPAIR
35041 Florence
Westland, MI 48185, USA
FFL Firearms Dealer

LUIS PEREZ
2310 Mckinley Street
Ypsilanti, MI 48197, USA
FFL Firearms Dealer

WILLIAM FAGAN
P.O. Box 425
Fraser, MI 48206, USA
Antique arms.
Catalogue $5.00 for 6 issues.

RICHARD J. SCHIEVENIN
19288 Eastborne
Harper Woods, MI 48225, USA
FFL Firearms Dealer

DAVID B. SMITH
5125 East Farrand Road
Clio, MI 48420, USA
FFL Firearms Dealer

GERRY L. CHEELY
9176 West Carpenter Road
Flushing, MI 48433, USA
FFL Firearms Dealer

J. B. GUN SHOP
7677 Martin Road
Imlay City, MI 48444, USA
FFL Firearms Dealer

JOHNSTON ENTERPRISES
1093 West Downey Avenue
Flint, MI 48505, USA
FFL Firearms Dealer

DAVID J. SCOTT
1850 Alcott Road
Saginaw, MI 48604, USA
FFL Firearms Dealer

THE L.M. GUN RACK
301 Twin Lake Road
Beaverton, MI 48612, USA
(517)-689-3132
Firearms, ammo & accessories.
S1,S4; 4;
CONTACT: Leo & Loraye McMahon

CHARLES MIRA
3320 West Shaffer Road
Coleman, MI 48618, USA
FFL Firearms Dealer

SMITH'S SHOOTING SUPPLIES
4949 N. Hemlock Road
Hemlock, MI 48626, USA
FFL Firearms Dealer

DOUGLAS E. HOLTZ
5132 South Dehmel Road
Frankenmuth, MI 48734, USA
FFL Firearms Dealer

STEPHEN HART
6982 West Tyler Road
Alma, MI 48801, USA
FFL Firearms Dealer

KEELER'S TRADING POST
5062 Freeman Road
Eaton Rapids, MI 48827, USA
( )-663-8777
Firearms
CONTACT: Roy Keeler

TEN GREENWOOD
2900 Fisk Road
Howell, MI 48843, USA
FFL Firearms Dealer

KIRNBERGER & ROSE
881 Wildemere Drive
Mason, MI 48854, USA
FFL Firearms Dealer

CLASSIC ARMS COMPANY
1600 Lake Lansing Road
Lansing, MI 48912, USA
517-484-6112 517-485-6063
Firearms
CONTACT: Carl Evanoff

L & K GUN & SUPPLIES
71 South McKinley
Battle Creek, MI 49017, USA
FFL Firearms Dealer

GILL DEVERNAY
6974 Little Paw Lake Road
Coloma, MI 49038, USA
FFL Firearms Dealer

WALT'S WORKSHOP
10119 East ML Avenue
Galesburg, MI 49053, USA
FFL Firearms Dealer

ROBERT G. HEATH
4766 -102 Avenue, 48th Street
Grand Junction, MI 49056, USA
FFL Firearms Dealer

GUARANTEED DISTRIBUTORS
County Road 687
Hartford, MI 49057, USA
Firearms

MIDWEST ARMS
P.O. Box 92
Niles, MI 49120, USA
FFL Firearms Dealer

THOMAS J. PAQUIN
2440 Lincoln Street
Muskegon, MI 49441, USA
FFL Firearms Dealer

MARK'S GUNS
2443 East Shettler Road
Muskegon, MI 49444, USA
FFL Firearms Dealer

THE LAST HOMESTEAD
Rt. 3, Box 205
Reed City, MI 49677, USA
FFL Firearms Dealer

LEROY A. RABIDEAU
10429 Chickagami Trail
Brutus, MI 49716, USA
FFL Firearms Dealer

ANTRIM MUNITIONS
RR 1, Box 35
Ellsworth, MI 49729, USA
FFL Firearms Dealer
CONTACT: Mark Groenink

PEACE OF MIND SALES
3405 Valdez Drive
Des Moines, IA 50310, USA
FFL Firearms Dealer

DOUGLAS R. CARLSON
1379 73rd Street, Dept. SGN
Des Moines, IA 50311, USA
Antique American firearms.
Catalogue 4.00.

P & M
1108 Mechanic
Osage, IA 50461, USA
FFL Firearms Dealer
CONTACT: Scott Putzier

DENNY'S SHOOTERS SUPPLY
124 East 18th Street
Cedar Falls, IA 50613, USA
Firearms & accessories.
List available.

KERN GUN SHOP
1200 Lilac Lane
Cedar Falls, IA 50613, USA
FFL Firearms Dealer

LORENZ OUTDOOR SUPPLY
600 5th Avenue, S.W.
Independence, IA 50644, USA
FFL Firearms Dealer

MILITARY FIREARMS

KROEGER KUSTOM STOCKS
144 Collins Avenue
Evansdale, IA 50707, USA
FFL Firearms Dealer

DOUG'S SHOOTERS SUPPLY
302 N. Brooks
Lennox, IA 50851, USA
FFL Firearms Dealer

DEAN FRITZ
P.O. Box 117
Castana, IA 51010, USA
FFL Firearms Dealer

AL'S GUNS & AMMO
Rt. 1
Milford, IA 51351, USA
FFL Firearms Dealer
CONTACT: Al Joyner

PAUL R. RATHMAN
3011 - 9th Avenue
Council Bluffs, IA 51501, USA
FFL Firearms Dealer

JD's TRADIN' POST
Rt. 1, Box 219-E
Honey Creek, IA 51542, USA
FFL Firearms Dealer

LYNN ADAMS
1202 Corning
Red Oak, IA 51566, USA
FFL Firearms Dealer

BARBAROSSATAL
RR 2, Box 73-A
Lansing, IA 52151, USA
FFL Firearms Dealer

K & L ENTERPRISES
3215 Big Bend Road
Ely, IA 52227, USA
Firearms
CONTACT: Kendall L. Homrighausen

THE AMMO BEARER
600 Highway One West
Iowa City, IA 52240, USA
FFL Firearms Dealer

EGLI GUN REPAIR
530 South Avenue B
Washington, IA 52353, USA
FFL Firearms Dealer

DAVE'S GUNS & AMMO
RR 5, Box 202
Ottumwa, IA 52501, USA
FFL Firearms Dealer

BILL'S DEFENSIVE FIREARMS
1304 Fairlane Drive
Bettendorf, IA 52722, USA
FFL Firearms Dealer

HAMILTON'S HOBBY
402 - 5th Street
Princeton, IA 52768, USA
FFL Firearms Dealer

JIM'S GUNS & AMMO

Rt. 1, Box 74
Eden, WI 53019, USA
FFL Firearms Dealer
CONTACT: James Koffman

JACK DEXTER
P.O. Box 435
Germantown, WI 53022, USA
414-251-8853
FFL Firearms Dealer

WORTHINGTON'S GUN SHOP
1307 Woodside Street
Hartland, WI 53029, USA
FFL Firearms Dealer

LARRY'S GUNS
Rt. 2, Box 116-A
Malone, WI 53049, USA
FFL Firearms Dealer

ROBERT L. DeWITT, JR.
211 Park Avenue
Pewaukee, WI 53072, USA
FFL Firearms Dealer

JAMES C. HANSON
1316 N. 11th Avenue
West Bend, WI 53095, USA
FFL Firearms Dealer

R. S. SUPPLIES
S47W35991 Meadows Drive
Dousman, WI 53118, USA
FFL Firearms Dealer

PIDCOCK'S GUN SHOP
13 West Geneva Street
Elkhorn, WI 53121, USA
FFL Firearms Dealer

SHERRON & SONS GUN & COIN SHOP
7201 - 60th Street
Kenosha, WI 53142, USA
FFL Firearms Dealer

GLENN WOLLENZIEHN
14825 W. Small Road
New Berlin, WI 53151, USA
414-422-0927
Antique ordnance.
Send SASE for list.

CASANOVA'S GUNS, INC.
1601 W. Greenfield Avenue
Milwaukee, WI 53204, USA
Collector books & firearms.

LLOYD'S GUNS
1946 S. 12th Street
Milwaukee, WI 53204, USA
FFL Firearms Dealer

INVESTMENT ARMS
1243 N. 29
Milwaukee, WI 53208, USA
(414)-933-7947
Ship & receive firearms, Retail sales, Remington "51" parts (1919 to 1929).
S1, S11, S12, S13; 11;
CONTACT: Brian M. Czerwinski
SALES DEPT: Brian M. Czerwinski
PURCHASING DEPT: Brian M. Czerwinski

RICHARD SEEBACH
2647 N. 47
Milwaukee, WI 53210, USA
FFL Firearms Dealer

MICHAEL MARSHALL
2736-B South 60th Street
Milwaukee, WI 53219, USA
FFL Firearms Dealer

RJB SALES SHOP
10824 W. Fiebrantz Avenue
Wauwatosa, WI 53222, USA
FFL Firearms Dealer

MANNING'S X-RING RELOADS
8610 N. 51st Street
Brown Deer, WI 53223, USA
FFL Firearms Dealer

R & G SUPPLIES
9984 W. Fond du Lac Avenue
Milwaukee, WI 53224, USA
FFL Firearms Dealer
CONTACT: Greg Barillari

STEVEN STUPAR
5534 Randal Lane
Racine, WI 53402, USA
FFL Firearms Dealer

JOHN FALLON
1012 Elm Street
Beloit, WI 53511, USA
FFL Firearms Dealer

JOHN F. BURN
P.O. Box 73
Beloit, WI 53511, USA
FFL Firearms Dealer

KORTH'S GUN SHOP
451 Ollie Street
Cottage Grove, WI 53527, USA
FFL Firearms Dealer
CONTACT: Karl E. Korth

STANLEY E. DENISON
Route 1, Box 174
Evanville, WI 53536, USA
FFL Firearms Dealer

U-LOAD
3787 Hwy 151 Rt 1
Sun Prairie, WI 53590, USA
FFL Firearms Dealer

SHANE ENTERPRISES
820 West Main Street
Sun Prairie, WI 53590, USA
FFL Firearms Dealer
CONTACT: Ernest Shane

NORM SANNES
666 S. Thompson Road
Sun Prairie, WI 53590, USA
FFL Firearms Dealer

WATERLOO WELDING & MACHINE SHOP
203 South Monroe Street
Waterloo, WI 53594, USA
FFL Firearms Dealer

MILITARY FIREARMS

GARY W. HENDRICKSON
2759 Ledgemont Street
Madison, WI 53711, USA
FFL Firearms Dealer

FRONTIER FIREARMS
350 Broadway
Platteville, WI 53818, USA
FFL Firearms Dealer
CONTACT: Jerald Stanton

GUS'S GUN SHOP
P.O. Box 693
Lake Delton, WI 53940, USA
FFL Firearms Dealer

JEROME K. DOOLITTLE
Rt. 2, Box 47A
Amery, WI 54001, USA
FFL Firearms Dealer

PAULSON BROS. ORDNANCE CORP.
P.O. Box 121
Clear Lake, WI 54005, USA
715-263-2112 715-263-2101
Rifled cannon work & supplies.
Write for information.
CONTACT: Bernie & Bruce Paulson

SYLTES HUNTING & FISHING
725 Mary Avenue
New Richmond, WI 54017, USA
FFL Firearms Dealer

J. K. SHOOTERS SUPPLY
132 East Ryan
Brillion, WI 54111, USA
FFL Firearms Dealer

SILVER CREEK SPORTSHOP
Star Route 1
Pembine, WI 54156, USA
FFL Firearms Dealer

RON BUCHMANN
Rt. 3, Box 293
Seymour, WI 54165, USA
FFL Firearms Dealer

ED'S GUN SHOP
Rt. 4, Box 320
Shawano, WI 54166, USA
FFL Firearms Dealer

GUNS UNLIMITED
1236 Bluebird Street
Sturgeon Bay, WI 54235, USA
FFL Firearms Dealer
CONTACT: K. M. Yedica

DAVID B. MUELLER
1944 University Avenue
Green Bay, WI 54302, USA
FFL Firearms Dealer

AL'S SPORTS
111 North Central
Marshfield, WI 54449, USA
FFL Firearms Dealer
CONTACT: David R. Knauf

BOB'S GUN SHOP
Rt. 1, Box 233, Cty. C
Wabeno, WI 54566, USA
FFL Firearms Dealer

RICHARD KLUG
912 Cass Street
LaCrosse, WI 54601, USA
FFL Firearms Dealer

LOUIS J. MANE
Rt. 4, Box 38
Black River Falls, WI 54615, USA
FFL Firearms Dealer

BRAD HANSON
204 Anderson St., Box 4
Coon Valley, WI 54623, USA
FFL Firearms Dealer

POOR BOY'S GUN SHOP
Rt. 2, Box 26
Hillsboro, WI 54634, USA
FFL Firearms Dealer

GOODVIEW GUNS
PO Box 476
Onalaska, WI 54650, USA
FFL Firearms Dealer
CONTACT: Charles E. Roberge

HAHN-KEE, INC.
840 Water Street
Eau Claire, WI 54701, USA
FFL Firearms Dealer

RUDY'S RELOADING & GUN REPAIR
Rt. 1, Box 77-A
Cadott, WI 54727, USA
FFL Firearms Dealer

BOB'S SHOOTING SUPPLIES
218 7th Avenue East
Ashland, WI 54806, USA
FFL Firearms Dealer

SCOTT WARWICK
Rt. 1, Box 148
Cumberland, WI 54829, USA
FFL Firearms Dealer

OLSEN ELECTRIC & HARDWARE
Northern Lights Road
Rt. 3, Box 869
Hayward, WI 54843, USA
FFL Firearms Dealer
CONTACT: Rodger L. Olsen

BIG FOOT ENTERPRISES
Rt. 8, Box 8299
Hayward, WI 54843, USA
FFL Firearms Dealer

RS GUN SALES, INC.
Rt. 2, Box 34
Neshkoro, WI 54960, USA
FFL Firearms Dealer
CONTACT: R. F. Scherbarth, Sr.

JAMES E. SCHEMA
528 Short Street
Faribault, MN 55021, USA
FFL Firearms Dealer

BILL SCHNEIDER SUPPLY COMPANY
970 W. 2nd Street
Rush City, MN 55069, USA
FFL Firearms Dealer

TODD BATES, III
P.O. Box 794
Willernie, MN 55090, USA
FFL Firearms Dealer

ALEX J. LAKE
1370 Osceola Ave.
St. Paul, MN 55105, USA
FFL Firearms Dealer

MINNESOTA FIREARMS EXCHANGE
310 - 167th Avenue N.E.
Anoka, MN 55303, USA
FFL Firearms Dealer

KEVIN SMUDER
Route 1, Box 104-J
Big Lake, MN 55309, USA
FFL Firearms Dealer

LAPLANT SERVICES
11490 122 Street
Cologne, MN 55322, USA
FFL Firearms Dealer
CONTACT: Ralph Laplant

GALE'S GUN SHOP
330 E. 2nd Street
Litchfield, MN 55355, USA
FFL Firearms Dealer

DOUG GILBERTSON
300 Dakota Ave. S.
Golden Valley, MN 55416, USA
FFL Firearms Dealer

NORTHWEST ASSOCIATED
CONSULTANTS
4820 Minnetonka Blvd.
Minneapolis, MN 55416, USA
FFL Firearms Dealer
CONTACT: David R. Licht

THE ARMORY
2911 Benjamin St., N.E.
Minneapolis, MN 55418, USA
FFL Firearms Dealer
CONTACT: Cliff McMeekin, Prop.

HEBRINKS GUNSHOP
480 - 67th Ave., N.E.
Fridley, MN 55432, USA
FFL Firearms Dealer
CONTACT: Walter Hebrink

BILL HICKS & COMPANY LTD.
3700 Annapolis Lane
Minneapolis, MN 55441, USA
(612)-559-4703
Firearms

STEVEN A. HOLT
56 Edison Blvd.
Silver Bay, MN 55614, USA
FFL Firearms Dealer

MILITARY FIREARMS

DAVID J. & ELAINE M. HAGEN
HCR-2, Box 378
Bovey, MN 55709, USA
FFL Firearms Dealer

JOHN P. JUGOVICH
222 S.W. 8th Street
Chisholm, MN 55719, USA
FFL Firearms Dealer

NORTHLAND SPORTS CENTER
P.O. Box 128
Floodwood, MN 55736, USA
FFL Firearms Dealer

SOUP TO NUTS STORE
2403 Splithand Road
Grand Rapids, MN 55744, USA
FFL Firearms Dealer
CONTACT: Russell G. Heavirland

JACK & JEAN KENNING
3610 - 20th Ave., N.W.
Rochester, MN 55901, USA
FFL Firearms Dealer

ED'S GUNS HOUSE
Rt. 1, Box 62
Minnesota City, MN 55959, USA
507-689-2925
Firearms.
List $1.00.

BOB GRAHAM
211 Floral Ave.
Mankato, MN 56001, USA
FFL Firearms Dealer

JOERG BROS. GUNSHOP
193 Mary Circle
North Mankato, MN 56001, USA
FFL Firearms Dealer
CONTACT: Carrol H. Joerg

HARVEY'S TRADING POST
113 N. Cedar
Luverne, MN 56156, USA
FFL Firearms Dealer

NORSTEN SUPPLY
1821 S. 1st
Willmar, MN 56201, USA
FFL Firearms Dealer
CONTACT: Gary Norsten

GUNS, ETC.
Route 3
St. Cloud, MN 56301, USA
FFL Firearms Dealer
CONTACT: Bruce Huebert

NORTH COUNTRY
315 - 20th Ave., N.
St. Cloud, MN 56301, USA
FFL Firearms Dealer
CONTACT: Rick Stevens

CONDON'S GUNS
State Road 77
Pine River, MN 56474, USA
FFL Firearms Dealer
CONTACT: Pat Condon

JOHN FINNEGAN
Rt. 3, Box 181
Wadena, MN 56482, USA
FFL Firearms Dealer

MDH SALES
RR 1
Elbow Lake, MN 56531, USA
FFL Firearms Dealer
CONTACT: Mark Huseth

JIM PIXLEY
23 7th Ave., N.E.
Pelican Rapids, MN 56572, USA
FFL Firearms Dealer

AMERICAN BUILDERS
P.O. Box 847
Bemidji, MN 56601, USA
FFL Firearms Dealer
CONTACT: Jerry Smith

JOSEPH R. WITTER
Route 8, Box 584
Bemidji, MN 56601, USA
FFL Firearms Dealer

TIESZEN'S GUN SHOP
57 N. Broadway
Marion, SD 57043, USA
FFL Firearms Dealer

DAKOTA GUNS
3800 S. Louise - Box 899
Sioux Falls, SD 57101, USA
FFL Firearms Dealer
CONTACT: H. B. Lawrence

BILL J. BLACKMAN
1701 Comet Road
Sioux Falls, SD 57103, USA
FFL Firearms Dealer

KAHLES GUNS
140 11th Street SW
Watertown, SD 57201, USA
FFL Firearms Dealer
CONTACT: William K. McClintic

BRADLEY D. JOHNSON
RR 2, Box 27
White, SD 57276, USA
FFL Firearms Dealer

FRONTIER ARMS
Rt. 4
Huron, SD 57350, USA
FFL Firearms Dealer
CONTACT: Larry R. Hulsted

UULEMANS
2524 - 3rd Ave., S.E.
Aberdeen, SD 57401, USA
FFL Firearms Dealer

PARTS UNLIMITED
HCR 37 #169A
Pierre, SD 57501, USA
FFL Firearms Dealer
CONTACT: Don McGregor

PETE'S GUN SHOP
Box 508
Ft. Pierre, SD 57532, USA
FFL Firearms Dealer

GALEN C. WHIPPLE
210 E. St. Charles
Rapid City, SD 57701, USA
FFL Firearms Dealer

DAKOTA LOG
108 E. Texas
Rapid City, SD 57701, USA
FFL Firearms Dealer
CONTACT: Jim Walstrom

LEWIS B. NUNNALLY, JR.
1402 E. Franklin Street
Rapid City, SD 57701, USA
FFL Firearms Dealer

WESTSIDE ARMS COMPANY
4907 Baldwin Street
Rapid City, SD 57702, USA
FFL Firearms Dealer

DECKER ARMS
Rt. 1, Box 102-E
Hot Springs, SD 57747, USA
FFL Firearms Dealer

HAROLD CROWE
1614 N. 5th St.
Fargo, ND 58102, USA
FFL Firearms Dealer

HE-MART
112 Main, Box 66
Cavalier, ND 58220, USA
FFL Firearms Dealer

LESLIE G. VOLLRATH
Rt. 1, Box 56
Pembina, ND 58271, USA
FFL Firearms Dealer

BOLEY GUN REPAIR & SALE
1414 S.W. 10th
Minot, ND 58701, USA
(701)-838-4735
FFL Firearms Dealer
S4,S7,S10,S11,S12; 8;
CONTACT: James H. Boley
SALES DEPT: Jim Boley
PURCHASING DEPT: Donna Boley

JAMES FLAHERTY
Box 61
Berthold, ND 58718, USA
FFL Firearms Dealer

SQUIRES ENTERPRISES
P.O. Box 61
210 Granite Street
Gardiner, MT 59030, USA
FFL Firearms Dealer

JIM'S GUN ROOM
735 Enfield Street
Billings, MT 59101, USA
FFL Firearms Dealer

MILITARY FIREARMS

ROBERT N. JOHNSON, JR.
1418 Mading Drive
Billings, MT 59105, USA
FFL Firearms Dealer

WILLIAM SCHIENO
1606 Janie Street
Billings, MT 59105, USA
FFL Firearms Dealer

BORNSCREEK RELOADING
Rt. 2, Box 346
Savage, MT 59262, USA
FFL Firearms Dealer

JERRY RONNING
Tongue River Stage
Niles City, MT 59301, USA
FFL Firearms Dealer

FORT YELLOWSTONE
194 & Mt. 16
Glendive, MT 59330, USA
FFL Firearms Dealer

GARY FARIS
P.O. Box 446
Jordan, MT 59337, USA
FFL Firearms Dealer

JOHN MAETZOLD
P.O. Box 267
Jordan, MT 59337, USA
FFL Firearms Dealer

NORTHWEST ARMS
634 - 23rd Street
Black Eagle, MT 59414, USA
FFL Firearms Dealer

JAMES R. MARTINSON
317 6th Avenue SE
Cut Bank, MT 59427, USA
FFL Firearms Dealer

M & L ENTERPRISES
P.O. Box 351
Shelby, MT 59474, USA
FFL Firearms Dealer

MUSSELSHELL LUMBER CO.
P.O. Box 277
5 Miles West of Melstone
Melstone, MT 59504, USA
FFL Firearms Dealer

JACK HALTER
P.O. Box 85
Big Sandy, MT 59520, USA
FFL Firearms Dealer

RICHARD L. FRANZ
524 Tamarack
Helena, MT 59601, USA
FFL Firearms Dealer

HUNT'N SHACK WEST
8375 Diamond Springs Drive
Helena, MT 59601, USA
FFL Firearms Dealer

A. J. OLLILA
6124 Hwy. 12 W.
Helena, MT 59601, USA
FFL Firearms Dealer

LEE'S SHOOTERS SUPPLY
P.O. Box 9048
Helena, MT 59604, USA
FFL Firearms Dealer
CONTACT: Oscar A. Lee

SOUTHSIDE SHOOTERS SUPPLY
Rt. 1, Box 8-F
Townsend, MT 59644, USA
FFL Firearms Dealer
CONTACT: M. D. Burtch

THOMAS E. BRADY
1330 South Church
Bozeman, MT 59715, USA
FFL Firearms Dealer

MICHAEL D. MITCHELL
P.O. Box 178
Virginia City, MT 59755, USA
FFL Firearms Dealer

TOED ENTERPRISES
127 East Front, #203
Missoula, MT 59802, USA
FFL Firearms Dealer

EDWIN C. TROESTER
P.O. Box 345
46 Overlook Road
Frenchtown, MT 59834, USA
FFL Firearms Dealer

TONY R. ARMOUR
638 Totem Road
Stevensville, MT 59870, USA
FFL Firearms Dealer

JOHN TINGLEY
P.O. Box 298
4 Miles East of Thompson Falls
Thompson Falls, MT 59873, USA
FFL Firearms Dealer

DOUGLAS C. LARSON
35 Amdahl Lane
Kalispell, MT 59901, USA
FFL Firearms Dealer

PHIL SCHREIBER
Rt. 1, Box 2
Eureka, MT 59917, USA
FFL Firearms Dealer

PARELLEX
1090 Fargo Avenue
Elk Grove Village, IL 60007, USA
312-228-0080
Survival & firearms supplies.
Catalogue available.

CREATIVE FIREARMS, LTD.
521 Crest Avenue
Elk Grove Village, IL 60007, USA
(312)-593-3535
Firearms & accessories.
S3,S6,S7,S11,S12; 4;
CONTACT: Tony Esposito
SALES DEPT: Tony Esposito
PURCHASING DEPT: Tony Esposito

Firearms, accessories, knives, etc. Old & new. Any & all. Plus custom work. Fast service & reasonable prices.

F & S SPORTING GOODS
7012 West Hillcrest Drive
Crystal Lake, IL 60014, USA
FFL Firearms Dealer
CONTACT: Tom Dowdy

RON'S FIREARMS SALES
18646 Westwood Plaza
Gurnee, IL 60031, USA
FFL Firearms Dealer

NESARD
P.O. Box 56
Lake Zurich, IL 60047, USA
312-381-7629
Firearms

McHENRY COUNTY FIREARMS
3632 West McCullom Lake Road
McHenry, IL 60050, USA
FFL Firearms Dealer

TEMPO COMPANY
422 Emerson Lane
Mundelein, IL 60060, USA
Guns, knives, and ammo.
S3; 20;
CONTACT: Dale Johnson
SALES DEPT: Dale Johnson
PURCHASING DEPT: Dale Johnson
Well stocked dealer. Approximately 200 different revolvers, auto pistols, rifles, shotguns & Class III weapons.

KENCO GUN & BOW
938 Whitfield Road
Northbrook, IL 60062, USA
FFL Firearms Dealer

MODCO
1215 Prairie Brook Drive
Palatine, IL 60067, USA
FFL Firearms Dealer

ROBERT H. HOLLOWAY, JR.
639 Fulton Street
Waukegan, IL 60085, USA
FFL Firearms Dealer

C. M. HILLMANN
3148 B Arizona
Great Lakes, IL 60088, USA
FFL Firearms Dealer

NORTHERN SPECIALTY PRODUCTS
P.O. Box 937
Wheeling, IL 60090, USA
(312)-459-6194
Firearms & edged weapons.
S1; 8;

B. G. SHOOTERS SUPPLY
219 Raupp Blvd.
Buffalo Grove, IL 60090, USA
FFL Firearms Dealer

MILITARY FIREARMS

HEWES LIMITED-ARMS DIVISION
10537 Crown Road, St. #9
Franklin Park, IL 60131, USA
FFL Firearms Dealer
CONTACT: Glenn J. Wehe

AARDVARK ENTERPRISES
P.O. Box 1046
Glendale Heights, IL 60139, USA
(312)-653-0133
Military vehicles & Jeep parts.
S6,S12; 6;
CONTACT: Dan Rhame
SALES DEPT: Dan Rhame
PURCHASING DEPT: Dan Rhame
Manufacturer of replica M60 MGs. Inf., A/B, patrolboat, & lightweight configurations; also limited quantities of early style models. Models are full-size, fit original GI mounts & incorporate aluminum and original GI parts, , but cannot fire and are completely legal. Prices from $500. - $1,000.

POPPEN ENTERPRISES
Rt. 3, 357 Highland Avenue
Hampshire, IL 60140, USA
FFL Firearms Dealer

THE GUN EMPORIUM
29 Hillside #4
Hillside, IL 60162, USA
FFL Firearms Dealer

SILHOUETTE SPORTING GOODS
2340 Laurel Lane
Sycamore, IL 60178, USA
FFL Firearms Dealer

ENFORCERS OUTFITTING,INC.
P.O. Box 1
Villa Park, IL 60181, USA
312-941-0615 EXT.714
Stun guns.

R. D. ENTERPRISES
P.O. Box 332
Winfield, IL 60190, USA
FFL Firearms Dealer

BILL'S SPORTING GOODS
3483 Challas Road
Crete, IL 60417, USA
FFL Firearms Dealer

W & S GUN SHOP
409 Cayuga
Joliet, IL 60432, USA
FFL Firearms Dealer

RICH'S SHOOTER SUPPLY
5400 West 157th Street
Oak Forest, IL 60452, USA
FFL Firearms Dealer

CATALOG ARMS, INC.
6000 West 79th Street
Burbank, IL 60459, USA
FFL Firearms Dealer

ROBERT J. SHACHTER
12033 South 73rd Court
Palos Heights, IL 60463, USA
FFL Firearms Dealer

PIONEER GUNS
1047 Riverview Drive
South Holland, IL 60473, USA
FFL Firearms Dealer

RAY WIKTOR
3506 Ashland Avenue
Steger, IL 60475, USA
FFL Firearms Dealer

THE GUN LODGE
991 Aurora Avenue, Rt. 25
Aurora, IL 60505, USA
FFL Firearms Dealer

RELOADER SERVICE
631 Watson Street
Aurora, IL 60505, USA
FFL Firearms Dealer

GARY D. STRASSER
4312 South Elm Avenue
Brookfield, IL 60513, USA
FFL Firearms Dealer

GOLDEN FEATHER & GEM SHOP
113 Grant Street
North Aurora, IL 60542, USA
FFL Firearms Dealer

NANCY CRANE
14231 Clark
Riverdale, IL 60627, USA
FFL Firearms Dealer

LESTER E. WESNER
4130 West 57 Street
Chicago, IL 60629, USA
FFL Firearms Dealer

CUSTOM ARMAMENT SYSTEMS CO.
5617-23 West Howard Street
Niles, IL 60648, USA
FFL Firearms Dealer

DONOHO'S GUN SHOP
RR 4
Rochelle, IL 61068, USA
FFL Firearms Dealer

CHUCK'S GUNS & AMMO
1112 Hamburg Road
Roscoe, IL 61073, USA
FFL Firearms Dealer

ROCK RIVER GUNS & AMMO
1000 Windsor Road
Loves Park, IL 61111, USA
FFL Firearms Dealer
CONTACT: Jeff Matthews

SPRINGFIELD ARMORY,INC.
420 West Main Street
Genesco, IL 61254, USA
Firearms & accessories.
$3.00 catalogue.

ROCK ISLAND ARMORY,INC.
111 E. Exchange
Genesco, IL 61254, USA
309 944-2109
Firearms & accessories

INDEPENDENT ENTERPRISES
124 South Monroe
Streater, IL 61364, USA
FFL Firearms Dealer

JAY E. THOMAS
RR 1, Box 160
Galesburg, IL 61401, USA
FFL Firearms Dealer

J. B. WASSI
14338 Grandview Drive
Chillicothe, IL 61523, USA
FFL Firearms Dealer

3-D SHOOTING SUPPLY
RR 1, Box 43
Hanna City, IL 61536, USA
FFL Firearms Dealer

WARD J. CAIN
601 Edwards
Henry, IL 61537, USA
FFL Firearms Dealer

GOLDSBOROUGH SHOOTERS SUPPLY
1004 East Lincoln
Bloomington, IL 61701, USA
FFL Firearms Dealer

F. J. VOLLMER & COMPANY,INC.
#3 Towanda Access Road
Bloomington, IL 61701-2485, USA
309-663-9494 309-663-9495
Firearms

MODERN GUNS AND ROCKS
P.O. Box 820
Bloomington, IL 61702, USA
FFL Firearms Dealer

CLARENCE CLAFLIN
204 West Clark
Heyworth, IL 61745, USA
FFL Firearms Dealer

DAVID C. HEATH
P.O. Box 15
Champaign, IL 61820, USA
FFL Firearms Dealer

SCHUTZ'S SHOOTERS
1718 Hedge Road
Champaign, IL 61821, USA
FFL Firearms Dealer
CONTACT: Terry Schutz

HOLYCROSS GUN SHOP
707 East 9th
Georgetown, IL 61846, USA
FFL Firearms Dealer

BUD'S GUNS & AMMO
RR 1, Box 169
Indianola, IL 61850, USA
FFL Firearms Dealer

FLASHBACK FIREARMS
1511 Central Avenue
Alton, IL 62002, USA
FFL Firearms Dealer
CONTACT: L. D. Chesnut

MILITARY FIREARMS

JIM SPRINGER
P.O. Box 116
Columbia, IL 62236, USA
FFL Firearms Dealer

IKE'S GUN REPAIR
RR 1, Box 273
Evansville, IL 62242, USA
FFL Firearms Dealer

HERBERT A. REINHARDT
Rt. 3, Box 54-A
Pinckneyville, IL 62274, USA
FFL Firearms Dealer

ERIC S. THIELE
PO Box 14
Blue Mound, IL 62513, USA
FFL Firearms Dealer

DENNIS GEESMANN
323 South Broad
Carlinville, IL 62626, USA
FFL Firearms Dealer

FRONTIER INDOOR RANGE
RR 1, Box 389B
Chatham, IL 62629, USA
FFL Firearms Dealer
CONTACT: Kenneth M. Burks

VERNON MEDLOCK
RR 1, Sandusky Road
Jacksonville, IL 62650, USA
FFL Firearms Dealer

ABLE DETECTIVE AGENCY, INC.
Firearms Department
320 North 5th Street
Springfield, IL 62701, USA
FFL Firearms Dealer

FRONTIER GUNS & AMMO
847 East North Avenue
Flora, IL 62839, USA
FFL Firearms Dealer

RICK'S GUN SUPPLY
Route 1, Box 412-A
Johnston City, IL 62951, USA
FFL Firearms Dealer

AKEN BROTHERS
RR 4, Box 192
Murphysboro, IL 62966, USA
FFL Firearms Dealer

TOM'S GUNS
12 Village Plaza
Arnold, MO 63010, USA
(314)-464-8667
Military firearms, edged weapons, ammo.
S1; 6;
CONTACT: Tom F. Runzi
SALES DEPT: Tom Runzi
PURCHASING DEPT: Tom Runzi

YELLOW ROSE FIREARMS
269 Village Mead
Ballwin, MO 63011, USA
FFL Firearms Dealer

THOR DISTRIBUTING

2945 Wedde Road
Barnhart, MO 63012, USA
FFL Firearms Dealer

JARVIS & JARVIS GUNSMITH
212 William
DeSoto, MO 63020, USA
FFL Firearms Dealer

S & S FIREARMS
P.O. Box 1137
Ballwin, MO 63022, USA
FFL Firearms Dealer

SCOTT R. CAMPBELL
576 Talon Court
Florissant, MO 63031, USA
FFL Firearms Dealer

INTERNATIONAL FIREARMS
4912 Orange Blossom Court
Hazelwood, MO 63042, USA
FFL Firearms Dealer

TROPHIES
Rt. 1, Box 131-A
St. Claire, MO 63077, USA
FFL Firearms Dealer

CHARLES V. VAN DYKE
2837 S. Jefferson
St. Louis, MO 63118, USA
FFL Firearms Dealer

PINE LAWN LOAN & JEWELRY CO.
6155 Natural Bridge Road
St. Louis, MO 63120, USA
FFL Firearms Dealer

AL ALLEN GUNS
3832 Buckley Road
St. Louis, MO 63125, USA
FFL Firearms Dealer

DONALD BAYER
117 East Velma
St. Louis, MO 63125, USA
FFL Firearms Dealer

WAFFENHAUS
731 Fieldcrest
Crestwood, MO 63126, USA
(314)-248-1517
Military firearms & accessories.
S11,S12; 3;
CONTACT: Jon H. Luer
SALES DEPT: Jon H. Luer
PURCHASING DEPT: Jon H. Luer
Waffenhaus is a small business specializing primarily in pre-1945 military small arms and occasionally in uniforms and technical manuals when available.

COMMENT FIREARMS
1553 Marbella Street
St. Louis, MO 63138, USA
FFL Firearms Dealer

VILLAGE SPORTS
1628 Clocktower Drive
Spanish Lake, MO 63138, USA
FFL Firearms Dealer

TOWER GUN SHOP
6 Spiede Lane
Creve Coeur, MO 63141, USA
FFL Firearms Dealer
CONTACT: Jack D. Lite

NATIONAL AUTOMATIC PISTOL
COLLECTORS ASSOCIATION
Box 15738 Tower Grove Station
St. Louis, MO 63163, USA
(314)-771-1160
Automatic pistols collectors association
S10; 19;
CONTACT: Thompson D. Knox

J. L. SABASKI
824 Fox Hill Road
St. Charles, MO 63301, USA
FFL Firearms Dealer

FRONTIER FIREARMS
P.O. Box 231
O'Fallon, MO 63366, USA
FFL Firearms Dealer
CONTACT: Bart Drake

R & R FIREARMS
18 Harborview Drive
Lake St. Louis, MO 63367, USA
FFL Firearms Dealer

KEVIN POWERS
15 Del Ray Court
St. Peters, MO 63376, USA
FFL Firearms Dealer

R & V GUNS & SUPPLIES
#25 Carpenter
St. Peters, MO 63376, USA
FFL Firearms Dealer

KENNETH S. & DOROTHY J. BOOTH
Rt. 3, Box 130
New London, MO 63459, USA
FFL Firearms Dealer

FRANK MALLIN
304 Hudson
Bucknor, MO 64016, USA
FFL Firearms Dealer

SCHWORER MOTOR COMPANY, INC.
11900 Food Lane
Grandview, MO 64030, USA
FFL Firearms Dealer

R & S CUSTOM
3826 Trailridge
Independence, MO 64055, USA
FFL Firearms Dealer

GEORGE ROGERS GUNS
5745 NO. Lakeview Drive
Kansas City, MO 64118, USA
FFL Firearms Dealer

HOEF ENGINEERING
P.O. Box 23971
Kansas City, MO 64141, USA
FFL Firearms Dealer

MILITARY FIREARMS

RON'S NORTHLAND GUNS & SUPPLIES
12106 N.W. Yukon Road
Kansas City, MO 64152, USA
FFL Firearms Dealer

WILBUR L. ADAMS
116 N. Laura Street
Maryville, MO 64468, USA
FFL Firearms Dealer

FARMERS PETROLEUM CO.
P.O. Box 162
Ridgeway, MO 64481, USA
FFL Firearms Dealer

HUNTERS FIREARMS & SUPPLIES
Rt. 4, Box 122
Joplin, MO 64801, USA
FFL Firearms Dealer

GARY D. NEELEY
RR 3, Box 311-A
Eldon, MO 65026, USA
FFL Firearms Dealer

JOHN W. BOWEN
313 Benton
Jefferson City, MO 65101, USA
FFL Firearms Dealer

AUNT HARRI'S PISTOL PARLOR
Rt. 6, Box 31
Columbia, MO 65202, USA
FFL Firearms Dealer

DAVE'S GUN SHOP
811 E. High Street
Boonville, MO 65233, USA
FFL Firearms Dealer

CARL D. FARRES
2502 Plays Avenue
Sedalia, MO 65301, USA
FFL Firearms Dealer

BILL'S AUTO SALES & SALVAGE
Wisdom Star Rt., Box 101
Warsaw, MO 65355, USA
FFL Firearms Dealer

CAHILL'S TV & APPLIANCE
Star Rt. 1, Box 49-B
Vienna, MO 65582, USA
FFL Firearms Dealer

THE GUN RUNNER
203 W. Walnut
Nixa, MO 65714, USA
FFL Firearms Dealer

LEE W. WALTERS
Rt. 2, Box 12
Norwood, MO 65717, USA
FFL Firearms Dealer

GUNSMOKE GUN GALLERY
1500 N. Glenstone
Springfield, MO 65803, USA
FFL Firearms Dealer

STEVE'S GUN SHOP
806 Crawford
Lawrence, KS 66044, USA

FFL Firearms Dealer

R. D. MIDDLEKAMP
15916 Brougham Court
Olathe, KS 66061, USA
FFL Firearms Dealer

RICHARD B. HINZE
Route 3, Box 134-2A
Tonganoxie, KS 66086, USA
FFL Firearms Dealer

FRANK MALLIN
2215 West, 103 South
Mission, KS 66206, USA
FFL Firearms Dealer

KENT M. SMITH
P.O. Box 12920
Overland Park, KS 66212, USA
FFL Firearms Dealer

WEST SIDE BAIT SHOP
512 West Street
Iola, KS 66749, USA
FFL Firearms Dealer

MEEK'S CROSSROADS STORE
RR #2
1 Mi. S. & 5 Mi. E. on Hwy 16
McCune, KS 66753, USA
FFL Firearms Dealer

JAMES L. MALLON
1710 Coronado Avenue
Emporia, KS 66801, USA
FFL Firearms Dealer

MELVIN MILITARY WEAPONS SALES
RR 2, Box 111
Lebo, KS 66856, USA
FFL Firearms Dealer
CONTACT: Melvin Uehling

MILLER AND SON ARMS SUPPLY
227 Pineview
Andover, KS 67002, USA
FFL Firearms Dealer
CONTACT: Bill Miller

DAVID L. BRENTLINGER
16171 W. Hwy. 54 #120
Goddard, KS 67052, USA
FFL Firearms Dealer

KANSAS MECHANICAL, INC.
1924 Arkansas
Wichita, KS 67203, USA
FFL Firearms Dealer

THE BRADSHAW'S
1855 North Richmond
Wichita, KS 67203, USA
FFL Firearms Dealer

JOHN H. BUSH
820 N. Parkwood
Wichita, KS 67208, USA
FFL Firearms Dealer

HUCKSTADT SUMMERS
2029 N. Woodlawn #508
Wichita, KS 67208, USA

FFL Firearms Dealer

CHARLES A. CROUSE
3120 S. Hillside
Wichita, KS 67216, USA
FFL Firearms Dealer

HAYNESS SERVICES
P.O. Box 9072
Wichita, KS 67277, USA
FFL Firearms Dealer

GOLD & SILVER EXCHANGE
213 West 8th
Coffeyville, KS 67337, USA
FFL Firearms Dealer

HAROLD GULLICK
412 North Parkview
Coffeyville, KS 67337, USA
FFL Firearms Dealer

S & T TRADING POST
921 Osage
Salina, KS 67401, USA
FFL Firearms Dealer
CONTACT: Tom Miller

FRIGON GUNS
627 West Crawford Street
Clay Center, KS 67432, USA
913-632-5607
Firearms

FLOYD L. SPRAGUE, JR.
115 North Cherry Street
Lindsborg, KS 67456, USA
FFL Firearms Dealer

REED'S HARDWARE & ELECTRIC
P.O. Box 188
Logan, KS 67646, USA
FFL Firearms Dealer

PEE WEE'S PISTOLS
1701 3rd Avenue
Dodge City, KS 67801, USA
FFL Firearms Dealer
CONTACT: Grady R. Phipps

ZACHARY BASKIN
1116 Colorado Street
Bellevue, NE 68005, USA
FFL Firearms Dealer

KEN W. NISLEY
6628 Franklin Street
Omaha, NE 68104, USA
FFL Firearms Dealer

PETERSON GUN & CLOCK SERVICE
2519 South 41st Street
Omaha, NE 68105, USA
FFL Firearms Dealer

KEISER'S KUSTOM KARTRIDGES
P.O. Box 27306
Ralston, NE 68127, USA
FFL Firearms Dealer

MILITARY FIREARMS

GOFF'S GUNS & SUPPLIES
7901 Edgewood Blvd.
Omaha, NE 68128, USA
FFL Firearms Dealer

THE GUNRUNNER
1426 South 167th Street
Omaha, NE 68130, USA
FFL Firearms Dealer
CONTACT: R. H. Disbrow

SSA ENTERPRISES
6303 Kentucky Road
Papillion, NE 68133, USA
FFL Firearms Dealer
CONTACT: Victor J. Samuel,Jr.

MATHER PHOTOGRAPHY
P.O. Box 404
217 Eldora Avenue
Weeping Water, NE 68463, USA
FFL Firearms Dealer
CONTACT: Wyman R. Mather

MIKE BAUR
5100 Adams
Lincoln, NE 68504, USA
FFL Firearms Dealer

O. C. BLACK
1220 N. 79th Street
Lincoln, NE 68505, USA
FFL Firearms Dealer

ARCHER ARMS
33rd & A Street
Lincoln, NE 68510, USA
FFL Firearms Dealer

R. HENRY
RR 1, Box 61-C
Norfolk, NE 68701, USA
FFL Firearms Dealer

DALE'S GUNS & SUPPLIES
P.O. Box 446
200 East 13th
South Sioux City, NE 68776, USA
FFL Firearms Dealer

THE SPECIAL WEAPONS SHOP
RR 2, Sheen's Tri. Ct. #4
Kearney, NE 68847, USA
FFL Firearms Dealer
CONTACT: Walter B. Kamp,II

DANIEL WARD
P.O. Box 275
Shelton, NE 68876, USA
FFL Firearms Dealer

LEONARD L. FATE
Rt. 1, Box 71
Clay Center, NE 68933, USA
FFL Firearms Dealer

LOCK, STOCK & BARREL GUN SHOP
P.O. Box B
Valentine, NE 69201, USA
402-376-2202
Firearms

WEST NEBRASKA GUNS & AMMO
Box 1-404 So. Park
Lyman, NE 69352, USA
FFL Firearms Dealer

JOHN M. FISCHER
401 Phosphor Avenue
Metairie, LA 70005, USA
FFL Firearms Dealer

JOHN'S HOUSE OF GUNS
3405 Rose Avenue
Chalmette, LA 70043, USA
FFL Firearms Dealer

ST. BERNARD PLATING
2119 Paris Road
Chalmette, LA 70043, USA
FFL Firearms Dealer

ROY L. JUNCKER
609 Fairfield Avenue
Gretna, LA 70053, USA
FFL Firearms Dealer

JOHNNIE E. PARKER
754 Glencove Lane
Gretna, LA 70053, USA
FFL Firearms Dealer

H & R GUNS AND SUPPLIES
308 Bannerwood Drive
Gretna, LA 70053, USA
FFL Firearms Dealer
CONTACT: Herbert G. Robin,Jr.

DEL-GULF SUPPLY CO., INC.
P.O. Box 484
3925 Peters Road
Harvey, LA 70059, USA
FFL Firearms Dealer
CONTACT: Chip Anderson

SIMMONS ARMS
4420 Toby Lane
Kenner, LA 70065, USA
FFL Firearms Dealer
CONTACT: Lawrence Craig Simmons

ERROL L. STEWARD
Route 1, Box 464
Saint Bernard, LA 70085, USA
FFL Firearms Dealer

C. G. SAVOIS
343 Willard Plaza
Westwego, LA 70094, USA
FFL Firearms Dealer

LOUIS F. BICOCCHI
217 Mouton Street
New Orleans, LA 70124, USA
FFL Firearms Dealer

F. PAUL FIFE,JR.
1000 North Cherry Street
Hammond, LA 70401, USA
FFL Firearms Dealer

PHELCO ARMS
1789 Highway, 22 East
Pouchatoula, LA 70454, USA
FFL Firearms Dealer

THE SHOOTIST
1789 Highway, 22 East
Pouchatoula, LA 70454, USA
FFL Firearms Dealer
CONTACT: Dennis Fontenot

TERRY M. WILLIAMS
838 Pine Tree Street
Slidell, LA 70458, USA
FFL Firearms Dealer

LEE CASON
Route 7, Box 7003
Slidell, LA 70461, USA
FFL Firearms Dealer

WILLIAM C. GIBSON
1007 East 12th
Crowley, LA 70526, USA
FFL Firearms Dealer

THE GUN RACK
1321 West Magnolia
Eunice, LA 70535, USA
FFL Firearms Dealer

MARTIN J. BIENVENU
126 Larue Blanc
Scott, LA 70583, USA
FFL Firearms Dealer

DOUG COUSINS
201 Broad
Scott, LA 70583, USA
FFL Firearms Dealer

FERDINAND B. McGEE
904 Alamo
Lake Charles, LA 70601, USA
FFL Firearms Dealer

SCOTT WEST
713-B Bradley
Lake Charles, LA 70605, USA
FFL Firearms Dealer

THE RANGE
3220 Jefferson Street
Baker, LA 70714, USA
FFL Firearms Dealer

GREAT SOUTHERN GUN & KNIFE
Rt. 4, Box 468
Denham Springs, LA 70726, USA
504-664-8653
FFL Firearms Dealer

A. Z. 'MICK' ABED
Rt. 13, Box 129, Bldg. 'B'
Denham Springs, LA 70726, USA
FFL Firearms Dealer

FOXWORTH GUNS
27550 Highway 441
Holden, LA 70744, USA
FFL Firearms Dealer
CONTACT: E. Allen Foxworth

WADE HANKS
43770 Hodgeson Road
Prairieville, IL 70769, USA
504-622-1072
FFL Firearms Dealer

MILITARY FIREARMS

DANIEL A. CHUSTZ
P.O. Box 16
Rougon, LA 70773, USA
FFL Firearms Dealer

MILLET INTERPRISES
Route 1, Box 195-A
Saint Amant, LA 70774, USA
FFL Firearms Dealer

DICKIE GUNS & ACCESSORIES
P.O. Box 98
Watson, LA 70786, USA
FFL Firearms Dealer

GENE'S PAWN & COIN SHOP
3345 Plank Road
Baton Rouge, LA 70805, USA
FFL Firearms Dealer

CLASSIC CAJUN ARMS
1950 Sherwood Forest Drive
Baton Rouge, LA 70815, USA
FFL Firearms Dealer
CONTACT: T. E. Millican

RONALD K. BABINS
10675 White Oak Drive
Baton Rouge, LA 70815, USA
FFL Firearms Dealer

JAD WOLF, BROKER
3021 Woodbrook Drive
Baton Rouge, LA 70816, USA
(504)-291-1696
Antique & modern military firearms.
S1,S4,S6,S12; 15;
CONTACT: Jad Allah Wolf,Jr.
SALES DEPT: Jad Wolf
PURCHASING DEPT: Jad Wolf
Collector of antique and modern military weapons, uniforms & memorabilia.

MAPIASO ENTERPRISES
221 Chimney Lane
Haughton, LA 71037, USA
FFL Firearms Dealer
CONTACT: Gary W. Smith

ROGER L. HARP
Route 5, Box 5309
Keithville, LA 71047, USA
FFL Firearms Dealer

SOUTHERN ARMS COMPANY
40 Tealwood
Shreveport, LA 71104, USA
FFL Firearms Dealer

THE GUN & KNIFE TRADER
204 East 70th
Shreveport, LA 71106, USA
FFL Firearms Dealer

THE FISH-N-RIG
1819 Jimmie Davis Highway
Bossier City, LA 71112, USA
FFL Firearms Dealer
CONTACT: Robert G. Bryant

GEORGE E. STADTER
202 Sandel Drive
Monroe, LA 71203, USA
FFL Firearms Dealer

F. F. GREGORY
515 Tulip Street
Rayville, LA 71269, USA
FFL Firearms Dealer

STEVE CARRICO
Route 1, Box 2048
Ruston, LA 71270, USA
FFL Firearms Dealer

BOLGAR ENTERPRISES
309 Mark Street
Alexandria, LA 71301, USA
FFL Firearms Dealer

DUTCHMAN'S ECONOMY GUN REPAIR
311 North Gayle
Bunkie, LA 71322, USA
FFL Firearms Dealer

J. C. PAWN SHOP
405 Tunica Drive East
Marksville, LA 71351, USA
FFL Firearms Dealer

GREG McMILLON
P.O. Box 1026
Columbia, LA 71418, USA
FFL Firearms Dealer

BILLY R. AINSWORTH
Rt. 1, Box 25
Kelly, LA 71441, USA
FFL Firearms Dealer

THE SWAP SHOP
Main & Church
Warren, AR 71671, USA
FFL Firearms Dealer

JAMES F. LANDERS
1155 Charley's Loop Road
Camden, AR 71701, USA
FFL Firearms Dealer

McMAHAN GUN SHOP
753 Dennis Street
Camden, AR 71701, USA
FFL Firearms Dealer

L. L. BASTON COMPANY
P.O. Box 1995
2101 North College
El Dorado, AR 71730, USA
501-863-5659 800-643-1564
Firearms accessories

DAVID STEVENS
3010 Ferrand Lane
El Dorado, AR 71730, USA
FFL Firearms Dealer

DEAL'S PLACE
919 North Washington
Murfreesboro, AR 71958, USA
FFL Firearms Dealer

WILLIAM V. KNOX
6 Beechwood
Cabot, AR 72023, USA
FFL Firearms Dealer

TATES MILL SHOOTING SUPPLIES
Rt. 1, Box 119-K
Cabot, AR 72023, USA
FFL Firearms Dealer

HARRIS SWAP & SPAWN SHOP
Rt. 3, Box 31
Clinton, AR 72031, USA
FFL Firearms Dealer

JAMES D. HARRISON
P.O. Box 4
Malvern, AR 72104, USA
FFL Firearms Dealer

KEN'S GUN SHOP
2211 Vancouver
Little Rock, AR 72204, USA
FFL Firearms Dealer

THE GUNNER'S MATE
6100 Evergreen Drive
Little Rock, AR 72207, USA
FFL Firearms Dealer
CONTACT: Richard Howell

JERRY HUMBRECHT
c/o J & J Mold & Die
9811 Interstate 30
Little Rock, AR 72209, USA
FFL Firearms Dealer

R. A. ALLEN
616 North 11th
Blytheville, AR 72315, USA
FFL Firearms Dealer

LOKER ENTERPRISES
Rt. 8, Box 375
Harrison, AR 72601, USA
FFL Firearms Dealer

HAMMERS SUPPLY
Rt. 3, Box 333
Eureka Springs, AR 72632, USA
FFL Firearms Dealer

KIRBY G. CARLTON
Rt. 1, Box 67
Jasper, AR 72641, USA
FFL Firearms Dealer

I-DEAL PAWN SHOP
P.O. Box 225
Prairie Grove, AR 72753, USA
FFL Firearms Dealer
CONTACT: Richard D. Skelton

TOMMY GARRETT
Rt. 1, Box 439
Alma, AR 72921, USA
FFL Firearms Dealer

WILLIAM K. ORMESBEE
Rt. 2, P.O. Box 394
Alma, AR 72921, USA
FFL Firearms Dealer

FREDERICK HOP ENTERPRISES
Rt. 1, Box 64
Greenwood, AR 72936, USA
FFL Firearms Dealer

MILITARY FIREARMS

OAK BOWER GUN SHOP
Rt. 1, Box 83
Mulberry, AR 72947, USA
FFL Firearms Dealer
CONTACT: Keith Gordon

RALPH L. BRYANT
2913 River Street
Ozark, AR 72949, USA
FFL Firearms Dealer

JOHN'S GUN SHOP
Rt. 1, Box 22
Marshall, OK 73056, USA
FFL Firearms Dealer

RICHARD BRITT
CB 1562
Mustang, OK 73064, USA
FFL Firearms Dealer

HDS INDUSTRIES
P.O. Box 1431
Oklahoma City, OK 73101, USA
FFL Firearms Dealer

K & G SPECIALTY ARMS
1400 Woodland Way
Oklahoma City, OK 73127, USA
FFL Firearms Dealer
CONTACT: K. S. Milanic

CHRIS QUERRY
8620 N.W. 91
Oklahoma City, OK 73132, USA
FFL Firearms Dealer

JIM'S TRADING POST
P.O. Box 235
Medicine Park, OK 73557, USA
FFL Firearms Dealer
CONTACT: James Cast

ONE MORE TIME SHOPPE
409 E. Maine
Enid, OK 73701, USA
Militaria & firearms
S3; 2;
CONTACT: C. J. Purcell
SALES DEPT: C.J. or Ross Purcell

STEVEN'S FIREARMS & ACCESSORIES
P.O. Box 107
Enid, OK 73702, USA
FFL Firearms Dealer

ALAN JUHL
Rt. 2, Box 8
Rosston, OK 73855, USA
FFL Firearms Dealer

ROY W. CRITCHNAU
210 S. Maple, Box 377
Oologah, OK 74053, USA
FFL Firearms Dealer

JOHN L. CRAWFORD
7603 N. 175th East Avenue
Owasso, OK 74055, USA
FFL Firearms Dealer

BOBBY L. KIRCHNER
RR 2, Box 110
Pawhuska, OK 74056, USA
FFL Firearms Dealer

JACKRABBIT GUNS
1321 E. Dewey
Sapulpa, OK 74066, USA
FFL Firearms Dealer
CONTACT: Robert M. Moon

JERI R. OSBORNE
Rt. 2, Box 11
Panhandle, TX 74068, USA
FFL Firearms Dealer

BOB DUNCAN GUNS OR AMMO
Rt. 1, Box 390
Skiatook, OK 74070, USA
FFL Firearms Dealer

BRAD OLIVER
Rt. 4, Box 584
Stillwater, OK 74074, USA
FFL Firearms Dealer

VEDA POINT BAIT & TACKLE
Rt. 1
Talala, OK 74080, USA
FFL Firearms Dealer
CONTACT: Don Ogle

MARSHALL ARMS
1511 S. Delaware Place
Tulsa, OK 74104, USA
FFL Firearms Dealer

RICHARD W. JONES
1210 S. 83 E. Avenue
Tulsa, OK 74112, USA
FFL Firearms Dealer

PHIL BUSH
505 Fairfax
Muskogee, OK 74403, USA
FFL Firearms Dealer

MYERS TRADING POST
710 S. 5th
Morris, OK 74445, USA
FFL Firearms Dealer

THE GUN BARREL
1301 W. Choctaw
Tahlequah, OK 74464, USA
FFL Firearms Dealer

GENE A. ERVIN
Box 237
Hartshorne, OK 74547, USA
FFL Firearms Dealer

FRANK O. ROBINSON
P.O. Box 32
McLoud, OK 74851, USA
FFL Firearms Dealer

CHRIS B. GRODHAUS
505 Post Oak Lane
Allen, TX 75002, USA
FFL Firearms Dealer

EICKS GUNSMITHING
2809 Arcadia Lane
Carrollton, TX 75006, USA
FFL Firearms Dealer

JAMES J. WALKER & ASSOCIATES
6400 Independence Pky #3602
Plano, TX 75023, USA
FFL Firearms Dealer

DOUBLE S SALES
3311 San Mateo
Plano, TX 75023, USA
FFL Firearms Dealer

MURANO ENTERPRISES
3824 Santiago Drive
Plano, TX 75023, USA
FFL Firearms Dealer
CONTACT: Robert R. Murano

LUNDSFORD'S FIREARMS & SUPPLY
8546 Santa Clara
Frisco, TX 75034, USA
FFL Firearms Dealer

B & C SUPPLY
1900 West Kingley, #203
Garland, TX 75041, USA
FFL Firearms Dealer

SILVER DOLLAR GUN & PAWN
2608 W. Walnut
Garland, TX 75042, USA
FFL Firearms Dealer

E. L. TABER
3317 O'Henry
Garland, TX 75042, USA
FFL Firearms Dealer

DANNY R. JANES
709 Park Forest
Garland, TX 75042, USA
FFL Firearms Dealer

SEMI-TECH,INC.
P.O. Box 402147
Garland, TX 75046, USA
Firearms accessories

ROBERT J. MOCK
P.O. Box 81808
5228 Knox Drive
Lewisville, TX 75056, USA
FFL Firearms Dealer

ARNOLD H. FULLER
1128 Highbridge Drive
Wylie, TX 75098, USA
FFL Firearms Dealer

RON COOPER
Rt. 1, L.C.E. 33
Malakoff, TX 75148, USA
FFL Firearms Dealer

THRIFTY SHOOTING SUPPLY
301 Elm
Waxahachie, TX 75165, USA
FFL Firearms Dealer
CONTACT: J. H. Pruett

MILITARY FIREARMS

WILEY & SONS
Rt. 1, Box 303
Wills Point, TX 75169, USA
FFL Firearms Dealer

JOSEPH A. BARBKNECHT
3219 Sharpview Lane
Dallas, TX 75228, USA
FFL Firearms Dealer

JAMES ABBOT
3219 Sharpview Lane
Dallas, TX 75228, USA
FFL Firearms Dealer

JAMES L. BRIGHT
8442 Hunnicut Road
Dallas, TX 75228, USA
FFL Firearms Dealer

DAVE'S HOUSE OF GUNS
9130 Viscount Row
Dallas, TX 75247, USA
Firearms

LEE BROWN
Rt. 10, Box 105
Paris, TX 75460, USA
FFL Firearms Dealer

POLICE EQUIPMENT COMPANY
Rt. 10, Box 150, Hwy. 195
Paris, TX 75460, USA
FFL Firearms Dealer

BOB SPLAWN
1124 Texas Blvd.
Texarkana, TX 75501, USA
FFL Firearms Dealer

FREDS GUN SHOP
P.O. Box 3393
Highway 718
Texarkana, AR 75504, USA
FFL Firearms Dealer
CONTACT: Fred Beeler

STEVEN'S GUNS & AMMO
Rt. 1, Box 156
New Boston, TX 75570, USA
FFL Firearms Dealer
CONTACT: Steven J. Oglesby

THE GUN DOCTOR, INC.
700 Glencrest Lane, Suite C
Longview, TX 75601, USA
FFL Firearms Dealer

C & S SPORTING GOODS
Rt. 10, Box 1165
Tyler, TX 75707, USA
FFL Firearms Dealer
CONTACT: Norman L. Collum

PRIDGEN'S SPORTING GOODS
Rt. 1, #6 Linden Lea Lane
Flint, TX 75762, USA
FFL Firearms Dealer

CORRIE RUPAR
Route 1, Box 234/G
Lindale, TX 75771, USA
FFL Firearms Dealer

ANTHONY W. JACKSON
Rt. 1, Box 324
Troup, TX 75789, USA
FFL Firearms Dealer

TOMMY RHOLES
P.O. Box 638
Van, TX 75790, USA
214-963-7758
Winchester firearms

SANFORD'S GUN SHOP
Rt. 1, Box 263, Hwy. 1252
Winona, TX 75792, USA
FFL Firearms Dealer
CONTACT: B. Z. Sanford

ANATOMY TARGETS COMPANY
Rt. 1, Box 166B
Lovelady, TX 75851, USA
FFL Firearms Dealer
CONTACT: Samuel S. Hallman

THE ARMORY
Rt. 10, Box 7740
Lufkin, TX 75901, USA
FFL Firearms Dealer
CONTACT: Daryl Gilbert

S & S SHOOTER'S SUPPLY
5609 Euclid Drive
Arlington, TX 76013, USA
FFL Firearms Dealer

THE WILD BUNCH WEAPONS COMPANY
2313 Roosevelt Drive, Suite A
Arlington, TX 76016, USA
(817)-860-2275 (817)-654-GUNS
Firearms & accessories.
CONTACT: Gregory J. Rainone

MICHAEL DOUTY
5906 Willow Branch
Arlington, TX 76017, USA
FFL Firearms Dealer

ALLEN M. HELD
5304 Vincennes Court
Arlington, TX 76017, USA
FFL Firearms Dealer

HUNTING WORKS
1805 Arthur
Colleyville, TX 76034, USA
FFL Firearms Dealer
CONTACT: Jim Martin

SMITH GUNS
P.O. Box 1590
Granbury, TX 76048, USA
FFL Firearms Dealer

DEFENSIVE ARMS
428 S. Main Street
Grapevine, TX 76051, USA
FFL Firearms Dealer
CONTACT: John A. Weil

THOMAS J. PUCCERELLA
1624 Campus Drive #151
Hurst, TX 76054, USA
FFL Firearms Dealer

LOUIS COKER
7601 Kevin Drive
Ft. Worth, TX 76118, USA
FFL Firearms Dealer

PATRICK C. HYDE
5700 Rio De Janeiro
N. Richland Hills, TX 76118, USA
FFL Firearms Dealer

DYLAN SALES COMPANY
P.O. Box 79253
Saginaw, TX 76179, USA
FFL Firearms Dealer

CHUCK FARMER
429 Ridgecrest Drive
Saginaw, TX 76179, USA
FFL Firearms Dealer

JOHN E. HARVEY
Twin Lakes # 136
Denton, TX 76201, USA
FFL Firearms Dealer

JOE D. IVY
Rt. 1, Box 212-A
Whitesboro, TX 76273, USA
FFL Firearms Dealer

RONALD M. BARR
1205 Warren Street
Wichita Falls, TX 76304, USA
FFL Firearms Dealer

WILLIAM VARDELL
4731 Eden Lane
Wichita Falls, TX 76306, USA
FFL Firearms Dealer

JAMES D. USELTON
Rt. 2, Box 124
Iowa Park, TX 76367, USA
FFL Firearms Dealer

CIRCLE 3 GUN SHOP
2012 Gordon
Vernon, TX 76384, USA
FFL Firearms Dealer
CONTACT: Neal Welch

GEM PAWN SHOP
112 South First
Temple, TX 76501, USA
FFL Firearms Dealer

TEMTEX SPORTING GOODS
1014 N. 13th Street
Temple, TX 76501, USA
FFL Firearms Dealer
CONTACT: William W. Hurley

CHARLES GOLDSTEIN
15 E. Wichita (M.P.R.)
Belton, TX 76513, USA
FFL Firearms Dealer

BROCKS GUN SHOP
1600 E. Hwy. 190
Copperas Cove, TX 76522, USA
FFL Firearms Dealer

MILITARY FIREARMS

KEMPNER TRUCK CENTER
Rt. 1, Box 185, FM 2808
Kempner, TX 76539, USA
FFL Firearms Dealer

RIVENDELL ARMS TRADERS
Rt. 6, Box 108-5
Killeen, TX 76541, USA
FFL Firearms Dealer

CENTEX GUN SHOP
P.O. Box 1673
Killeen, TX 76541, USA
FFL Firearms Dealer
CONTACT: Edwin Velez

GLYNN ALEWINE
P.O. Box 1475
2324 John Road
Killeen, TX 76543, USA
FFL Firearms Dealer

THE HARGIS COMPANY, INC.
P.O. Box 526
1205 Washington
Waco, TX 76701, USA
FFL Firearms Dealer

BRAD WHIDDON
311 Central
Waco, TX 76705, USA
FFL Firearms Dealer

AMERICAN DERRINGER CORPORATION
P.O. Box 8983
Waco, TX 76714, USA
(817)-799-9111
Manufacturer of firearms.
; 7; ; 4,000; ; 100.00
Catalogue available. Annually.
CONTACT: Robert Saunders, President
SALES DEPT: Bob Saunders
Manufacturer of stainless steel Derringers in 26 calibers, including 9mm, .30 Special, .357 Mag., 44 Mag., .45 Auto, .45 Colt/.410. Also imports .410 Buckstot.

TONY KRISCHKE
201 Center Avenue
Brownwood, TX 76801, USA
FFL Firearms Dealer

JAMES W. LEDBETTER
P.O. Box 130
1014 Fisher Street
Goldthwaite, TX 76844, USA
FFL Firearms Dealer

DANNY W. RIPPLE
P.O. Box 60043
San Angelo, TX 76906, USA
FFL Firearms Dealer

H DOUBLE O HEADQUARTERS
206 1/2 11th Street
Ozone, TX 76943, USA
FFL Firearms Dealer

JOHN ABATTE GUNSMITHING
1608 Bonnie Brae
Houston, TX 77006, USA
FFL Firearms Dealer

D. SCOTT PRINGLE
2619 Tannehill
Houston, TX 77008, USA
FFL Firearms Dealer

T.V.'s RELOADING SHOP
119 Maxey Road
Houston, TX 77013, USA
FFL Firearms Dealer

FRED TOWERY
1304 Forest Oaks Drive
Houston, TX 77017, USA
FFL Firearms Dealer

THOMAS F. 'TOM' SHAW, III
10822 Longshadow lane
Houston, TX 77024, USA
FFL Firearms Dealer

MASSEY'S GUNSHOP
3619 Wood Valley Drive
Houston, TX 77025, USA
FFL Firearms Dealer
CONTACT: Oly D. Massey

NAVAS ARMS, INC.
1804 Harrington
Houston, TX 77026, USA
FFL Firearms Dealer
CONTACT: Benjamin & Richard Nava

SPEARS FIREARMS
9106 Nook Court
Houston, TX 77040, USA
FFL Firearms Dealer

MARO'S GUNS
11410 Galbreath
Houston, TX 77066, USA
FFL Firearms Dealer

GARRETT'S FIREARMS
11103 Somerford
Houston, TX 77072, USA
FFL Firearms Dealer

THE GUNNERY
2033 Peppermill
Houston, TX 77080, USA
FFL Firearms Dealer

JAMES R. VOGT
15607 Midridge
Houston, TX 77084, USA
FFL Firearms Dealer

TEXAS AMERICAN ARMS
3826-B Greenhouse Road
Houston, TX 77084, USA
FFL Firearms Dealer
CONTACT: Johnny Broyles

D & A GUN COMPANY
7226 Barton Oaks
Houston, TX 77095, USA
FFL Firearms Dealer

EUGENE LLOYD
291 Scarborough Drive, #801
Conroe, TX 77304, USA
FFL Firearms Dealer

MASTER SHOOTERS SUPPLY
1811 Spruce Knob
Kingwood, TX 77339, USA
FFL Firearms Dealer

RAMSHACK SHOOTERS SUPPLIES
Rt. 2, Box 540
Huntsville, TX 77340, USA
FFL Firearms Dealer
CONTACT: Victor W. Pears

ALAN HOLK
4806 Locust
Bellaire, TX 77401, USA
FFL Firearms Dealer

VICTOR F. SLANINA, JR.
Rt. 6, Box 6138
Brazoria, TX 77422, USA
FFL Firearms Dealer

RICKY P. KRAMER
P.O. Box 365
813 Clubside Drive
E. Bernard, TX 77435, USA
FFL Firearms Dealer

J. P. RACIDE
5322 Finsbury Field
Katy, TX 77449, USA
FFL Firearms Dealer

DONALD D. HEMME
P.O. Box 95
2214 Fall Meadow
Missouri City, TX 77459, USA
FFL Firearms Dealer

DAVID A. STANBROUGH
6534 S.F. Austin
Freeport, TX 77541, USA
FFL Firearms Dealer

LEIGH L. ADAMS
5222 Appleblossom
Friendswood, TX 77546, USA
FFL Firearms Dealer

JOHN ALLEN
105 Bayou Vista Drive
Hitchcock, TX 77563, USA
FFL Firearms Dealer

MUSKET ARMS COMPANY
1632 E. Main
League City, TX 77573, USA
FFL Firearms Dealer

I & J INTERESTS
614 Pine Circle Drive
Seabrook, TX 77586, USA
FFL Firearms Dealer

STEVE KOSH
P.O. Box 1227
Bridge City, TX 77611, USA
FFL Firearms Dealer

ROBERT C. PASTORELLA, SR.
2701 Magnolia
Groves, TX 77619, USA
FFL Firearms Dealer

MILITARY FIREARMS

JAMES E. JORDAN
5401 Gulfway Drive #310
Groves, TX 77619, USA
FFL Firearms Dealer

PEOPLES GUNS
2013 11th Street
Port Neches, TX 77651, USA
FFL Firearms Dealer
CONTACT: Barry Vance

ROBERT G. MILLER
1313 Block Street
Port Neches, TX 77651, USA
FFL Firearms Dealer

HARRY GARZA
2211 Ave. A
Beaumont, TX 77701, USA
FFL Firearms Dealer

MONTY'S
5275 Gladys
Beaumont, TX 77706, USA
FFL Firearms Dealer

BENCHREST
P.O. Box 7901
Beaumont, TX 77706, USA
FFL Firearms Dealer

THE OUTPOST/SHOOTERS SUPPLY
2811 Willhelm
Bryan, TX 77801, USA
FFL Firearms Dealer

MICHAEL PAULUS
2601 Rountree #7
Bryan, TX 77801, USA
FFL Firearms Dealer

GREG'S GUNS
3001 Meadowlane
Victoria, TX 77901, USA
FFL Firearms Dealer

BLAYLOCK'S GUNWORKS
Rt. 3, Box 103-A
Victoria, TX 77901, USA
(512)-573-2744
Firearms & accessories
S3,S4,S6,S12; 5;
CONTACT: Dale Blaylock
SALES DEPT: Dale Blaylock
PURCHASING DEPT: Dale Blaylock
Military & general firearms dealers

DON L. BUEHNER
1802 E. Loma Vista
Victoria, TX 77901, USA
FFL Firearms Dealer

RODS & GUNS
P.O. Box 990
10th & Adams Street
Port O'Connor, TX 77982, USA
FFL Firearms Dealer
CONTACT: Harlem M. Terry

C. B. GUNS & AMMO
P.O. Box 95
Weesatche, TX 77993, USA
FFL Firearms Dealer

CONTACT: Charles F. Borgfeld

BOERNE ARMS COMPANY
Rt. 5, Box 5623
Boerne, TX 78006, USA
FFL Firearms Dealer
CONTACT: Jim Clarke

BROWN ACCESSORIES
P.O. Box 344
Devine, TX 78016, USA
FFL Firearms Dealer

MURIS CUSTOM FIREARMS
805 Andrews
Laredo, TX 78041, USA
FFL Firearms Dealer

GREG SKALOMENOS
632 Hwy. 46 S
New Braunfels, TX 78130, USA
FFL Firearms Dealer

BIG BILLS GUNS
264 Kerlick Lane
New Braunfels, TX 78130, USA
FFL Firearms Dealer
CONTACT: William F. Kerlick,Jr.

FIRST TEXAS PAWN
1301-A Pat Booker Road
Universal City, TX 78148, USA
FFL Firearms Dealer

WINGSHOOTERS LTD
6104 Broadway
San Antonio, TX 78209, USA
FFL Firearms Dealer

ABRAHAM & JOELINE MOGERMAX
8015 Peale
San Antonio, TX 78239, USA
FFL Firearms Dealer

KENT D. GERSTNER
5603 Cary Grant
San Antonio, TX 78240, USA
FFL Firearms Dealer

BARRY V. TIPPITT
12103 Ridge Summit
San Antonio, TX 78247, USA
FFL Firearms Dealer

LEON SPRINGS ARMORER
7803 Summit Circle
San Antonio, TX 78256, USA
FFL Firearms Dealer
CONTACT: c/o Richard W. Strimel

RICHARD N. PANNELL
210 Sabine
Portland, TX 78374, USA
FFL Firearms Dealer

ROGER W. BENNETT
111 Granby Place
Portland, TX 78374, USA
FFL Firearms Dealer

KENNETH MICHAEL KOONCE
Rt. 1, Box 125
Taft, TX 78390, USA

FFL Firearms Dealer

TEXAS CENTURION PRODUCTS
3419 Ayers
Corpus Christi, TX 78415, USA
FFL Firearms Dealer

WILSON'S GUNS
1726 Triple Crown Drive
Corpus Christi, TX 78417, USA
FFL Firearms Dealer

R. M. CANO
913 Slabaugh
Mission, TX 78572, USA
FFL Firearms Dealer

J. HILL CORPORATION
1402 S. Cage, Box 46
Pharr, TX 78577, USA
FFL Firearms Dealer

CACTUR ROSE GUN SHOP
P.O. Box 162
Cedar Creek, TX 78612, USA
FFL Firearms Dealer

TIM WITHERS
Rt. 4, Box 83
Gonzales, TX 78629, USA
FFL Firearms Dealer

COLLECTIBLE FIREARMS
Rt. 2, Box 338-C
San Marcos, TX 78666, USA
FFL Firearms Dealer
CONTACT: M. J. Pierson

BILL KEELS
4716 Duval Road #0-40
Austin, TX 78727, USA
FFL Firearms Dealer

M. L. MAUND
7805 Lindenwood Circle
Austin, TX 78731, USA
FFL Firearms Dealer

CEDAR VALLEY ASSOC.
Rt. 27, Box 74W
Austin, TX 78737, USA
FFL Firearms Dealer

JIM TRAWEEK
7107 Sir Gawain
Austin, TX 78745, USA
FFL Firearms Dealer

DETECTION,INC.
6717 Burnet Road
Austin, TX 78757, USA
FFL Firearms Dealer

SCOTT MAGIC COMPANY
1007 Kramer Lane
Austin, TX 78758, USA
FFL Firearms Dealer

LARRY W. VARNER
11 Donna Drive
Uvalde, TX 78801, USA
FFL Firearms Dealer

MILITARY FIREARMS

ARMADILLO ARMS
1102 Etta
Friona, TX 79035, USA
FFL Firearms Dealer

BOB'S GUN SHOP
406 W. 14th
Friona, TX 79035, USA
FFL Firearms Dealer

KEN DANNER
Rt. 1, Box 73-B
Happy, TX 79042, USA
FFL Firearms Dealer

DAN SELLERS
2020 Christine
Pampa, TX 79065, USA
FFL Firearms Dealer

ALAN PETE COOMBES
8215 Goodnight Tr.
Amarillo, TX 79110, USA
FFL Firearms Dealer

MIKE KENDALL
5106 Arden Road
Amarillo, TX 79110, USA
FFL Firearms Dealer

S & J GUNS
124-A North, Main Box 360
Deaver City, TX 79323, USA
FFL Firearms Dealer

LEWIS B. NICKENS
P.O. Box 933
802 7th Street
Wolfforth, TX 79382, USA
FFL Firearms Dealer

DANNY GREGG
Rt. 4, Box 717
Lubbock, TX 79424, USA
FFL Firearms Dealer

MELTON'S SPORTING GOODS
1900 W. Drive
Snyder, TX 79549, USA
FFL Firearms Dealer

ED'S GUN SHOP
702 Sayles Drive
Abilene, TX 79605, USA
FFL Firearms Dealer

NEAL'S GUNS
133 Portland
Abilene, TX 79605, USA
FFL Firearms Dealer

PHIL M. DOLBOW
616 W. Nobles
Midland, TX 79701, USA
FFL Firearms Dealer

THOMAS J. MOTYCKA
1604 West Dengar
Midland, TX 79705, USA
FFL Firearms Dealer

THIELEN FIREARMS & AMMO CO.
P.O. Box 356
Ft. Stockton, TX 79735, USA
FFL Firearms Dealer

T. C. HALL
413 W. 10th Street
Monahans, TX 79756, USA
FFL Firearms Dealer

TOMMY SWAFFORD
2002 E. 10th
Odessa, TX 79761, USA
FFL Firearms Dealer

THE GUN SHACK
4613 McKnight
Odessa, TX 79762, USA
FFL Firearms Dealer
CONTACT: Travis Dunn

RICHARD RANDALL
513 Avenue A Circle
Odessa, TX 79763, USA
FFL Firearms Dealer

DAVE HERTEL
6944 Ramada
El Paso, TX 79912, USA
FFL Firearms Dealer

EDWARD M. D'AREY
10061 Chick-A-Dee
El Paso, TX 79924, USA
FFL Firearms Dealer

DACO INDUSTRIES
9513 Carnegie
El Paso, TX 79925, USA
FFL Firearms Dealer
CONTACT: David J. Davis

RUSSELL W. DENHAM
2328 Octubre Drive
El Paso, TX 79935, USA
FFL Firearms Dealer

HARTMUT MUELLER
10535 Montwood Drive #177
El Paso, TX 79935, USA
FFL Firearms Dealer

BRYANT ENTERPRISES, INC.
1961 Ratner Circle
El Paso, TX 79936, USA
FFL Firearms Dealer

DAVID P. MARSHALL
1533 Paul Harney
El Paso, TX 79936, USA
FFL Firearms Dealer

DAVE QUEVEDO
P.O. Box 370383
El Paso, TX 79937, USA
FFL Firearms Dealer

DALE S. DORTCH
6878 West 74th Avenue
Arvada, CO 80003, USA
FFL Firearms Dealer

G & S SPORTS
7069 Swadley Court
Arvada, CO 80004, USA
FFL Firearms Dealer

GENERAL SERVICES
10891 E. Lowry Place
Aurora, CO 80010, USA
(303)-367-2326
Firearms consultant
S1,S6,S11,S12; 14;
CONTACT: John W. Jackson
SALES DEPT: John W. Jackson
PURCHASING DEPT: John W. Jackson
I act as a consultant for those who need help concerning the operation, history, and repair of all types of firearms, including Class 3 weapons, also finder's service, & seller's service.

GUN BOX
3188 Zion Street
Aurora, CO 80011, USA
FFL Firearms Dealer

B. C.'s
1568 South Elkhart Street
Aurora, CO 80012, USA
FFL Firearms Dealer
CONTACT: Bret M. Clausen

STEVE CALZADA
16409 East Evans Avenue
Aurora, CO 80013, USA
FFL Firearms Dealer

RON REESOR
12200 East Iliff, #208
Aurora, CO 80014, USA
FFL Firearms Dealer

R & R ENTERPRISES
21517 East Powers Lane
Aurora, CO 80015, USA
FFL Firearms Dealer
CONTACT: Ken Roberts

DAVID R. CRUCKSHANKS
1310 Fresno Court
Broomfield, CO 80020, USA
FFL Firearms Dealer

TED A. ROCKENHAUS
2690 Ridge Drive
Broomfield, CO 80020, USA
FFL Firearms Dealer

DENNIS M. SEARLES
8485 West 38th Avenue
Wheat Ridge, CO 80033, USA
FFL Firearms Dealer

SOUTHWEST SPORTS SPECIALTIES
8893 West Star Avenue
Littleton, CO 80123, USA
FFL Firearms Dealer
CONTACT: Bob Gates

DONALD A. REHLING
7641 South Jellison Street
Littleton, CO 80127, USA
FFL Firearms Dealer

'A' AMMO & GUN SUPPLY
41243 Frontier Road
Parker, CO 80134, USA
FFL Firearms Dealer

MILITARY FIREARMS

GUNSLINGER GUN SHOP
7821 East Ponderosa Drive
Parker, CO 80134, USA
FFL Firearms Dealer
CONTACT: Joseph M. Skalisky

RLE GUNS
10646 South State Highway 67
Sedalia, CO 80135, USA
FFL Firearms Dealer

FRENCHS GUN SHOP
258 Broadway
Denver, CO 80203, USA
FFL Firearms Dealer

GLENN E. MILLER
2254 Dahlia Street
Denver, CO 80207, USA
FFL Firearms Dealer

JOHN BROWN
4037 North Newton Street
Denver, CO 80211, USA
FFL Firearms Dealer

SCOTTIES GUNS & MILITARIA
5606 East Colfax
Denver, CO 80220, USA
(303)-399-GUNS
Firearms & general militaria.
S1,S4,S12; 12;
CONTACT: David Jewell
SALES DEPT: Scottie
PURCHASING DEPT: Scottie

SCOOTIES GUN & MILITARIA
5606 East Colfax Avenue
Denver, CO 80220, USA
FFL Firearms Dealer

MEADOWLARK SPORTS
1460 Brentwood
Lakewood, CO 80228, USA
FFL Firearms Dealer
CONTACT: Bill Meneely

WILLIAM CONYERS
7845 East Cedar Avenue
Denver, CO 80231, USA
FFL Firearms Dealer

FEATHER ENTERPRISES
2300 Central Avenue, #K
Boulder, CO 80301, USA
303-442-7021
Firearms accessories

C & L ENTERPRISES
5271 Spotted Horse Trail
Boulder, CO 80301, USA
FFL Firearms Dealer

VEGA TOOL COMPANY
2756 North 47th Street
Boulder, CO 80301, USA
FFL Firearms Dealer
CONTACT: Thomas R. Rosa

THE KING'S KID
3595 Darley Avenue
Boulder, CO 80303, USA
FFL Firearms Dealer

ELIZABETH MITCHELL
405 South 40th
Boulder, CO 80303, USA
FFL Firearms Dealer

ERO DISTRIBUTING COMPANY
P.O. Box 3431
Boulder, CO 80307, USA
(303)-499-1169
Firearms & ammunition
S1,S11; 3;
CONTACT: Vic Amenti
SALES DEPT: Vic Ament
PURCHASING DEPT: Vic Ament
ERO Distributing Co. is a supplier of fine firearms and ammunition to the discriminating buyer. Special orders are our specialty.

RAM LINE,INC.
406 Violet Street
Golden, CO 80401, USA
Firearms accessories

PESTINGER CONSTRUCTION
1453 Ponderosa Drive
Evergreen, CO 80439, USA
FFL Firearms Dealer

JAMES DONALD
P.O. Box 38
Kremmling, CO 80459, USA
FFL Firearms Dealer

DON RULLE ENTERPRISES
1128 Meadow Street
Longmont, CO 80501, USA
FFL Firearms Dealer

B.A.S. WHOLESALE FIREARMS
1429 Sumac Street
Longmont, CO 80501, USA
FFL Firearms Dealer

MUTTHERSBOUGH GUNSMITHING
9661 North 89th Street
Longmont, CO 80501, USA
FFL Firearms Dealer

F. P. GRESKY
2351 Ridgecrest Road
Fort Collins, CO 80524, USA
FFL Firearms Dealer

RPN GUN SHOP
316 West 51st Street
Loveland, CO 80537, USA
FFL Firearms Dealer

ARTHUR S. MINNIER
P.O. Box 170
Lyons, CO 80540, USA
FFL Firearms Dealer

PINEWOOD INDUSTRY
P.O. Box 950
Lyons, CO 80540, USA
FFL Firearms Dealer

GOOSE HAVEN HUNTING SUPPLIES
36615 Highway 257
Windsor, CO 80550, USA
FFL Firearms Dealer

RANDY CASS GUN SHOP
4351 WCR 77
Briggsdale, CO 80611, USA
FFL Firearms Dealer

RL'S GUN SALES
712 Phelps Street
Sterling, CO 80751, USA
FFL Firearms Dealer

BIG T. GUNS
Rt. 1, Davis Road
Peyton, CO 80831, USA
FFL Firearms Dealer
CONTACT: Ron Townes

DOUGLAS WOZNY
P.O. Box 1037
Woodland Park, CO 80863, USA
FFL Firearms Dealer

JAN'S 4 GUNS
306 Montclair Street
Colorado Springs, CO 80910, USA
FFL Firearms Dealer

DEAN BYRNE
P.O. Box 13382
Fort Carson, CO 80913, USA
FFL Firearms Dealer

JIM DUNN
1840 Mineola Street
Colorado Springs, CO 80915, USA
FFL Firearms Dealer

MONITOR ARMS
100 Tyndall Street
Colorado Springs, CO 80916, USA
FFL Firearms Dealer

TOM BRADY
5136 Pony Soldier Drive
Colorado Springs, CO 80917, USA
FFL Firearms Dealer

FRANK V. LOUCHTE
004 Appaloosas
Trujille Creek-Middle Fork
Aguilar, CO 81020, USA
FFL Firearms Dealer

BILL ARMSTRONG
P.O. Box 632
Rocky Ford, CO 81067, USA
FFL Firearms Dealer

DUSTY COVER USED BOOKS
309 West Main
Trinidad, CO 81082, USA
FFL Firearms Dealer

CARROL JOE CARTER
97 Sierra
Alamosa, CO 81101, USA
FFL Firearms Dealer

ECONOMY ARMS & AMMO
0657 Leprechaun Lane
Howard, CO 81233, USA
FFL Firearms Dealer

MILITARY FIREARMS

SIDNEY C. PHIPPS
2202 Hawthorne Avenue
Grand Junction, CO 81501, USA
FFL Firearms Dealer

ROULETTE ENTERPRISE
P.O. Box 2957
Grand Junction, CO 81502, USA
FFL Firearms Dealer

ORDNANCE PARK CORPORATION
657-20 1/2 Road
Grand Junction, CO 81503, USA
Civil War artillery supplies.
Catalogue $2.00.

MAVERICK TRADING COMPANY
2798 Oxford Avenue
Grand Junction, CO 81503, USA
FFL Firearms Dealer

RICHARDSON'S SPORTS SUPPLY
2704 Rincon Drive
Grand Junction, CO 81503, USA
FFL Firearms Dealer

SHOOTERS SUPPLY
3246 Front Street
Clifton, CO 81520, USA
FFL Firearms Dealer

STATE STREET SPORTS
19 West State Marbie
Carbondale, CO 81623, USA
FFL Firearms Dealer
CONTACT: Lee Haugen

GLEN V. NORTON
645 Grace Drive
Carbondale, CO 81623, USA
FFL Firearms Dealer

DONALD L. KITSMILLER
1133 Barclay
Craig, CO 81625, USA
FFL Firearms Dealer

ARMAGH, LTD.
P.O. Box 1110
Eagle, CO 81631, USA
FFL Firearms Dealer

MARV'S PLACE, INC.
223 W. 16th Street
Cheyenne, WY 82001, USA
FFL Firearms Dealer

L & S SPORT SHOP
1519 Otto Road
Cheyenne, WY 82007, USA
FFL Firearms Dealer

HUCK'S FIREARMS SUPPLIES
4746 E. Skyline Dr. #48
Laramie, WY 82070, USA
FFL Firearms Dealer

COAST TO COAST
1918 East A
Torrington, WY 82240, USA
FFL Firearms Dealer
CONTACT: Charles Huff

BUCKBOARD BILL'S
P.O. Box 332
Ten Sleep, WY 82442, USA
FFL Firearms Dealer

TOFFLEMIRE'S ARMS & OIL PRODUCTS
108 South 11th E.
Riverton, WY 82501, USA
FFL Firearms Dealer

HORSE SHOE HEART TRADERS
604 N. Broadway
Riverton, WY 82501, USA
FFL Firearms Dealer
CONTACT: Tom Boll

J. M. DUNSWORTH
845 South 9th
Lander, WY 82520, USA
FFL Firearms Dealer

NATRONA HOPLOLOGICAL SERVICES
3210 Bella Vista Drive
Casper, WY 82601, USA
FFL Firearms Dealer

CODY S. FOSTER
6493 Village Drive
Casper, WY 82604, USA
FFL Firearms Dealer

ROBERT E. FAUNCE
209 N. 3rd, Box 1143
Glenrock, WY 82637, USA
FFL Firearms Dealer

MIKE DAVIS
Box 482
Glenrock, WY 82637, USA
FFL Firearms Dealer

POWDER HORN GUNS
501 S. Douglas
Gillette, WY 82716, USA
307-682-7254
Firearms & accessories
S1,S4,S7,S12; 3;
CONTACT: Mike Carver
Powder Horn - Guns. Guns & ammo; modern, antique, military. We buy, sell, trade, and pawn and type weapon.

METAL WORKS
4111 Teepee
Gillette, WY 82716, USA
FFL Firearms Dealer
CONTACT: Tom E. Howard

J. R.'s ENTERPRISES
#88 Coyote Trail
Gillette, WY 82716, USA
FFL Firearms Dealer

RON HUCKINS
P.O. Box 171
288 Noble
Osage, WY 82723, USA
FFL Firearms Dealer

RICHARD J. LACEY
1300 New Hampshire #41
Rock Springs, WY 82901, USA
FFL Firearms Dealer

DARRELL SLATON
106 Mountain Road
Rock Springs, WY 82901, USA
FFL Firearms Dealer

THE DOC'S
736 Pilot Butte Avenue
Rock Springs, WY 82901, USA
FFL Firearms Dealer

PINEDALE GUN & PAWN SHOP
34 N. Franklin
Pinedale, WY 82941, USA
FFL Firearms Dealer

CHUCK BIRKEMEYER
Box 527
Jackson, WY 83001, USA
FFL Firearms Dealer

RELOADING BENCH OF WYOMING
Rt. 1
Afton, WY 83110, USA
FFL Firearms Dealer

HOKANSON GUNS
P.O. Box 247
Freedom, WY 83120, USA
FFL Firearms Dealer
CONTACT: Lin Dee Hikanson

J. T.'s GUN SHOP
27 North, 600 West
Blackfoot, ID 83221, USA
FFL Firearms Dealer

BEATTIE'S GUNS & ANTIQUES
P.O. Box 531
454 N. State Street
Shelley, ID 83274, USA
FFL Firearms Dealer

C. G. COX
171 E. Oak
Shelley, ID 83274, USA
FFL Firearms Dealer

BLAKE A. PARKER
c/o Jerry Davis
P.O. Box 852
Soda Springs, ID 83276, USA
FFL Firearms Dealer

JERRY L. McMULLIN
727 Jensen
Idaho Falls, ID 83401, USA
FFL Firearms Dealer

S & D SUPPLY
1175 First Street
Idaho Falls, ID 83401, USA
FFL Firearms Dealer

JON L. WILKINSON
747 E. 13th Street
Idaho Falls, ID 83401, USA
FFL Firearms Dealer

SC ENTERPRISES
Rt. 1, Box 27
North Fork, ID 83466, USA
FFL Firearms Dealer
CONTACT: Joseph Syceylo

MILITARY FIREARMS

7 DEVILS SPORTSHOP
P.O. Box 169
312 S. Main Street
Riggins, ID 83549, USA
FFL Firearms Dealer
CONTACT: Fred J. Hess

DHB ENTERPRISES
P.O. Box 433
220A E. Park Street
Emmett, ID 83617, USA
FFL Firearms Dealer

SCOTT GOODMAN
P.O. Box 1715
McCall, ID 83638, USA
FFL Firearms Dealer

K/BAR FIREARMS
P.O. Box 856
Nampa, ID 83651, USA
(208)-465-7537
Military firearms
S4,S6,S12,S13; 4;
CONTACT: Ken Smith
SALES DEPT: Ken Smith
PURCHASING DEPT: Ken Smith
All types of collectible and current military small arms. Automatic weapons, small launchers, grenades and ammo for qualified license holders.

BIGHORN GUN SHOP
337 South 14th Street
Payette, ID 83661, USA
(208)-642-3066
Gunsmithing
S1,S11,S12,S13; 3;
CONTACT: John F. Shuke,Jr.
SALES DEPT: John Shuke
PURCHASING DEPT: John Shuke
Gunsmithing of all types, custom wood and metal work also, Guns, Scopes, Cases, Ammo, etc. I will custom order anything for 10% over cost.

DELUXE SWAP SHOP
110 South 11th
Boise, ID 83701, USA
(208)-343-0666
Firearms and accessories
S1;
CONTACT: Lewis R. Pierce

MIKE CRANE
5298 Turret
Boise, ID 83703, USA
FFL Firearms Dealer

CHARLES DAVIDSON
2417 South Pond
Boise, ID 83705, USA
FFL Firearms Dealer

SALMON RIVER ARMS
3216 Centennial Avenue
Boise, ID 83706, USA
FFL Firearms Dealer
CONTACT: Rayner C. Howard

HERN IRON WORKS
N 1900 C Millview Lane
Coeur D'Alene, ID 83814, USA
208-765-3115
Cast iron cannon barrel foundry.

Send SASE for catalogue.

M.D. ENTERPRISES
913 W. Palouse R. Dr. #11
Moscow, ID 83843, USA
FFL Firearms Dealer
CONTACT: Marck Cramer

ANTIQUE GUN SHOPPE
E. 2640 Ponderosa Blvd.
Post Falls, ID 83854, USA
(208)-773-1320 (208)-773-0106
General militaria, museum, antique artillery.
S6,S8,S11,S12; 28;
CONTACT: Glen E. Mattox
SALES DEPT: Glen E. Mattox
We offer a free museum showing development of U.S. fighting arms - For sale - A complete line of antique arms and military items including cannons and Gatling Gun.

BOBBY L. HARTLEY
92 North 775 B
American Fork, UT 84003, USA
FFL Firearms Dealer

C. E. PERSINGER
64 West 600 North
Kaysville, UT 84037, USA
FFL Firearms Dealer

BOBS GUNS
1558 E. Oakridge Drive
Layton, UT 84041, USA
FFL Firearms Dealer
CONTACT: Robert L. Sprenkle

LARS - ARMS
1500 N. 1200 #58
Layton, UT 84041, USA
FFL Firearms Dealer

M. N. MASCARO
618 E. 7200 South
Midvale, UT 84047, USA
FFL Firearms Dealer

TED H. DAVIS
P.O. Box 835
655 S. 400 E
Orem, UT 84058, USA
FFL Firearms Dealer

PAPER PUNCHER
5897 South 2600 West
Roy, UT 84069, USA
FFL Firearms Dealer

THE HIRED GUN
937 East 10715 South
Sandy, UT 84070, USA
FFL Firearms Dealer
CONTACT: Terry L. Tippets

BOB'S GUNS
Box 836
Vernal, UT 84078, USA
FFL Firearms Dealer
CONTACT: Robert L. Hurley

DEEP CREEK MOUNTAIN RANCH
Callao Star Rt., Box 380
Wendover, UT 84083, USA
FFL Firearms Dealer
CONTACT: George Douglass

SANDY SHOOTERS SUPPLY
8466 Strato Drive
Sandy, UT 84092, USA
FFL Firearms Dealer
CONTACT: Robert N. Westover

DON E. OLSEN
333 North 300 West
Salt Lake City, UT 84103, USA
FFL Firearms Dealer

STEVENS SALES-SERVICE
2562 Lambourne Avenue
Salt Lake City, UT 84109, USA
FFL Firearms Dealer

OLYMPIC ARMS
4204 Rowland Drive
Salt Lake City, UT 84124, USA
FFL Firearms Dealer

SEAMONS FIREARMS
614 N. Main
Lewiston, UT 84320, USA
FFL Firearms Dealer
CONTACT: Robert Seamons

THE GIFT HOUSE
120 - 25th
Ogden, UT 84401, USA
FFL Firearms Dealer

WATKINS SPORTING GOODS
875 North 900 East
Price, UT 84501, USA
FFL Firearms Dealer
CONTACT: Dennis Watkins

BRAND PORTERS GUNS & AMMO
1004 North Covecrest Drive
Price, UT 84501, USA
FFL Firearms Dealer

VAN WAGENEN FINANCE
75 North 500 West
Provo, UT 84601, USA
FFL Firearms Dealer
CONTACT: Norman Van Wagenen

DON WEIDINGER
2267 West 710 North
Provo, UT 84601, USA
FFL Firearms Dealer

GARY DELSIGNORE
3675 Cottonwood Drive
Cedar City, UT 84720, USA
801-586-2505
Firearms

ALGER ENTERPRISES
P.O. Box 241
Enterprise, UT 84725, USA
FFL Firearms Dealer

DELOY V. EVANS
P.O. Box 799
145 W. 300 N.
Hurricane, UT 84737, USA
FFL Firearms Dealer

MILITARY FIREARMS

JOSEPH K. HARWOOD
100 South 60 West Central
Monroe, UT 84754, USA
FFL Firearms Dealer

INTERPORT
Box 1796
St. George, UT 84770, USA
Machine-guns & parts.
Catalogue $2.00.

LEE WEBB
1944 West 1700 North
St. George, UT 84770, USA
FFL Firearms Dealer

JEWEL BOX, INC.
601 North Central
Phoenix, AZ 85004, USA
FFL Firearms Dealer

J. CURTIS EARL
5512 N. Sixth Avenue
Phoenix, AZ 85012, USA
(602)-264-3166
Automatic weapons & parts.
S6; 21;
Catalogue $5.00. Annually.
SALES DEPT: Curt Earl
PURCHASING DEPT: Curt Earl
Specializing in the classic machineguns and submachineguns of the world, as well as modern automatics. Send $5.00 for large 44 page photo illustrated brochure listing guns, ordnance, parts and accessories, laws and facts of interest of my $2,000,000.00 inventory.

ARIZONA AMMO SUPPLY
4008 West Maryland
Phoenix, AZ 85019, USA
FFL Firearms Dealer

MAJ. CHARLES W. DUNDEE
3538 West McLellan Blvd.
Phoenix, AZ 85019, USA
FFL Firearms Dealer

PRIME TARGETS
8935 N. 2nd Way, Suite 125-A
Phoenix, AZ 85020, USA
Targets

T & M GUNS
2906 West Michigan Avenue
Phoenix, AZ 85023, USA
FFL Firearms Dealer

NORTHWEST ARMS & AMMO
4001 West Corrine
Phoenix, AZ 85029, USA
FFL Firearms Dealer
CONTACT: Dave Jensen

H & H MAGNUM SALES
4805 West Thomas Road
Phoenix, AZ 85031, USA
FFL Firearms Dealer

EARL EDWARDS
6539 West Colledge Street
Phoenix, AZ 85033, USA
FFL Firearms Dealer

BNJ WEAPONS ENTERPRISES
P.O. Box 20424
Phoenix, AZ 85036, USA
FFL Firearms Dealer

THOMAS P. BUGIE
5040 East Shomi Street
Phoenix, AZ 85044, USA
FFL Firearms Dealer

SIGNUS
Box 33712-K15
Phoenix, AZ 85067, USA
Smoke generating pyrotechnics

WAYNE PADEN
2544 East Fairbrook Street
Mesa, AZ 85203, USA
FFL Firearms Dealer

KENS GUNS
2800 West Whitely
Apache Junction, AZ 85220, USA
FFL Firearms Dealer
CONTACT: Kenneth C. Allped

CASA GRANDE GUN SHOP
P.O. Box 157
108 West 4th St., Suite #6
Casa Grande, AZ 85222, USA
FFL Firearms Dealer
CONTACT: Steven O. Shanafelt

CUSTOM GUN CARE
5339 E. Orchid Lane
Paradise Valley, AZ 85253, USA
FFL Firearms Dealer

JAMES E. SANDERS
4910 East Pershing
Scottsdale, AZ 85254, USA
FFL Firearms Dealer

SPECIAL ORDER
1855 East Don Carlos #2
Tempe, AZ 85281, USA
FFL Firearms Dealer

HUGH J. SCHMAHL
1038 West 23rd Street
Tempe, AZ 85282, USA
FFL Firearms Dealer

JAMES S. STUART, SR.
5618 North 63rd Avenue
Glendale, AZ 85301, USA
FFL Firearms Dealer

DELBERT BROWN
4639 West Becken Lane
Glendale, AZ 85304, USA
FFL Firearms Dealer

GUNS 'N' SUTFF
14251 North 50 Drive
Glendale, AZ 85306, USA
FFL Firearms Dealer

LAST CHANCE GUN SHOP
4925 West Crocus
Glendale, AZ 85306, USA
FFL Firearms Dealer
CONTACT: Davin Hooker

CAINS SHOOTER SUPPLY
Route 1, Box 584
Buckeye, AZ 85326, USA
FFL Firearms Dealer

ED BURTON
19845 West Van Buren
Buckeye, AZ 85326, USA
FFL Firearms Dealer

WORLD OF LUGERS
19044 North 98th Lane
Peoria, AZ 85345, USA
Firearms.
List $2.00, 5 issues for $5.00.
CONTACT: Ralph E. Shattuck

THE SHOOTER'S SHOP
2053 S. 11th Avenue
Yuma, AZ 85364, USA
(602)-782-6494
Military firearms, ammo & accessories.
S1,S4,S11,S12; 2;
CONTACT: Joseph M. Smyth
SALES DEPT: Joe E. Smyth
PURCHASING DEPT: Joe E. Smyth
Specializing in sales of military arms, ammunition and accessories, also sporting arms and sporting goods. Weapons for sale and trade. Special orders accepted.

RAYMOND G. DAVIS
P.O. Box 58-4
Benson, AZ 85602, USA
FFL Firearms Dealer

HERITAGE ARMS GALLERY
P.O. Box 5843
310 Arizona Street
Bisbee, AZ 85603, USA
FFL Firearms Dealer

DANIEL CHAN
849 South 9th Avenue
Tucson, AZ 85701, USA
FFL Firearms Dealer

PRECISION SHOOTING EQUIPMENT
P.O. Box 5487
Tucson, AZ 85703, USA
Firearms accessories

THE ARMORY
One North Paseo Pedro
Tucson, AZ 85710, USA
FFL Firearms Dealer

THE SHOOTIST
3946 East Speedway
Tucson, AZ 85712, USA
FFL Firearms Dealer

ED'S
7750 East Clarance Place
Tucson, AZ 85715, USA
FFL Firearms Dealer

ARTHUR H. RAHM
1905 N. Country Club Drive
Tucson, AZ 85716, USA
FFL Firearms Dealer

MILITARY FIREARMS

M. D. MARTINEZ
909 East Mesquite Drive
Tucson, AZ 85719, USA
FFL Firearms Dealer

FIREARM CATALOG SHOWROOM
P.O. Box 13664
Tucson, AZ 85732, USA
FFL Firearms Dealer

DUNN'S GUNS
7080 West Bopp Road
Tucson, AZ 85746, USA
FFL Firearms Dealer

MICHAEL THOMAS
P.O. Box 763
Lakeside, AZ 85929, USA
FFL Firearms Dealer

JEFF HAMMOND
P.O. Box 1102
St. Johns, AZ 85936, USA
FFL Firearms Dealer

JONES GUN SHOP
P.O. Box 537
114 Rolling Hills
Taylor, AZ 85939, USA
FFL Firearms Dealer

JACK'S SPORTING GOODS
3563 North Walker
Flagstaff, AZ 86001, USA
FFL Firearms Dealer

DONALD JANSEN
170 Chaco Trail
Flagstaff, AZ 86001, USA
FFL Firearms Dealer

RED ROCK FIREARMS
75 Broken Lance Way
Sedona, AZ 86336, USA
FFL Firearms Dealer
CONTACT: Mike Roberts

F & M SALES
3737 Roosevelt Street
Kingman, AZ 86401, USA
FFL Firearms Dealer

J. W. GALBRAITH
3051 Star Drive
Lake Havasu City, AZ 86403, USA
FFL Firearms Dealer

CHARLES M. HETON
P.O. Box 2328
Window Rock, AZ 86515, USA
FFL Firearms Dealer

HENRY L. MAURINO
4301 4th St. N.W.
Albuquerque, NM 87107, USA
FFL Firearms Dealer

RICHARD E. SPROWL
6710 Cochiti S.E. #21
Albuquerque, NM 87108, USA
FFL Firearms Dealer

QUARTERMASTER SALES
7201 Avenida la Costa N. E.
Albuquerque, NM 87109, USA
505-821-4946
Firearms

R. ARREDONDO
512 General Patch S.E.
Albuquerque, NM 87123, USA
FFL Firearms Dealer

INTERNATIONAL CULTURAL EXCHANGE
600 Vista Abajo N.E.
Albuquerque, NM 87123, USA
FFL Firearms Dealer
CONTACT: Kenneth M. O'Keefe

GARY L. CUNNINGHAM
1725 Fornax Road
Rio Rancho, NM 87124, USA
FFL Firearms Dealer

JOHN B. VAN WAGNER
3300 May Circle
Rio Rancho, NM 87124, USA
FFL Firearms Dealer

AL'S SERVICE
609 E. Apache at Butler
Farmington, NM 87401, USA
FFL Firearms Dealer

SOFTWORLD, INC.
1267 46th Street
Los Alamos, NM 87544, USA
FFL Firearms Dealer

STEVE WORK
P.O. Box 2172, Campus Station
Socorro, NM 87801, USA
FFL Firearms Dealer

MILLER FIREARMS SUPPLY
P.O. Box 432
Truth or Consequence, NM 87901, USA
FFL Firearms Dealer

SLICKS GUN SHOP
102 W. Hill, Box 588
Hatch, NM 87937, USA
FFL Firearms Dealer

RUDY ARREY
916 S. Ruby
Deming, NM 88030, USA
FFL Firearms Dealer

DON RUDD
1405 Florida Street
Silver City, NM 88061, USA
FFL Firearms Dealer

SACRAMENTO FIREARMS
2784 Sacramento Court
Holloman AFB, NM 88330, USA
FFL Firearms Dealer
CONTACT: Bob Ehmen

'ACES' GUN SHOP
441 Sunburst Drive
Henderson, NV 89015, USA
FFL Firearms Dealer
CONTACT: George Abernathy

RANDEL ENTERPRISES
P.O. Box 1089
Henderson, NV 89015, USA
FFL Firearms Dealer

GADDY'S GUNS
Foote Trailer Ct. #13
Silver Peak, NV 89047, USA
FFL Firearms Dealer

HIGH DESERT SHOOTERS EMPORIUM
P.O. Box 70210
Las Vegas, NV 89170, USA
702-796-1133
Firearms

K.I.S. GUNSMITH
1830 Ruby Street
Ely, NV 89301, USA
FFL Firearms Dealer

JERRY'S GUNS & AMMO
P.O. Box 126
300 N. Stateline Drive
McDermitt, NV 89421, USA
FFL Firearms Dealer

FRANK'S FIREARMS & SUPPLIES
6th & 'A' Streets
Mina, NV 89422, USA
FFL Firearms Dealer

THE GUN EXCHANGE
363 Roberts Street
Reno, NV 89502, USA
FFL Firearms Dealer
CONTACT: Robert F. Enewold

RENO PROSPECTORS SUPPLY
1385 S. Wells
Reno, NV 89502, USA
FFL Firearms Dealer

CATCO
316 California Avenue, #341
Reno, NV 89509, USA
707-253-8338
Firearms accessories

L & L GUN SALES
1325 Trainer Way
Reno, NV 89512, USA
FFL Firearms Dealer

STEWART M. RUTTENBERG
892 East Williams Street
Carson City, NV 89701, USA
FFL Firearms Dealer

MARK MARSHALL
1669 N. Edmonds Drive
Carson City, NV 89701, USA
FFL Firearms Dealer

RAYMOND G. THOMPSON
2158 Norman Lane, #121-5
Battle Mountain, NV 89820, USA
FFL Firearms Dealer

BRASS RAIL
1136 North La Brea
Los Angeles, CA 90038, USA
FFL Firearms Dealer

MILITARY FIREARMS

MECCA HOBBIES
11024 Washington Boulevard
Culver City, CA 90232, USA
FFL Firearms Dealer

LEBLANC'S GUN SERVICE
15126 Dunston Drive
Whittler, CA 90604, USA
FFL Firearms Dealer

GUERIN INDUSTRIES
P.O. Box 2298
La Habra, CA 90631, USA
FFL Firearms Dealer
CONTACT: Firearms Division

HERITAGE CLASSICS
16415 Alexander Place
Cerritos, CA 90701, USA
FFL Firearms Dealer
CONTACT: Melvin R. Cummings

TIANIAI HU
12539 East Corellian Court
Cerritos, CA 90701, USA
FFL Firearms Dealer

NIDA RADIO & ELEC.
15719 Texaco Avenue
Paramount, CA 90723, USA
FFL Firearms Dealer
CONTACT: Kevin R. Nida

JIM WALKER-SPECIALTIES
4225 Lime Ave., Bixby Knolls
Long Beach, CA 90807, USA
FFL Firearms Dealer

RANDCO
P.O. Box 1981
Monrovia, CA 91016, USA
Firearms parts.
Catalogue $1.00.

THE GUN SHOP
2629 East Foothill Blvd.
Pasadena, CA 91107, USA
FFL Firearms Dealer

INTERSTATE ARMS
18526 Parthenia Street
Northridge, CA 91324, USA
818-349-8430 617-667-9877
Firearms

SID JETTELSON
18925 Vose Street
Reseda, CA 91335, USA
FFL Firearms Dealer

THE GOLDEN INGOT
20929 Ventura Blvd., #37
Woodland Hills, CA 91364, USA
FFL Firearms Dealer

INCO
P.O. Box 3111
Burbank, CA 91504, USA
(818)-842-4094
Exotic weapons dealer.
S6; 30;
Catalogue available.
CONTACT: S. M. Kessler

SALES DEPT: S. M. Kessler
PURCHASING DEPT: W. E. Little

CARPERTIER'S GUNSMITHING
11100 Cumpston, Unit 40
North Highlands, CA 91601, USA
FFL Firearms Dealer

STALO M. LINARES
7614 Morella Avenue
North Highlands, CA 91605, USA
FFL Firearms Dealer

BUMBLE BEE WHOLESALE, INC.
12521 - 3 Oxnard Street
North Hollywood, CA 91606, USA
818-985-2939
Firearms

RON GASTON
18209 East Newburgh
Azusa, CA 91702, USA
FFL Firearms Dealer

FEDERAL ORDNANCE, INC.
P.O. Box 6050
1443 Potrero Avenue
S. El Monte, CA 91733, USA
(818)-350-4161 (800)-423-4552
Firearms, ammunition & accessories.
S3,S5; 20;
Catalogue available. Quarterly.
CONTACT: Burton Brenner
SALES DEPT: Rose Murillo
PURCHASING DEPT: Richard Siegel
Importer/wholesaler of surplus military weapons & ammunition. Manufacturer of shooting accessories.

MICHAEL COX
2114 Sierra Leone Avenue
Rowland Heights, CA 91748, USA
FFL Firearms Dealer

FIRST CABIN LTD.
5137 Montclair Plaza Lane
Montclair, CA 91763, USA
FFL Firearms Dealer
CONTACT: Skip Anderson

J.R.'s SERVICE
331 North Gerona Avenue
San Gabriel, CA 91775, USA
FFL Firearms Dealer

GARY SOLIS
2918 Lancaster Road
Carlsbad, CA 92008, USA
FFL Firearms Dealer

MOKTAR KOURDA TRADING POST
951 Eton Court
Chula Vista, CA 92010, USA
FFL Firearms Dealer

DENNIS L. BOND
622 Stanley Court
El Cajon, CA 92020, USA
FFL Firearms Dealer

WEAVER ARMS
P.O. BOX 3316
Escondido, CA 92025, USA

Firearms

THE SHOOT SHOPPE
1631 Gary Lane
Escondido, CA 92026, USA
FFL Firearms Dealer
CONTACT: David Osburn

SCULPTURES IN SAND
2008 East Washington
Escondido, CA 92027, USA
FFL Firearms Dealer

KEN ADAMS
1442 South Mission Road
Fallbrook, CA 92028, USA
FFL Firearms Dealer

MICHAEL P. MURPHY
11621 Valley Vista Road
Lakeside, CA 92040, USA
FFL firearms dealer & communications equipment.

MARVIN S. BERGER
1770 Terraza Street
Oceanside, CA 92054, USA
FFL Firearms Dealer

JO-MAC TRADING POST II
108 South Freeman Street
Oceanside, CA 92054, USA
FFL Firearms Dealer

WILLIAM FLOURNOY
14260 Garden Road, Suite 25-A
Poway, CA 92064, USA
FFL Firearms Dealer

ERNIE M. WEBB
12411 Paseo Colina
Poway, CA 92064, USA
FFL Firearms Dealer

W. A. SUTTON
P.O. Box 4
San Luis Rey, CA 92068, USA
FFL Firearms Dealer

FRONTIER GUN SHOP
3445 University
San Diego, CA 92100, USA
Firearms

G. T. IMPORTS
3818 4th Avenue
San Diego, CA 92103, USA
Gun leather & holsters.
CONTACT: George Schmall

BULLSEYE ARMS
701 North 20th Street
Banning, CA 92220, USA
FFL Firearms Dealer

JOHN F. LENDERMAN
1401 La Brucherie Road, #15
El Centro, CA 92243, USA
FFL Firearms Dealer

MILITARY FIREARMS

JIM'S GUNSHOP
54-622 Herrera
La Quinta, CA 92253, USA
FFL Firearms Dealer
CONTACT: James Langford

C. M. BROYHILL
18217 Danbury Avenue
Hesperia, CA 92345, USA
FFL Firearms Dealer

MYRON W. STICKLEY
10208 Choiceana Avenue
Hesperia, CA 92345, USA
FFL Firearms Dealer

YOUR BEST SHOT
P.O. Box 704
Pinon Hills, CA 92372, USA
FFL Firearms Dealer
CONTACT: Earl Curtis

H. R. GALLANT
27891 Covington Way
Sun City, CA 92381, USA
FFL Firearms Dealer

RIVERSIDE GUN
4146 Melrose #6
Riverside, CA 92504, USA
FFL Firearms Dealer
CONTACT: Bradley L. Dischner

JAMES HUTCHISON
6836 Weaver Street
Riverside, CA 92504, USA
FFL Firearms Dealer

WILBUR C. ALBERGE
6530 Baker Street
Riverside, CA 92509, USA
FFL Firearms Dealer

JAMES R. HUFFMAN
813 Worthington Street
Brea, CA 92621, USA
FFL Firearms Dealer

ROBERT LEE
P.O. Box 1462
Garden Grove, CA 92642, USA
FFL Firearms Dealer

ALCO WHOLESALE
16462 Gothard St., Suite E
Huntington Beach, CA 92647, USA
FFL Firearms Dealer

BULLHEAD ENTERPRISES
6582 Dohrn Circle
Huntington Beach, CA 92647, USA
FFL Firearms Dealer

SMITH FIREARMS
16052 Giarc Lane
Huntington Beach, CA 92647, USA
FFL Firearms Dealer

TOM BAKER
P.O. Box 305
Huntington Beach, CA 92648, USA
FFL Firearms Dealer

MIDWAY GUNS
15056 Midway Place #4
Midway City, CA 92655, USA
FFL Firearms Dealer

N. AMERICAN SCHOOL OF FIREARMS
4400 Campus Drive
Newport Beach, CA 92660, USA
Firearms instruction school

THOMAS GUNS WHOLESALE
511 East 1st, Suite C
Tustin, CA 92680, USA
FFL Firearms Dealer

NORMAN HOWER
P.O. Box 1211
Tustin, CA 92681, USA
Collects Pedersen & British cartridges.

SOUTHERN CALIF. JEWELRY & LOAN
12522 Westminster Blvd.
Santa Ana, CA 92706, USA
FFL Firearms Dealer

WEBER'S FIREARMS
222 South Primrose Street
Anahein, CA 92804, USA
FFL Firearms Dealer

DON BROWN
2515 West Winston, #25
Anaheim, CA 92804, USA
FFL Firearms Dealer

DUFFEYS GUNS
1764 North Woodwind Lane
Anaheim, CA 92807, USA
FFL Firearms Dealer

CODE 3
1630 Callens Road
Ventura, CA 93003, USA
FFL Firearms Dealer

BERNARD M. WITTKINS
244 Petit Avenue
Ventura, CA 93003, USA
FFL Firearms Dealer

C OF C
1244 El Monte
Simi Valley, CA 93065, USA
FFL Firearms Dealer

HILLDALE SALES
1625 Tierra Rejada Road
Simi Valley, CA 93065, USA
FFL Firearms Dealer

HAROLD A. PARKS
2098 Kasten Street
Simi Valley, CA 93065, USA
FFL Firearms Dealer

LINDSAY AUTO SUPPLY
P.O. Box 70
Lindsay, CA 93247, USA
FFL Firearms Dealer

DANNY C. HOPKINS
711 First Street
Taft, CA 93268, USA

FFL Firearms Dealer

J. S. HANDGUN WORKS
2203 Watson Street
Bakersfield, CA 93308, USA
FFL Firearms Dealer

ROBERT C. MEHCIZ
5501-12 Lennox Avenue
Bakersfield, CA 93309, USA
FFL Firearms Dealer

B & A SALES
23 Squire Road
San Luis Obispo, CA 93401, USA
FFL Firearms Dealer

RICHARD MILLER
1398 Paso Robels Avenue
Los Osus, CA 93402, USA
FFL Firearms Dealer

PINE TREE GUNS & SUPPLIES
821 Cornwall Street
Cambria, CA 93428, USA
FFL Firearms Dealer
CONTACT: Michael Molnar, Jr.

THOMAS B. GOLDING
Star Route, Box 210
Santa Margarita, CA 93453, USA
FFL Firearms Dealer

JIM SMITH
4845 Kenneth Avenue
Santa Maria, CA 93455, USA
FFL Firearms Dealer

JERRY YOUNG
1025 West Upjohn
Ridgecrest, CA 93555, USA
FFL Firearms Dealer

MARK E. MALLORY
1667 Purvis Avenue
Clovis, CA 93612, USA
FFL Firearms Dealer

STIDSTON'S FIREARMS
1456 Wrenwood
Clovis, CA 93612, USA
FFL Firearms Dealer

GEORGES GUN SUPPLY
5782 North 4th Street
Fresno, CA 93710, USA
FFL Firearms Dealer

VALLEY ARMS, INC.
4974 North Fresno Street, #204
Fresno, CA 93726, USA
FFL Firearms Dealer

GOLDEN WEST SPORTS
2747 North Clovis Avenue
Fresno, CA 93727, USA
FFL Firearms Dealer

TOM CAUDLE
2826 West Locust
Fresno, CA 93771, USA
FFL Firearms Dealer

MILITARY FIREARMS

CALIFORNIA WEAPONS
250 Del Medio, #301
Mountain View, CA 94040, USA
FFL Firearms Dealer

PAUL L. DISTERDICK
31 Saw Mill Court
Mountain View, CA 94043, USA
FFL Firearms Dealer

DENNIS HENSON
920 Round Hill Road
Redwood City, CA 94061, USA
FFL Firearms Dealer

HELLER ARMS LTD.
P.O. Box 578
San Bruno, CA 94066, USA
Firearms parts
CONTACT: Thomas W. Heller

G. F. GRESS & L. L. WONG
1307 Grandview Drive
South San Francisco, CA 94080, USA
FFL Firearms Dealer
CONTACT: c/o Chessmen Enterprises

DENNIS PINKE
1225 Vienna Drive #427
Sunnyvale, CA 94086, USA
FFL Firearms Dealer

SPAULDING ARMOURY
801 Foster City Blvd., #10
Foster City, CA 94404, USA
FFL Firearms Dealer

THE GUNSLINGER
2382 Ironwood Place
Alamo, CA 94507, USA
FFL Firearms Dealer

JIM B. BARNARD
P.O. Box 42
Benicia, CA 94510, USA
FFL Firearms Dealer

CAL ARMS COMPANY
P.O. Box 21143
Concord, CA 94521, USA
FFL Firearms Dealer

HEASTONS FIREARMS & GUNSMITHING
225 Ilene Drive
Pleasant Hill, CA 94523, USA
FFL Firearms Dealer

PEARCE'S GIFT SHOP
885 Loring Avenue
Crockett, CA 94525, USA
FFL Firearms Dealer

OLD WEST GUN ROOM
3509 Carlson Blvd.
El Cerrito, CA 94530, USA
Firearms

DUNCAN'S SHOOTING SUPPLY
4938 Wheler Court
Fremont, CA 94538, USA
FFL Firearms Dealer

THE GUN PRO
280 West 'A' Street, #56
Hayward, CA 94541, USA
FFL Firearms Dealer
CONTACT: Ronald E. Kinyon

COMBAT ARMS
3636 Castro Valley Blvd. #8
Castro Valley, CA 94546, USA
(415)-538-6544
Firearms, ammo, gunsmithing and accessories.
S1; 4; 1,000; $28.00
CONTACT: Richard M. Bash
SALES DEPT: Richard M. Bash
PURCHASING DEPT: Richard M. Bash
We provide hardware & training for para-military, survivalists & civilians interested in home defense. Stocking HK, FN, Steyr, Colt, AK, & knowledgeable about them also.

DEAD EYE SUPPLY
259 Morello Avenue
Martinez, CA 94553, USA
FFL Firearms Dealer
CONTACT: Kevin Smith

R.J.'s GUN STOP
3596 Woodbrook
Napa, CA 94558, USA
FFL Firearms Dealer

ANDY'S ARMS & ACCESSORIES
4950 Croydon Place
Newark, CA 94560, USA
FFL Firearms Dealer

TEDFORD P. EADS
5654 Geranium Court
Newark, CA 94560, USA
FFL Firearms Dealer

FRANK A. JOHNSON
36531 Cedar Blvd.
Newark, CA 94560, USA
FFL Firearms Dealer

REPPAS SPORTING GOODS
6635 Dublin Blvd.
Dublin, CA 94568, USA
FFL Firearms Dealer

SANCHES SALES
880 Tradewind Lane
Rodeo, CA 94572, USA
FFL Firearms Dealer

NOR-CAL SHOOTING SPORTS
2004 Elliott Drive
Vallejo, CA 94590, USA
FFL Firearms Dealer

THE ARMORY
1305 Whipple Ave
Redwood City, CA 94602, USA
FFL Firearms Dealer
CONTACT: Joan Voelzke

CLYDE V. JACKSON
968 Park Lane
Oakland, CA 94610, USA
FFL Firearms Dealer

BRUCE L. YOW
407 Orange St., Suite 412
Oakland, CA 94610, USA
FFL Firearms Dealer

BILL McELROY
38 Amicita Avenue
Mill Valley, CA 94941, USA
FFL Firearms Dealer

E.D.S.
311 Bonita Drive
Aptos, CA 95002, USA
FFL Firearms Dealer

RADER'S JEWELRY & LOAN CO.
1209 Pacific Avenue
Santa Cruz, CA 95060, USA
FFL Firearms Dealer

PERFORMANCE MACHINE ENGINEERING
2640 #2, 17th Avenue
Santa Cruz, CA 95065, USA
FFL Firearms Dealer

CLARENCE B. RAMSEY
1408 Keoncrest Avenue
San Jose, CA 95110, USA
FFL Firearms Dealer

ALL AMERICAN SALES
6395 El Paseo
San Jose, CA 95120, USA
FFL Firearms Dealer
CONTACT: George Coughraw

RAYMOND RAPP
6335 Mayo Drive
San Jose, CA 95123, USA
FFL Firearms Dealer

D & J SHOOTING SUPPLY
3259 Parkhaven Drive
San Jose, CA 95132, USA
FFL Firearms Dealer
CONTACT: R. I. Otwell

JR SYSTEMS
3200 Knightswood Way
San Jose, CA 95148, USA
FFL Firearms Dealer

X.O.'s GUNS
1660 West Flora Street
Stockton, CA 95203, USA
FFL Firearms Dealer

PHIL'S FIREARMS
327 Santa Barbara Court
Stockton, CA 95210, USA
FFL Firearms Dealer

THE GUN CABINET
P.O. Box 117
25571 East Elm
Farmington, CA 95230, USA
FFL Firearms Dealer
CONTACT: James Davis

RC PRODUCTS
413 Daisy Avenue
Lodi, CA 95240, USA
FFL Firearms Dealer

MILITARY FIREARMS

CENTRAL VALLEY ARMS
716 Evergreen
Manteca, CA 95336, USA
FFL Firearms Dealer
CONTACT: Edward McNeese

SECOND CHANCE PAWN SHOP
137 W 19th Street, Apt. 3
Merced, CA 95340, USA
FFL Firearms Dealer

WARNER ENTERPRISES
801 W. Roseburg, Suite 174
Modesto, CA 95350, USA
FFL Firearms Dealer

MOREY'S GUNS & AMMO
1400 North Carpenter Rd., #D
Modesto, CA 95351, USA
FFL Firearms Dealer

WOODWARD'S FIREARMS
549 East Hawkeye
Turlock, CA 95380, USA
FFL Firearms Dealer
CONTACT: Mark E. Woodward

BILL BLANC
46 Plymouth Way
Santa Rosa, CA 95401, USA
FFL Firearms Dealer

J & R OUTFITTERS
1530 Monterey Drive
Santa Rosa, CA 95405, USA
FFL Firearms Dealer

JACK B. AZEVEDO
24830 Ward Avenue
Ft. Bragg, CA 95437, USA
FFL Firearms Dealer

M3 SHOOTERS SUPPLY
P.O. Box 127
Garberville, CA 95440, USA
FFL Firearms Dealer

FAR WEST SHOOTERS SUPPLY
630 Gatliff Avenue
Eureka, CA 95501, USA
FFL Firearms Dealer
CONTACT: R. A. Messer

BILL GOLUB, FIREARMS BROKER
8716 Latimer Way
Fair Oaks, CA 95628, USA
(916)-967-8645
Special order arms & ammunition.
S3; 5;
CONTACT: William M. Golub
SALES DEPT: Bill Golub
PURCHASING DEPT: Bill Golub
Special order all Arms & Ammo. At cost plus 10%. Current & surplus military and all civilian arms are available in any quantity.

ROBERT L. JEWETT
8125 Sunset Ave., Suite 224
Fair Oaks, CA 95628, USA
FFL Firearms Dealer

DARRYL or STAN HANKINS
303-B Montrose Drive
Folsom, CA 95630, USA
FFL Firearms Dealer

CALABOOSE II
6521 Larchmont Drive
North Highlands, CA 95660, USA
FFL Firearms Dealer

JAMES E. RICE
10416 Croetto Way
Rancho Cordova, CA 95670, USA
FFL Firearms Dealer

SKYCLONES AVIATION
2285 Oakvale Drive
Shingle Springs, CA 95682, USA
FFL Firearms Dealer
CONTACT: Stoney Stonich

PACIFIC INTERNATIONAL
MERCHANDISING CORPORATION
2215 'J' Street
Sacramento, CA 95816, USA
916-446-2737
Firearms

THOMAS A. ALSTON
6291 Warehouse Way
Sacramento, CA 95826, USA
FFL Firearms Dealer

CLAUDE A. TOWNSEND
8953 Fashion Drive
Sacramento, CA 95826, USA
FFL Firearms Dealer

R & S FIREARMS
3209 Hugarth Drive
Sacramento, CA 95827, USA
FFL Firearms Dealer
CONTACT: Rob Nunes

DAN RUIZ
Star Route, Box 118, G-6
Berry Creek, CA 95916, USA
FFL Firearms Dealer

BIG G'S SPORTING GOODS
12671 Centerville Road
Chico, CA 95926, USA
FFL Firearms Dealer

R. C. WINCHESTER
P.O. Box 26
Yuba City, CA 95992, USA
FFL Firearms Dealer

SHOOTERS SUPPLY
6500 Doberman Drive
Redding, CA 96003, USA
FFL Firearms Dealer

G. D. STANKEY
1676 North Street
Anderson, CA 96007, USA
FFL Firearms Dealer

PAUL'S GUNS
810 Rio Street
Red Bluff, CA 96080, USA
FFL Firearms Dealer
CONTACT: Paul R. Moon

NIALL E. STALLARD
P.O. Box 824
Aiea, HI 96701, USA
FFL Firearms Dealer

BRAD'S GUN SHOP
1718 WILIKINA DRIVE
Wahlawa, HI 96786, USA
FFL Firearms Dealer

S. A. CALDWELL
P.O. Box 1713
Honolulu, HI 96806, USA
FFL Firearms Dealer

WALLCARD SPORTING EQUIPMENT
17885 S.W. TV Highway
Aloha, OR 97006, USA
FFL Firearms Dealer

ELDEN DUNSMOOR
P.O. Box 323
Gresham, OR 97030, USA
FFL Firearms Dealer

GRESHAM GUN WORKS
52 N.W. 2nd
Gresham, OR 97030, USA
FFL Firearms Dealer

JAMES R. EBERT
16236 S. Hunter Avenue
Oregon, OR 97045, USA
FFL Firearms Dealer

THOMAS & STEVEN CADDY
4601 Victory Road
Troutdale, OR 97060, USA
FFL Firearms Dealer

VIKING GUN WORKS, INC.
P.O. Box 1725
Beaverton, OR 97075, USA
FFL Firearms Dealer

STUCK'S FIREARMS & SUPPLIES
200 S. Alpine Drive
Cornelius, OR 97113, USA
FFL Firearms Dealer
CONTACT: John D. Stuck

BRIAN GEHLEY
Rt. 1, Box 119
Gaston, OR 97119, USA
FFL Firearms Dealer

W & W SPORTING GOODS
11842 S.E. Ankeny
Portland, OR 97216, USA
FFL Firearms Dealer

JOHN DAVIS
4920 N.E. 55th
Portland, OR 97218, USA
FFL Firearms Dealer

TOM KELLY
5805 N.E. Alberta
Portland, OR 97218, USA
FFL Firearms Dealer

MILITARY FIREARMS

STEVE BURTON
7180 S.W. Taylors Ferry Road
Portland, OR 97223, USA
FFL Firearms Dealer

SHAWN K. WATE
4353 Northeast 133rd
Portland, OR 97230, USA
FFL Firearms Dealer

GUNS BY THE ORDER
9 S.E. 136th
Portland, OR 97233, USA
FFL Firearms Dealer

CRAIG SCHIMSCHOK
17028 S.E. McKinley Road
Portland, OR 97236, USA
FFL Firearms Dealer

MAURYS GUN RACK
5939 S.E. 82nd Avenue
Portland, OR 97266, USA
Firearms

DALE C. SCHMIDT
610 Howell Prairie Rd., S.E.
Salem, OR 97301, USA
FFL Firearms Dealer

HARRY HALVERSON
5090 Coloma Ct. S.E.
Salem, OR 97306, USA
FFL Firearms Dealer

OAKVILLE OUTFITTERS
1455 N.W. 13th Street
Corvallis, OR 97330, USA
FFL Firearms Dealer
CONTACT: Norman Pawlowski

MICK'S GUNS
37391 Kgal Drive
Lebanon, OR 97355, USA
FFL Firearms Dealer

GAUL'S RADIO SERVICE
313 N. Water Street
Silverton, OR 97381, USA
FFL Firearms Dealer

JOHN W. HART
25275 Rice Road
Sweet Home, OR 97386, USA
FFL Firearms Dealer

T & R SPORTING GOODS
P.O. Box 1413
Gold Beach, OR 97444, USA
FFL Firearms Dealer

JACK DAILEY
Rt. 3, Box 136
Reedsport, OR 97467, USA
FFL Firearms Dealer

RON GLENN
3324 Winona Road
Grants Pass, OR 97526, USA
FFL Firearms Dealer

MAUSER SPORT WHOLESALERS
19940 Juniper Lane
Bend, OR 97701, USA
FFL Firearms Dealer

MALCOLM'S GUNS
1343 S.W. 41st Street
Pendelton, OR 97801, USA
FFL Firearms Dealer

WM. JERRY WELCH
P.O. Box 4
Third & Amy
Haines, OR 97833, USA
FFL Firearms Dealer

COLEMANS GUNS
Rt. 1, Box 430C, 4th St. W.
Irrigon, OR 97844, USA
FFL Firearms Dealer

BOW HUNTERS HAVEN
2802 N. Cherry
La Grande, OR 97850, USA
FFL Firearms Dealer

D ENTERPRISES
400 Walnut
La Grande, OR 97850, USA
FFL Firearms Dealer

JIM & CAROL WIMER
P.O. Box 115
La Grande, OR 97850, USA
FFL Firearms Dealer

CANDY'S
P.O. Box 459
Union, OR 97883, USA
Waterproof fuse
S6; 2;
CONTACT: Robert Zinter
WATERPROOF FUSE. Finest available from leading U.S. manufacturer. Send S.A.S.E. and $1.00 for Sample Fuse.

ALLEN L. JONES
416 Ehrgood Avenue
Nyssa, OR 97913, USA
FFL Firearms Dealer

JAMES C. TAYLOR
32454 - 46th Place South
Auburn, WA 98001, USA
FFL Firearms Dealer

CHUCK SAVICKI
3409 C St., N.E. #11
Auburn, WA 98002, USA
FFL Firearms Dealer

DARRIGAN ENTERPRISES
4900 Highland Drive
Bellevue, WA 98006, USA
FFL Firearms Dealer

GET TRADING COMPANY
48620 - 284th Ave., S.E.
Enumclaw, WA 98022, USA
FFL Firearms Dealer
CONTACT: George Trepus

RICHARD CAVANAUGH
3833 S.W. 339th Street
Federal Way, WA 98023, USA
FFL Firearms Dealer

MARTY CURTIS
27251 144th S.E. Place
Kent, WA 98031, USA
FFL Firearms Dealer

THORA ROWLEY
13509 - 120 Ave. N.E.
Kirkland, WA 98034, USA
FFL Firearms Dealer

RICK INMAN
12608 N.E. 119th St. B-3
Kirkland, WA 98034, USA
FFL Firearms Dealer

ADAMS GUNWORKS
16410 28th Avenue W
Lynnwood, WA 98037, USA
FFL Firearms Dealer

RICK'S SUPPLY
22304 - 39th Ave. West
Mountlake Terrace, WA 98043, USA
FFL Firearms Dealer
CONTACT: Richard C. Mayer

THE ARMORED RECON SHOP
5801 -224TH Pl. S.W.
Mountlake Terrace, WA 98043, USA
FFL Firearms Dealer

BRUCE W. BOOTS
13248 Renton Avenue S
Renton, WA 98055, USA
FFL Firearms Dealer

K-C ENTERPRISES
16836 - 122 S.E.
Renton, WA 98055, USA
FFL Firearms Dealer

JERRY L. HONEA
12815 - S.E. 106th
Renton, WA 98056, USA
FFL Firearms Dealer

WAFFENFABRIK - WEST
2730 - 49th Ave. S.W.
Seattle, WA 98116, USA
FFL Firearms Dealer

BAZAR SPORTS & CUTLERY
6710 3rd Avenue Northwest
Seattle, WA 98117, USA
FFL Firearms Dealer
CONTACT: Allen Robinson

JOHN S. MILLER
8810 - 36th Ave. S.W.
Seattle, WA 98126, USA
FFL Firearms Dealer

ROOSEVELT SPORT & GUN
900 North 103rd
Seattle, WA 98133, USA
FFL Firearms Dealer

MILITARY FIREARMS

NEMCO
P.O. Box 45813
Seattle, WA 98145, USA
FFL Firearms Dealer
CONTACT: Edward Lowe

MO's GUN SHOP
12230 23rd Ave. South
Seattle, WA 98168, USA
FFL Firearms Dealer
CONTACT: Michael S. Palermo

REINER GRAUMANN
11033 Crestwood Dr. South
Seattle, WA 98178, USA
FFL Firearms Dealer

NORTHWEST SHOOTERS SUPPLY
628 - 58th Street
Everett, WA 98203, USA
FFL Firearms Dealer
CONTACT: Richard C. Birch

HOWARD D. LEONARD
9908 - 48th Avenue West
Everett, WA 98204, USA
FFL Firearms Dealer

MARTY'S GUN SALES
26306 Clear Creek Road
Darrington, WA 98241, USA
FFL Firearms Dealer

PETE'S PLACE
P.O. Box #1
Lakewood, WA 98259, USA
FFL Firearms Dealer

MARDO SALES
12824 48th Drive NE
Marysville, WA 98270, USA
FFL Firearms Dealer

CRESCENT BAY TRADING COMPANY
8166 850th East
Oak Harbor, WA 98277, USA
FFL Firearms Dealer

SNOHOMISH GUN WORKS
6729 - 187th St. S.E.
Snohomish, WA 98290, USA
FFL Firearms Dealer

J. MARTIN ENTERPRISES
5702 - 163rd Ave. S.E.
Snohomish, WA 98290, USA
FFL Firearms Dealer

R.A.I. SALES
26511 Sumner-Buckley Hwy.
Buckley, WA 98321, USA
FFL Firearms Dealer

ALAN WOSLEGER
321 - 158th KPS
Lakebay, WA 98349, USA
FFL Firearms Dealer

DON ROBERTS
821 Billy Smith Road
Port Angeles, WA 98362, USA
Firearms collector

MINTER CREEK SPORTING GOODS
2236 S.W. Pine Road
Port Orchard, WA 98366, USA
FFL Firearms Dealer

MICHAEL J. WILLIAMS
637 S.W. Birch Road
Port Orchard, WA 98366, USA
FFL Firearms Dealer

JIM'S PAWN HAUS
111 Anderson Parkway
Poulsbo, WA 98370, USA
FFL Firearms Dealer

HAROLD P. KNEPPER
120 Candlewyck Dr. West
Tacoma, WA 98467, USA
FFL Firearms Dealer

OLYMPIC ARMS, INC.
624 Old Pacific Hwy S.E.
Olympia, WA 98503, USA
206-456-3471
Firearms accessories

CHET'S CUSTOM CARTRIDGES
P.O. Box C
699 Sussex Avenue
Tenino, WA 98589, USA
FFL Firearms Dealer

A. D. HUGHES ENTERPRISES
12030 Bald Hill Rd. S.E.
Yelm, WA 98597, USA
(206)-458-7277
Firearms, ammo, knives, blackpowder guns & accessories, optics, outdoors supplies
S1; 11;
CONTACT: Alan D. Hughes
SALES DEPT: Alan D. Hughes
PURCHASING DEPT: Alan D. Hughes
Firearms, ammunition, knives, optics, blackpowder guns & accessories, reloading equipment & supplies, gun safes, leather goods, hunting & fishing equipment, etc. Full service firearms & related items store, specializing in unique & hard to get items; with low prices and friendly service.

BEAR ARMS
17905 Cole Witter Road
Battle Ground, WA 98604, USA
FFL Firearms Dealer

H & L SHOOTERS SUPPLY
482 B Highway 506
Castle Rock, WA 98611, USA
FFL Firearms Dealer

LARRY HEASLEY COMPANY
P.O. Box 575
Ilwaco, WA 98624, USA
FFL Firearms Dealer

EDWARD J. NORTH
P.O. Box 1312, MCAS
F.P.O. Seattle, WA 98764, USA
FFL Firearms Dealer

KENNETH L. RUTH
2413 Larson Road
Yakima, WA 98908, USA

FFL Firearms Dealer

BOB'S GUNS
P.O. Box 478
42 Lakeview Avenue
Electric City, WA 99123, USA
FFL Firearms Dealer

TINTSULATION GUN SALES
3813 N. Division
Spokane, WA 99207, USA
FFL Firearms Dealer

BRAD L. FULLER
E. 6813 Second #12
Spokane, WA 99212, USA
FFL Firearms Dealer

BIG JOHN'S GUNS
4240 Laurel Drive
West Richland, WA 99352, USA
FFL Firearms Dealer

VALLEY CUSTOM ARMS
1313 Perkins
Richland, WA 99352, USA
FFL Firearms Dealer

ONE TO ONE GUNS
1304 Roberdeau
Richland, WA 99352, USA
FFL Firearms Dealer

T & R FIREARMS & SHOOTERS
SRA #25
11661 Rockridge Drive
Anchorage, AK 99502, USA
FFL Firearms Dealer
CONTACT: Bob Sanders

STUARTS SPORT SPECIALTIES
7081 Chad Street
Anchorage, AK 99502, USA
FFL Firearms Dealer

CALVIN R. PAPPAS
P.O. Box 43033
Anchorage, AK 99509, USA
FFL Firearms Dealer

THOMAS M. PEETZ
P.O. Box 4-112
Anchorage, AK 99509, USA
German machinegun collector

EAGLE RIVER LOAN
P.O. Box 4903
Eagle River, AK 99577, USA
FFL Firearms Dealer

JAMES ALLEN'S
P.O. Box 51
Terrace Drive & 2nd Street
Glenallen, AK 99588, USA
FFL Firearms Dealer

ALASKA ARMS & AMMO
1652 E. Aliak
Kenai, AK 99611, USA
FFL Firearms Dealer

MILITARY FIREARMS

FUNNY RIVER OUTFITTERS
P.O. Box 3353
Kenai, AK 99611, USA
FFL Firearms Dealer
CONTACT: Joe W. Harrison

DON'S GUN ROOM
S.R.D. Box 9100
Palmer, AK 99645, USA
FFL Firearms Dealer

JOHNSON BROTHERS
P.O. Box 3774
Soldotna, AK 99669, USA
FFL Firearms Dealer

RIDGE RUNNER GUNSMITHING
P.O. Box 3575
Soldotna, AK 99669, USA
FFL Firearms Dealer

TOM'S GUN SHOP
Rt. 2, Box 257
81 Mi. Marker Sterling Hwy.
Sterling, AK 99672, USA
FFL Firearms Dealer
CONTACT: Thomas F. Vogel

THOMAS & ROSEMARY CRAIG
965 Dewood
Wasilla, AK 99687, USA
FFL Firearms Dealer

GEORGE BOICE
570 West Ponderosa Lp.
Wasilla, AK 99687, USA
FFL Firearms Dealer

RAINBOW PAWN & GUN
P.O. Box 870854
Mile 48.5 Parks Highway
Wasilla, AK 99687, USA
FFL Firearms Dealer

WASILLA PAWN SHOP
P.O. Box 870213
Wasilla, AK 99687, USA
FFL Firearms Dealer

ARCTIC SPORT FIREARMS
1096 O'Brien Street
Fairbanks, AK 99701, USA
FFL Firearms Dealer

DOWN UNDER GUNSMITHS
206 Driveway
Fairbanks, AK 99701, USA
FFL Firearms Dealer

FOX FIREARM SALES
358 Chatanika Trail
Fairbanks, AK 99701, USA
FFL Firearms Dealer

CHUCK'S ROD & GUN SHOP
218 Yale Way
Fairbanks, AK 99701, USA
FFL Firearms Dealer
CONTACT: Charles A. Griffey

TROY REYNOLDS
P.O. Box 403
Delta Junction, AK 99737, USA

FFL Firearms Dealer

MIKE'S GUNS
P.O. Box 178
Petersburg, AK 99833, USA
FFL Firearms Dealer
CONTACT: Michael C. Cooper

NORTHERN VENTURES
P.O. Box 1065
Sitka, AK 99835, USA
FFL Firearms Dealer
CONTACT: Marvin J. Krause

SOUTHWEST GUNSMITHING
P.O. Box 1814
Wrangell, AK 99929, USA
FFL Firearms Dealer

DOUGLAS HOULT
P.O. Box 153
Coaldale, Alberta T0K 0L0, Canada
Ordnance collector.

CITY GUNS (PTY) LTD.
67 Hout Street
Cape Town, 8001, South Africa
Antique & modern firearms & accessories.

TOKAI GUNSHOP
Blue Route Centre
P.O. Box 77
Tokai, 7965, South Africa
(021) 72-0747
Rifles & accessories.

HOFFMAN & REINHART WAFFEN
Stauffacherquai 56
Zurich, CH 8004, Switzerland
Military firearms & accessories.

MESSRS. HOFMANN & REINHART
56 Stauffacherquai
Zurich, 8004, Switzerland
01-2411632
Firearms dealer.

E. BROLL
Bergstrasse 38
Spiegelberg, 7151, West Germany
07194/228
Collector's munitions & militaria.
List available.

MILITARY FILM/VIDEO

MERCHANTS OF MEMORIES
P.O. Box 3625
Philadelphia, PA 19125, USA
Video-cassettes & tapes of radio shows.
Catalogue available.

VICTORY VIDEO
P.O. Box 87
Toney, AL 35773, USA
(205)-852-1098
Military history video cassettes.
S4,S5,S13; 4;
Catalogue available. Quarterly.
CONTACT: Earl Blair
SALES DEPT: Earl Blair
PURCHASING DEPT: Earl Blair

World War II military and aviation history video cassettes for sale. Quarterly catalog. $3.00 for one year catalog subscription.

HAMMER
P.O. Box 1393
Columbus, IN 47201, USA
German marches on cassettes.
S6,S11,S12; 6;
Catalogue available.
CONTACT: Karl Hammer
Manufacturer/Wholesaler: WW II German march cassettes, flags, pins, and posters. Custom capacity retailer: Videos and books. We buy pre-1945, 78 rpm German march records.

TIMKIN FILMS
P.O. Box 32
Tinley Park, IL 60477, USA
Military history video cassette.

EDWARD TOPOR
4313 S. Marshfield Avenue
Chicago, IL 60609, USA
Military newsreels and films.
List $1.00.

INTERNATIONAL HISTORICAL FILMS
P.O. Box 29035
Chicago, IL 60629, USA
(312)-436-8051 (312)-436-0038
Military history video-cassettes.
S6,S12; 9;
Write or phone for free catalogue. 2 or 3 times annually. CONTACT: Peter P. Bernotas
SALES DEPT: Peter Bernotas
PURCHASING DEPT: Peter Bernotas
VIDEOCASSETTES FOR SALE - WWI through the Falklands Campaign. Specializing in WW II German Newsreels and feature films reproduced from the worlds largest privately owned collection of original 35MM source materials, including contemporary Soviet and British videocassettes.

FILMS INCORPORATED
5547 N. Ravenswood
Chicago, IL 60640, USA
(312)-878-2600 (800)-323-4222
War-related films & video-cassettes.
S3,S4,S8,S9,S10,S13; 48;
Catalogue available. Annually.
CONTACT: Charles Benton
SALES DEPT: Dale Dohm
PURCHASING DEPT: Dale Dohm
Film and videocassettes on war-related topics, including 'VIETNAM: A TELEVISION HISTORY', 'RETURN TO IWO JIMA', and the 'WAR' series.

D.R.V.
2609 Loving
Fort Worth, TX 76106, USA
Nazi video-cassettes & posters.
Send SASE.

ARCHIVE VIDEO PRODUCTIONS
6951 Cobblestone Court
Ft. Worth, TX 76140, USA
817-478-8592
Military history video cassettes.

MILITARY FILM/VIDEO

TJE VIDEO PRODUCTIONS
P.O. Box 50141
Reno, NV 89513, USA
Military firearms & how-to video-cassettes.
S6; 2;
Catalogue available. Continuously updated.
CONTACT: John Seginski
SALES DEPT: John Seginski
Produces low cost how-to videos on firearms ass/disassy on guns, such as govt. .45, PPK, AR-15, etc., and outdoor subjects.

FFERDE GROFE FILMS
3100 Airport Drive
Santa Monica, CA 90405, USA
Films & video-cassettes.

AAF MUSEUM
P.O. Box 7325
Thousand Oaks, CA 91359, USA
Military history video-cassettes.

UDS
P.O. Box 1945
San Marcos, CA 92069, USA
Military history video-cassettes.

MAIL ORDER VIDEO
7888 Ostrow Street, Suite A
San Diego, CA 92111, USA
(619)-569-4000 (619)-569-4106
Military history video-cassettes.
S6,S13; 5;
Catalogue available.
CONTACT: Lenny Magill
SALES DEPT: Alan Creamer
PURCHASING DEPT: Alan Creamer
Largest distributor of sporting-gun tapes, documentaries of war and weaponry, self defense, gun competitions.

KEN STONE
P.O. Box 4304
Garden Grove, CA 92642, USA
Video-cassette producer & distributor.

N. F. VIDEO LIBRARY, INC.
177 Webster Street, #244
Monterey, CA 93940, USA
213-278-0378
Military aviation video-cassettes.
CONTACT: William J. Nagy

CAV
P.O. Box 47
Burlingame, CA 94010, USA
Military history video-cassettes.
Free catalogue.

MIKE AWBREY
9 Loma Mar Court
Sacramento, CA 95828, USA
Military video-cassettes & books.

FLAGS/BANNERS

BOB BANKS
18901 Gold MIne Court
Brookeville, MD 20833, USA
U.S. flag collector

HAMMER
P.O. Box 1393
Columbus, IN 47201, USA
German marches on cassettes.
S6,S11,S12; 6;
Catalogue available.
CONTACT: Karl Hammer
Manufacturer/Wholesaler: WW II German march cassettes, flags, pins, and posters. Custom capacity retailer: Videos and books. We buy pre-1945, 78 rpm German march records.

FLAGS, FLAGPOLES & SAUNAS
P.O. Box 2066
Palm Springs, CA 92263, USA
619-327-3735
Flag manufacturer.
52 page color catalogue $1.00.
CONTACT: Henry Untermeyer

MICHAEL JANKE
Postfach 50 10
Bad Salzuflen, 4902, West Germany
(0 52 22) 7 23 30
Military flags.

FRATERNAL ITEMS

WORLD EXONUMIA
P.O. Box 4143APB
Rockford, IL 61110, USA
(815)-226-0771
Military & civilian exonumia.
S6,S7; 12; 22,000; P.O.R.
Catalogue available. As needed.
CONTACT: Rich Hartzog
We buy, sell, auction (consignments wanted) military, other tokens, historical medals, G.A.R., other fraternal, CW dog tags (Wanted!), more. No decorations. Next auction catalog $5.00.

GENERAL MILITARIA

THE RUPTURED DUCK
P.O. Box 457
Hubbardston, MA 01452, USA
(617)-928-4495
U.S., German, & Japanese militaria.
CONTACT: Bill Shea

HOUSE OF BILLY YANK
22 Franconia Street
Worcester, MA 01602, USA
General American militaria.
12 lists for $5.00.

THE QUARTERMASTER
Box 903S, West Side Station
Worcester, MA 01602, USA
General militaria.
Send SASE for free List.

DONS MILITARY OUTPOST
15 Keans Road
Burlington, MA 01803, USA
General militaria.

JIM HALL
20 Bow Street
Beverly, MA 01915, USA
German militaria.

MILITARY & NAUTICAL COLLECT.
140 Derby Street (First Floor)
Salem, MA 01970, USA
(617)-745-6945
General militaria.
S3,S4,S5,S6,S10,S12; 18;
Catalogue available. As needed.
CONTACT: John J. Harty
SALES DEPT: John Harty
PURCHASING DEPT: John Harty
Retail, wholesale, mail-order, shop, shows - General militaria, surplus, uniforms, edged weapons, firearms, headgear, insignia, military equipment, optical equipment, accoutrements, medals, orders, decorations, military vehicles.

REX STARK - AMERICANA
49 Wethersfield Road
Bellingham, MA 02019, USA
(617)-966-0994
Fine quality mail auctions of all kinds of historical Americana circa 1750-1950.
S4,S7,S13; 10;
Catalogue available. Bi-annually.
CONTACT: Rex Stark
SALES DEPT: Rex Stark
PURCHASING DEPT: Rex Stark
Publish top quality mail auctions of historical Americana - political, advertising, military, toys, photography, sports, police & fire, etc. Subscription $5.00 for 3 issues of catalog.

C. EUGENE SWEZEY
3 Oxbow Road
Medfield, MA 02052, USA
General militaria.

SUTLERS WAGON
P.O. Box 5
Cambridge, MA 02139, USA
General militaria.

S. L. STEINBERG
P.O. Box 512
Malden, MA 02148, USA
General militaria.

JOHN GISETTO
390 Plymouth Street
Middleborough, MA 02346, USA
617-947-1546
Massachusetts militaria.

D. NICHOLSON
186 Elm Street
N. Attleboro, MA 02760, USA
General militaria.

JAMES B. KAHN, M.D.
Route 107
Deerfield, NH 03037, USA
General militaria.

NORTHSHORE MILITARY
P.O. Box 267
Hollis, NH 03049, USA
General militaria.

GENERAL MILITARIA

GLOBE MILITARIA, INC.
RFD 1, Box 269
Keene, NH 03431, USA
(603)-352-1961
General militaria.
S4; 17;
Catalogue available. Monthly.
CONTACT: Jack Meanen
SALES DEPT: Mary or Cheryl
PURCHASING DEPT: Mary or Cheryl

C. E. GUARINO
Box 49 Berry Road
Denmark, ME 04022, USA
General militaria.

GREGG DAVIS
P.O. Box 490
Rumford, ME 04276, USA
(207)-364-3190 (207)-364-7893
Antique through modern military relics, collectibles, mostly WWII.
S3,S6,S12; 3;
SALES DEPT: Greg Davis
PURCHASING DEPT: Greg Davis
Militaria - Buy-Sell-Trade Nazi, U.S., Jap. etc. bayonets, swords, daggers, medals, insignia, flags, helmets, buckles, patches, firearms, field gear, books, photos, other relics.

MacDONALD'S MILITARY
Eustis, ME 04936, USA
Memorabilia & Maine mementoes.

CENTURY ARMS
P.O. Box 714
St. Albans, VT 05478, USA
802-524-9541
General militaria.

JOHN DENEHY, JR.
P.O. Box 1231
Bristol, CT 06010, USA
General militaria.

JAMES H. HALLAS
1510 Main Street
Glastonbury, CT 06033, USA
General militaria.

BRACKIN'S MILITARIA
P.O. Box 23
Manchester, CT 06040, USA
General militaria.
Send $6.00 for 4 issues of list.
CONTACT: Mike Brackin

WILLIAM C. MOORE
P.O. Box 11048
Newington, CT 06111, USA
General militaria.

DAVID M. LAZARUS
66 Cameo Drive
Willimantic, CT 06226, USA
General militaria.

LEBANON SPORT CENTRE
Old Route 2, Box 302
Lebanon, CT 06249, USA
(203)-886-2477
General military collectables.
S3; 16;

PURCHASING DEPT: Ed Lipinski

W. VYE
P.O. Box 33
Durham, CT 06422, USA
German militaria.
Send SASE for List.

ORDNANCE CHEST
P.O. Box 905
Madison, CT 06443, USA
General militaria & books.
3 Lists for $5.00.

BURGESS
100 Hillcrest Avenue
Meriden, CT 06450, USA
General militaria.
Catalogue $2.00.

R. J. DRENZEK
593 Park Street
Bridgeport, CT 06608, USA
203-335-4208
General militaria.

RICHARD PINTO & COMPANY
P.O. Box 5
Trumbull, CT 06611, USA
General militaria.

ROBERT L. HARRIS
U.S Route 7, Kent Road
Cornwall Bridge, CT 06754, USA
203-868-2060 (HM) 203-672-6440 (WK)
Military antiquities Dealer & Exporter.

NORM FLAYERMAN
P.O. Box 1000
Squash Hollow Road
New Milford, CT 06776, USA
203-354-5567
General militaria.
Catalogue available.

GUTHMAN AMERICANA
P.O. Box 392
Westport, CT 06881, USA
203-259-9763
American militaria.

MILITARIA, INC.
138 Kearny Avenue
Kearny, NJ 07032, USA
201-998-7471
General militaria.

WILLIAM ILLENYE
128 Quincy Street
Passaic, NJ 07055, USA
General militaria.

THE NEWBEDFORD ARSENAL
P.O. Box 661
Holmdel, NJ 07733, USA
General militaria.

CONLY D. BROOKS
509 Garden Street
Mt. Holly, NJ 08060, USA
General militaria.

PIONEER & COMPANY

216 Haddon Avenue
Westmont, NJ 08108, USA
General militaria.

ARMIES OF THE PAST LTD.
P.O. Box 3311
Trenton, NJ 08619, USA
(609)-890-0142
General militaria.
S1,S5; 20;
Catalogue available. Three times per year, $5.00.
CONTACT: Wayne Thomas

D P LIMITED
P.O. Box 6247
Bridgewater, NJ 08807, USA
201-722-0182
Firearms & general militaria.
CONTACT: Daniel D. Powell, Jr.

JAMES P. SWEENEY
226 N. Main Street
Milltown, NJ 08850, USA
War of 1812 collector.

GEORGE A. SCHNEIDER
6 Dill Court
Old Bridge, NJ 08857, USA
Indian Wars collector.

M. C. WHITWORTH
HQ EUCOM, Box 793
APO New York, NY 09128, USA
General militaria.

DONALD SOKOL
USAF Regional Medical Center
APO New York, NY 09220, USA
General militaria.

HARVEY BELLOVIN
410 East 64th Street
New York, NY 10021, USA
General militaria.
Military oriented civilian items.

MONTE'S MILITARIA
2109 Broadway
New York, NY 10023, USA
General militaria dealer.

THE SOLDIER SHOP
1222 Madison Avenue
New York, NY 10028, USA
212-535-6788
General militaria & reproductions.
Catalogue available.

COLLECTORS ANTIQUITIES
60 Manor Road
Staten Island, NY 10310, USA
212-981-0973
General militaria.
Quarterly catalogue $5.00 per year.
CONTACT: Jacques N. Jacobsen, Jr.

GENERAL MILITARIA

GENE CHRISTIAN
3849 Bailey Avenue
Bronx, NY 10463, USA
(212)-KI8-0243
General elite military unit items.
S4,S8;
All items re: Foreign Legion; Devils Island; China (U.S., French, White Russian, Warlords, Shanghai Volunteers, Police, Fire); Military marked kukris.

GRANDE ARMEE
P.O. Box 395
Yonkers, NY 10710, USA
914-779-4156 EVE. 212-687-4055 EXT 446
General militaria.
Catalogue available.
CONTACT: Michael Morris

JOSEPH E. GARABRANT
14 Douglas Street
Pearl River, NY 10965, USA
General militaria dealer.

CYRUS GALLETTA
268 Cottage Road
Valley Cottage, NY 10989, USA
General militaria.
Free List, send SASE.

DOUGLAS ROBIN BARKER
47 Inwood Road
Port Washington, NY 11050, USA
General militaria.
S4;

SPARR'S MILITARIA STORE
41-15 Broadway
Long Island City, NY 11103, USA
718-728-6014
General militaria.

POSTAL HISTORY TRADERS, LTD.
P.O. Box 58, Van Brunt Station
Brooklyn, NY 11215, USA
General militaria.

ALPINE ARMS
6716 Ft. Hamilton Parkway
Brooklyn, NY 11219, USA
General militaria.

THEME PRINTS
P.O. Box 123
Bayside, NY 11361, USA
General militaria.
Catalogue $1.00.

DICK WEISLER
53-07 213th Street
Bayside, NY 11364, USA
U.S.M.C. collector.

S & S FIREARMS
74-11 Myrtle Avenue
Glendale, NY 11385, USA
General militaria & gun parts.
Catalogue available.

KEN DRAKE
Box 972
Amityville, NY 11701, USA
516-799-4694
Military weapons collector.

THE BROAD ARROW
407 Bernice Drive
Bayport, NY 11705, USA
General militaria.

ALBERT G. BEETAR
1875 Monroe Street
N. Bellmore, NY 11710, USA
General militaria.
Write for List.

HSE MILITARIA
3794 Dianne Street
Bethpage, NY 11714, USA
(516)-796-6376
General militaria, insignia & medals.
S4,S12; 20;
Catalogue available. Monthly.
CONTACT: Harvey S. Eisenberg
SALES DEPT: Harvey
PURCHASING DEPT: Harvey
We offer U.S. and foreign medals, WWII German medals, badges and insignia, world-wide aviation, para and elite forces wings and badges.

RICHARD HOLT & SON
4 Marion Street
Nesconset, NY 11767, USA
General militaria & medals.
List $1.00.
CONTACT: Richard & John Holt

JOHN D. RUPP
5 Beach Road, Eatons Neck
Northport, NY 11768, USA
General militaria.
List available.

FAIRFAX
P.O. Box 461
Sayville, NY 11782, USA
(516)-589-4403
General militaria, bayonets.
S4; 16;
Catalogue available. Annually, with free suppliments. CONTACT: M. Fairfax
SALES DEPT: M. Fairfax
PURCHASING DEPT: M. Fairfax
We list for sale all types of military items from Civil War through WWII with a specialty in bayonets.

GEORGE W. JUNO
765 Saratoga Road
Burnt Hills, NY 12027, USA
American Revolutionary War militaria.
S.A.S.E. for list.

AMERICAN MILITARY ANTIQUES, LTD
P.O. Box 3227
Schenectady, NY 12303, USA
518-356-4406
General militaria.
CONTACT: Joe Marino

SCHERER'S ANTIQUES
Rt. 149 & Bay Road
Lake George, NY 12845, USA
(518)-793-1628 (518)-793-0617
Headgear, uniforms, edged weapons, medals, photos, toy soldiers, 1745-1945.
S1,S4,S12; 30;
Catalogue available. Annually.
CONTACT: Bob Scherer
SALES DEPT: Bobby Scherer
Located on a main thoroughfare in Lake George, halfway between Fort Ticonderoga and Saratoga Battlefield (30 miles each way). Also fine antiques and furniture.

VICTOR ARCIDINO
21 Water Street
Baldwinsville, NY 13027, USA
General militaria.

EDWARD J. MONARSKI
1050 Wadsworth Street
Syracuse, NY 13208, USA
315-455-1716
General militaria & gun show promoter.
Catalogue available.

KRUPPER
P.O. Box 177
Syracuse, NY 13208, USA
General militaria

BAZ ANTIQUES
720 B Fountain Street
Clinton, NY 13323, USA
General militaria

THE SOLDIERS TRUNK
RD 2, Box 218
W. Winfield, NY 13491, USA
(315)-822-6388
Militaria mail-order auction.
S7; 2;
Catalogue available. 6 issues per year.
CONTACT: William D. Wise
SALES DEPT: William or Linda Wise
PURCHASING DEPT: William or Linda Wise

JACK DEAVEANS
1506 High Street
Utica, NY 13501, USA
315-732-7739
General militaria

MOHAWK ARMS, INC.
P.O. Box 399
Utica, NY 13503, USA
General militaria.
5 Catalogues per year $24.50.

RONALD W. BULLOCK
P.O. Box 312
9 North Washington Avenue
Oxford, NY 13830, USA
607-843-6266
General militaria & gun show promoter.

DER JAGER
P.O. Box 92
Orchard Park, NY 14127, USA
German militaria

THE CRACKED POT
P.O. Box 11114
Buffalo, NY 14211, USA
General militaria
CONTACT: Charles Scaglione

GENERAL MILITARIA

TRANSIT VALLEY MILITARIA
2160 Maple Road
Williamsville, NY 14221, USA
General militaria.
List available.

JERRY BACHMAN
1025 Chili Avenue
Rochester, NY 14611, USA
716-328-7870
Military surplus & militaria.
Prices & literature upon request.

GARDE DU CORPS MILITARIA
P.O. Box 222
Allentown, NY 14707, USA
General militaria.
Send $5.00 for latest catalogue.

JIM SCHERMERHORN
Rt. #1
Westfield, NY 14787, USA
Japanese militaria.
Catalogue available.

GEORGE B. HARRIS
P.O. Box 1136
5212 Elm Valley Road
Alfred, NY 14802, USA
U.S. militaria dealer.
Catalogue $5.00.

GARY P. ARMITAGE
P.O. Box 145
Rockwood, PA 15557, USA
Afrika Korps collector

ROBERT HICKMAN
Route 4, Rockford Road
Butler, PA 16001, USA
Imperial German shoulder strap collector

FORT CRAWFORD PRODUCTIONS, INC.
Box 476
Slippery Rock, PA 16057, USA
412-794-6840
Gun show promoter & general militaria.
CONTACT: George Klein

ROBERT KOMOROWSKI
P.O. Box 47
Conneautville, PA 16406, USA
General militaria

CLINTON J. BEBELL
P.O. Box 3201
Erie, PA 16508, USA
General militaria.
SASE for list.

KEMER H. RUNKLE
3027 N. 4th Street
Harrisburg, PA 17110, USA
General militaria

ROMAN'S ARMY STORE
4369 Sunset Pike
Chambersburg, PA 17201, USA
717-263-3700
General militaria
CONTACT: V. Roman

THE RELIC SHOP
3240 Davidsburg Road
Dover, PA 17315, USA
General militaria

ENFIELD ARSENAL
P.O. Box 272
Gettysburg, PA 17325, USA
General militaria
CONTACT: Hugh Cunningham

BATTLEFIELD MILITARY MUSEUM
Rt 140, Box 192
Gettysburg, PA 17325, USA
General militaria

HERITAGE TRAILS
P.O. Box 307
Turbotville, PA 17772, USA
General militaria.
Catalogue $1.00.
CONTACT: Bob Swope

AMERICAN ANTIQUES
P.O. Box 116
Allenwood, PA 17810, USA
General militaria

THE MUSKET & DRUM RELIC SHOP
RD 4, Box 148
Easton, PA 18042, USA
General militaria
CONTACT: Bill & Carol Weiss

TOYS FOR BIG BOYS
3400 Nazaraeth Pike
Easton, PA 18043, USA
General militaria & bayonets

JACK H. GUSS, II
RD. 2, Box 405
Northampton, PA 18067, USA
General militaria.
List available.

U.S. SPRINGFIELD MILITARIA
RD 2, Box 631
Slatington, PA 18080, USA
(215)-767-9371
General militaria, scabbard repros.
List $1 & SASE.
CONTACT: Douglas & Kathie Tietze

THOMAS PETERS
227 W. Federal Street
Allentown, PA 18103, USA
215-791-2552 215-434-6631
German militaria

THE CROOKED CANE
PO Box 49
Sterling, PA 18463, USA
General militaria
CONTACT: Coggins

ED WINN, SR.
1 Holiday Drive
Duryea, PA 18642, USA
General militaria & gun show promoter.
List $1.00.

QUARTERMASTER CORPS
P.O. Box 908
Buckingham, PA 18912, USA
(215)-794-5606
Toy soldiers in action poses from armies 1860-1918.
S4,S13; 3;
Catalogue available.
CONTACT: Adele Hoegermeyer
SALES DEPT: Carl Hoegermeyer
PURCHASING DEPT: Adele Hoegermeyer

J. DeCHRISTOPHER
P.O. BOX 457
FEASTERVILLE, PA 19047, USA
General militaria.
Catalogue available.

G. J. WATTS
801 Bullock Avenue
Yeadon, PA 19050, USA
General militaria.
Send SASE for List.

KURT STEIN
833 CRESTVIEW DRIVE, R 37
SPRINGFIELD, PA 19064, USA
General militaria

GOLDEN LION MILITARIA
5228 Arlington Street
Philadelphia, PA 19131-3203, USA
General militaria.
Catalogue available.

ANYTHING & EVERYTHING SHOP
36 Waterloo Avenue
Berwyn, PA 19312, USA
215-647-8186
Firearms & general militaria
CONTACT: Ed Hoch

BRANDYWINE BATTLEFIELD MUSEUM
P.O. Box 265
Chadds Ford, PA 19317, USA
General militaria

DANIEL GRIFFIN
205 Buck Run
Coatesville, PA 19320, USA
215-383-1210
General militaria

PHOENIX MILITARIA CORPORATION
P.O. Box 66
163 Troutman Road
Arcola, PA 19420, USA
(215)-933-0909
General militaria & publisher of militaria collectors books & magazines.
S6; 19; 23,000; $60.00
Catalogue $3.00. Semi-annually.
CONTACT: Terry Hannon, President
SALES DEPT: Irene Karas
PURCHASING DEPT: Terry Hannon
Send $3.00 or 3 pieces of military insignia for our latest catalogue of U.S. & world-wide militaria. Thousands of items listed including: books, manuals, medals, insignia, equipment, uniform components, badges, etc.

GENERAL MILITARIA

'THE PHOENIX EXCHANGE'
P.O. Box 66
163 Troutman Road
Arcola, PA 19420, USA
(215)-933-0909
Quarterly military collectors magazine.
; 2;
CONTACT: Terry Hannon, Publisher
SALES DEPT: W. R. Bendel
PURCHASING DEPT: Terry Hannon
AD MANAGER: William Bendel
'The Phoenix Exchange' functions as the military collector's & dealer's marketplace. Articles cover a wide spectrum of militaria, but attend to each subject in detail. Send $2.00 for a sample copy.

HERITAGE COLLECTORS' SOCIETY
P.O. Box 389
311 W. Main Street
Lansdale, PA 19446, USA
(215)-362-0976 (215)-362-2021
Historical artifacts and memorabilia.
S6; 10;
Catalogue available. 3 per year.
CONTACT: Thomas A. Lingenfelter
SALES DEPT: Tom Lingenfelter

THE FIFE & DRUM
P.O. Box 6
Valley Forge, PA 19481, USA
General militaria.
Catalogue available.

ANTIQUE ARMS & ARMOUR
P.O. Box 116
Ziegerville, PA 19492, USA
Antiques & German militaria
CONTACT: Gerhard Windbiel

JACK ROMIG
RD #1, Box 354
Barto, PA 19504, USA
General militaria & insignia

GREAT WAR MILITARIA
RD #2, Box 10
Kutztown, PA 19530, USA
215-683-8944
WWI militaria
CONTACT: DR. David L. Valuska

MORTON INSIGNIA
526 Blackbird Drive
Bear, DE 19701, USA
(302)-322-6188
Books & general militaria. Militaria show promoter.
S12;
CONTACT: Stan Blake

WAR RELICS SHOP
Rt #1, Box 154
Milford, DE 19963, USA
302 422-5487
General militaria
CONTACT: Jeff Millman

ANTIQUES & COLLECTIBLES
6310 Martina Terrace
Lanham, MD 20706, USA
General militaria.
Send SASE for list.

SNYDER'S TREASURES
3520 Mullin Lane
Bowie, MD 20715, USA
301-262-5735
General militaria & gun show promoter.
List available.
CONTACT: Charles E. Snyder

AMERICAN MILITARY ANTIQUES
8048 Main Street
Ellicott City, MD 21043, USA
301-465-6827
General militaria.
Write or call for details.
CONTACT: Courtney B. Wilson

ERWIN L. BELL
2629 Coon Club Road
Westminster, MD 21157, USA
301-876-1973
General militaria

AVALON FORGE
409 Gun Road
Baltimore, MD 21227, USA
301-242-8431
General militaria.
Catalogue available.

CECIL C. ENNIS
9510 Burton Avenue
Baltimore, MD 21234, USA
301-665-2397
General militaria

NATIONAL CAPITAL HISTORIC SALES
7721 'D' Fullerton Road
Springfield, VA 22153, USA
(703)-569-6663
General world militaria.
S3,S4,S5,S6; 6;
Catalogue available.
CONTACT: George A. Petersen
SALES DEPT: George Petersen
PURCHASING DEPT: George Petersen
Militaria. US & foreign. WWI, WWII, Korea, Vietnam. German, U.S., British, Russian. Buy, sell, rent, consulting, museum work. Call for information.

PAUL SINOR
P.O. Box 12230
Arlington, VA 22209, USA
General militaria.
Catalogue $1.00.

G. S. KNIGHT
7214 Regent Drive
Alexandria, VA 22307, USA
Pre-Indian Wars collector

THE COLLECTOR'S ARMOURY, INC.
P.O. Box 59
Alexandria, VA 22313, USA
(703)-549-0772
General militaria.
Free catalogue.

SUMMIT POINT TRADING COMPANY
P.O. Box 427
Edinburg, VA 22824, USA
(703)-984-9444 (703)-740-8065
All antique historical militaria.

S4,S12;
Catalogue available. 1 to 3 times per year.
SALES DEPT: J. M. Bracken

MOSS PUBLICATIONS
P.O. Box 729
Orange, VA 22960, USA
(703)-672-5921
U.S. Civil War book publisher.
S6,S8,S9,S10,S12; 9;
Catalogue available. Annually.
CONTACT: S. W. Sylvia & M. J. O'Donnell
SALES DEPT: Nancy Rossbacher
PURCHASING DEPT: Nancy Rossbacher
Moss Publications is dedicated to the preparation and distribution of information for the historian and collector. We specialize in the Civil War field.

STAHLHELM MILITARIA
P.O. Box I
West Point, VA 23181, USA
(804)-843-2232
General military collectables
S1,S12; 8;
CONTACT: Bob Kurfees or Will Addison
SALES DEPT: Bob Kurfees
Collector, dealers with a general selection of collectibles from the Civil War thru Vietnam. Specializing in German edged weapons, hand guns and U.S. Air Corps insignia.

EARL N. LEVITT
303 Indian Springs Road
Williamsburg, VA 23185, USA
General militaria collector.
Send for list of items wanted.

RON WOLIN
437 Bartell Drive
Chesapeake, VA 23320, USA
General militaria

JAMES BROWN
506 Village Drive
Vinton, VA 24179, USA
Collector of general militaria

STEVE'S GUNS & SURVIVAL EQUIPMENT
P.O. Box 12819
Winston-Salem, NC 27117, USA
General militaria & gun show promoter.
CONTACT: Steve R. Foster

MAYHEW-REECE AND ASSOCIATES
P.O. Box 20081
Greensboro, NC 27420, USA
(919)-294-3226
Original & replica 3rd Reich militaria.
S6,S12,S13; 14;
Catalogue available. Bi-monthly.
SALES DEPT: A. D. Reece, Jr.
PURCHASING DEPT: S. D. Mayhew
We specialize in Third Reich collectibles from World War II including postcards, photocards, cancelled envelopes, stamps, banknotes, and coins. Collector leaflets, medals, membership pins, armbands, belt buckles, and other miscellaneous items are occasionally offered. Send $1.00 for list.

TAR HEEL MILITARIA
Rt. #2, Box 479
Concord, NC 28025, USA
General militaria

GENERAL MILITARIA

218 ANTIQUES & ARMORY
P.O. Box 1408
218 Castle Street
Wilmington, NC 28402, USA
General militaria
CONTACT: Paul B. Berry

BILL BROOKS
4901 Indian Trail
Wilmington, NC 28403, USA
919-392-2961
General militaria

H. J. MILITARIA
P.O. Box 9992-0992
Columbia, SC 29290, USA
Foreign general militaria.
List $1.00, book list $.50.

ZIPPS MILITARIA
P.O. Box 223
Travelers Rest, SC 29690, USA
General militaria

PALMETTO COLLECTABLES
2606 Walnut Avenue
Barnwell, SC 29812, USA
(803)-259-5913
General militaria.
S1,S2,S6,S7,S12; 10;
Catalogue available. Annually.
CONTACT: Cindy Lancaster
SALES DEPT: Cindy Lancaster
PURCHASING DEPT: Cindy Lancaster
We carry one of the largest selections of Coca-Cola and German memorabilia. Shipping outside of the U.S. as well. We are basically a mail-order business.

MILITARY WORLD
4975 Jimmy Carter Blvd.
Norcross, GA 30093, USA
(404)-925-0875
General militaria & gun show promoter.
S1;
CONTACT: Don Bramblett

BRIGADE QUARTERMASTERS, LTD.
Dept. 119P
1025 Cobb International Blvd.
Kennesaw, GA 30144, USA
(404)-428-1234
Militaria, uniforms, survival gear, EMT supplies, clothing, optics, & tools.
S3,S6; 10;
Catalogue available. Semi-annually.
CONTACT: Mitchell L. WerBell, IV
SALES DEPT: Ken Southerland
PURCHASING DEPT: Carl Bricken
AD MANAGER: Budd Austeinov
Brigade Quartermasters is primarily a mail order catalogue house, but we do have a retail showroom in Kennesaw, GA and sell wholesale to authorized dealers.

STEVE GRIFFITH
P.O. Box 585
Kennesaw, GA 30144, USA
404-422-9924
General militaria

THE AIRBORNE SHOP
3086 Vinings Ferry Drive, N.W.
Atlanta, GA 30339, USA
General airborne militaria
CONTACT: William S. Edwards

THE CORPORAL
P.O. Box 420486
Atlanta, GA 30342, USA
General militaria.
Send SASE for catalogue. Quarterly.

MILITARY MARKETING, INC.
Box 720024
Atlanta, GA 30358, USA
General militaria.
List $1.00.

THE COUNTRY STORE
117 Beaver Creek Road
Kathleen, GA 31047, USA
General militaria.

SAN JUAN PRECIOUS METALS CORP.
4818 San Juan Avenue
Jacksonville, FL 32210, USA
904-387-2212
General militaria
Catalogue $2, 12 issues $20
CONTACT: Ron Gordon

WARTIME COLLECTABLES
2626-40 Mission Road
Tallahassee, FL 32304, USA
(904)-576-1552
General militaria.
S12;
CONTACT: Andrew Lipps
Dealer in all types of militaria with an emphasis on WWI-II U.S. and WWII German.

SIN LOI PRODUCTIONS, INC.
1105 N. Main St., Suite 2-A
Gainesville, FL 32601, USA
(904)-466-3595
Vietnam War commemorative items.
S4,S5,S12; 5;
Catalogue available. Annually.
CONTACT: Mark Johnson
SALES DEPT: M. Johnson
PURCHASING DEPT: M. Johnson
Sin Loi Productions - 50 different T-shirts commemorating the Vietnam War. Also patches, caps, sport shirts, DI's & more.

GLENTIQUES, LTD
P.O. Box 8807
Coral Springs, FL 33075, USA
German regimental steins. Mail-order auction.
Catalogue $3.00.
CONTACT: Gary Kirsner

JOE HAMILTON
530 N.W. 31 Avenue
Miami, FL 33125, USA
(305)-643-3764
General militaria.
S12;
Specializing in WWII Italian Facist and Nazi political, civil, government daggers, tunics, hats, flags. Also military swagger sticks and dress daggers world wide.

MUSEUM OF HISTORICAL ARMS
1038 Alton Road
Miami Beach, FL 33139, USA
General militaria.

HOLBROOK ANTIQUE MILITARIA
5296 S.W. 65 Avenue
Miami, FL 33155-6439, USA
General militaria

PAUL WESOLOWSKI
3702 N.E. 171 Street
N. Miami Beach, FL 33160, USA
Polish reference work & militaria.
S6,S8,S9,S10,S11,S12; ; 10,000; ; 100.00
CONTACT: Paul Wesolowski
Collector of Polish militaria for personal collection. Interested in medals, orders and militaria of other nations. Will trade with collectors. Have book for sale on Polish Orders, Medals, Badges and Insignia, 1705-1985, Cost $29.50 post paid in US, Canada and world add $5.00.

JUNK TO JEMS
P.O. Box 24483
Oakland Park, FL 33307, USA
General militaria
CONTACT: James Lorah

AMERICAN HERITAGE INVESTMENTS
1574 Pennwood Circle North
Clearwater, FL 33516, USA
General militaria.

LIBERTY ARMS
P.O. Box 72
Auburndale, FL 33843, USA
General militaria
CONTACT: John S. Rice

WWII RELICS
P.O. Box 682
Dunedin, FL 34296, USA
General militaria.
Catalogue $5.00.

MILITARY WAREHOUSE
Box 682
Dunedin, FL 34296, USA
German militaria.
Catalogue $5.00.

ARISAKA ARMORY
4001 Windermere Drive
Tuscaloosa, AL 35405, USA
(205)-556-0086
WWI & WWII militaria, modern hunting weapons.
S4,S8,S9,S10,S11,S12; 4;
Catalogue available. Bi-annually.
CONTACT: D. H. White, Jr.
SALES DEPT: Greg Taylor
PURCHASING DEPT: William White/George Taylor AD MANAGER: Kathleen Harper, Editor
Military rifle sales specializing in Japanese. Publisher "Japanese Type 99 Arisaka". Maintain museum of WWWI & II weapons of most countries.

CLARK'S
1524 Overlook Shopping Center
Mobile, AL 36618, USA
General militaria & collectables.
List available.

GENERAL MILITARIA

RELIQUIAE
4908 Pebble Creek Drive
Antioch, TN 37013, USA
Memorabilia, antiques & collectibles.
CONTACT: Lillian & George Haven

CUMBERLAND MARKETING INTERNATIONAL
P.O. Box 40461
Nashville, TN 37204, USA
(615)-298-1401
General militaria & video tape.
S4,S5;
CONTACT: Richard S. Gardner

LONE STAR
2998 Elm Hill Road
Nashville, TN 37214, USA
General militaria.

BRADLEY ENTERPRISES
P.O. Box 373-PMS
Highway 361
Elizabethton, TN 37644, USA
(615)-725-3610
Custom G.I. dog tag imprinter.
S6,S12; 3;
Catalogue available. Upon request.
CONTACT: Richard J. Bradley
SALES DEPT: Richard J. Bradley
PURCHASING DEPT: Richard J. Bradley
Make custom imprinted G.I. Dog Tags. Five lines with eighteen characters and spaces. Chains included. Single $3.50, Double $5.50, Postpaid. Also make pet ID tags.

MILITARY SURPLUS SUPPLY
5594 Airways
Memphis, TN 38116, USA
(901)-346-0090
General militaria & surplus.
S6;
$1.00 list of WWII items.
CONTACT: J. Brannan,Jr.

DER REICHGRUPPE
P.O. Box 4778
Louisville, KY 40204, USA
German militaria.
Catalogue $2.00.

ASHER'S MILITARY
P.O. Box 19103
3415 Bonaventure Blvd.
Louisville, KY 40219, USA
Vietnam War militaria.
S3,S6,S12; 7;
Catalogue available. Annually.
CONTACT: Gary Asher
SALES DEPT: G. Asher
PURCHASING DEPT: G. Asher
Buy-Sell-Trade Vietnam era weapons, clothing, field gear,and any related militaria pertaining to 'nam.

LARRY PITMAN
Rt. 5, Zanzibar Farm
Paris, KY 40361, USA
General militaria collector.

GARDNER MILITARIA
301 Meadow Lane
Frankfort, KY 40601, USA
(502)-223-5581
General militaria
S6,S12;
CONTACT: Phil Gardner
Specialists in American and Japanese militaria 1900-1945.Always purchasing U.S. bullion patches, Ike jackets and Japanese medals. Buying entire collections.

GHQ
106 Wallace
Covington, KY 41014, USA
General militaria
CONTACT: L. Perttula

L. GOODRIDGE
735 Ridgedale Court
Hebron, KY 41048, USA
German militaria.
Catalogue available.

ROGER S. STEFFEN HISTORICAL MILITARIA
14 Murnan Road
Cold Spring, KY 41076, USA
(606)-431-4499
Arms & armour, orders & medals, and military curios mail-auctions.
Catalogue available.
CONTACT: Roger S. Steffen

DAVID BINKLEY
25981 Willowbend Road
Perrysburg, OH 43551, USA
1-419-874-5627 1-419-259-5898
Identification models collector.

VAN R. ROGERS PRODUCTIONS
P.O. Box 147
8230 Fairmount
Novelty, OH 44072, USA
General militaria

JOHN M. KADUCK
P.O. Box 02152
Cleveland, OH 44102, USA
General militaria

AMERICAN MILITARY PRODUCTS
1818 Wilbur Road
Medina, OH 44256, USA
General militaria.
Catalogue available.

HERMAN A. WIESENSEE
99 Charles Street
Struthers, OH 44471, USA
General militaria

MT. WASHINGTON ANTIQUES
3742 Kellogg
Cincinnati, OH 45226, USA
General militaria
CONTACT: c/o Fergusons Antiques

QUEEN CITY MILITARY SALES
5443 Fox Road
Cincinatti, OH 45239, USA
General militaria.
Catalogue available.

ARMS DEPOT
702 Watervliet Avenue
Dayton, OH 45400, USA
General militaria

A-1 SALVAGE & MILITARY SURPLUS
132 1/2 East 3rd Street
Dayton, OH 45402, USA
513 252-4191
General militaria & WWII surplus.
List $2.00.

ROBERT PORTER
1939 N. Defiance, Lot 15
Ottawa, OH 45875, USA
General militaria.
Write for details.

INDIANAPOLIS ARMY-NAVY
6028 East 21st Street
Indianapolis, IN 46219, USA
(317)-356-0858
Military surplus & insignia.
S1,S4; 36;
CONTACT: Bob Reiter
SALES DEPT: Bob Reiter
PURCHASING DEPT: Bob Reiter
We offer military insignia and surplus of all kinds, field equipment, outerwear, camping, etc. Fast dependable service. Satisfaction guaranteed. Mail-orders, MasterCard, VISA accepted.

T. DeVONA
Box 242
Cedar Lake, IN 46303, USA
General militaria.
Send SASE with $.39 for catalogue.

M & M MEMENTOES
P.O. Box 245
Crown Point, IN 46307, USA
(219)-663-7105 (219)-663-2160
Research Indiana C.W., C.W. relics, all types other militaria RevWar to Vietnam.
S4,S12,S13; 6;
Catalogue available. Bi-annually.
CONTACT: Morris M. Wood,II
SALES DEPT: Morris M. Wood
PURCHASING DEPT: Morris M. Wood
List #4 of Civil War and other militaria S.A.S.E. Paper, books, swords, relics, medals, guns. Collecting Indiana and Michigan Civil War items. Buy-Sell-Trade.

HAMMER
P.O. Box 1393
Columbus, IN 47201, USA
German marches on cassettes.
S6,S11,S12; 6;
Catalogue available.
CONTACT: Karl Hammer
Manufacturer/Wholesaler: WW II German march cassettes, flags, pins, and posters. Custom capacity retailer: Videos and books. We buy pre-1945, 78 rpm German march records.

WILLIAM FAGAN
126 Belleview
Mount Clemens, MI 48043, USA
General militaria.
Catalogue $5.00.

GENERAL MILITARIA

GEORGE W. LIZENBY
182 Redwood Drive
Troy, MI 48084, USA
Firearms & general militaria collector.

WEBSTER DISPLAYS
843 Riverside Drive, Suite 3
Battle Creek, MI 49015, USA
(616)-968-2442
Uniform display supplies.
S5,S8,S12; 4;
Catalogue available.
CONTACT: Dave Webster
SALES DEPT: Dave Webster
Torso forms for military uniforms. $35.00 + $3.00 postage; styrofoam heads, fleshtone, male face, $5.50 ea. or $41.00 for 8. White faceless heads $3.10 ea. or $34.00 for 12. These heads are ideal for headgear display as they preserve and restore original shape. Minimum order $15.00.

JOSEPH BELL
5228 Larkin N.E.
Comstock Park, MI 49321, USA
(616)-784-6059
Japanese & general military relics.
S12,S13;

TAL LIEPA
P.O. Box 16
Clive, IA 50053, USA
Baltic states militaria collector.

ROGERS COIN & GUN
98 Ellison Road
Lisbon, IA 52253, USA
(319)-455-2957
Wholesale/retail military items. Specialty - custom ordering.
S12; 8;
CONTACT: Roger D. Curl
SALES DEPT: Roger D. Curl
PURCHASING DEPT: Roger D. Curl
F.F.L. dealer specializing in military items thru 'Nam. Special emphasis on WWII and M1 carbine variations.

R & R ENTERPRISES
P.O. Box 471
Springville, IA 52336, USA
Vietnam War militaria.
Illustrated list $1.00.

DISTRIBUTING COMPANY
P.O. Box 331
Cedarburg, WI 53012, USA
General militaria.
Catalogue $10.00 (refundable).
CONTACT: David Mathes

PIECES OF HISTORY
Rt. 3, Box 971
Salem, WI 53168, USA
General militaria

J. R. PRASSER
P.O. Box 829
Fond Dulac, WI 54935, USA
Vietnam militaria.

AL LANGE
P.O. Box 11574
Twin Cities Branch Station
St. Paul, MN 55111, USA
Vietnam War militaria collector.

BATTLEFIELD MILITARY BOOKSTORE
1600 B West Lake Street
Minneapolis, MN 55408, USA
612-823-3711
General militaria & bookseller.
Catalogue $1.00.

E. OHLGREN
10372 Columbus Circle
Bloomington, MN 55420, USA
General militaria.
List available.

HIGHLAND PUBLISHERS
9000 Tenth Avenue N.
Golden Valley, MN 55427, USA
(612)-788-2444
Military book publishers.
S2,S6,S9,S10; 7;
SALES DEPT: Karla Jones
PURCHASING DEPT: Karla Jones
Publisher -"Americas Medal of Honor".

CHARLES L. FEAY
2863 114th Lane N.W.
Coon Rapids, MN 55433, USA
(612)-421-3731
Uniforms, medals, & unit histories.

DOUGLAS A. ANDERSON
P.O. Box 63
Oslo, MN 56744, USA
WWI & WWII general militaria collector.

JAMES M. BISHOP
P.O. Box K
Warroad, MN 56763, USA
(218)-386-1535 (218)-681-8141
General militaria & weaponry.
S4,S12;
SALES DEPT: James M. Bishop
General interest in all things military - primary interest is weaponry.

PETER ERICKSON
917 18th Street So.
Fargo, ND 58103, USA
Western collectables.
Send S.A.S.E. for list.

GERALD R. MILLER
522 5th Avenue North
Great Falls, MT 59401, USA
General militaria

GERALD MILLER
714 9th Street South
Great Falls, MT 59405, USA
General militaria collector.

CANYON FERRY ARMS
5840 Canyon Ferry Road
Helena, MT 59601, USA
General militaria

HAYES OTOUPALIK
Rt. 2, Evaro Road
Missoula, MT 59801, USA
406 549-4817
American militaria 1833-1945.
4 catalogues for $6.00.

PARKER
210 Walnut
Elk Grove, IL 60007, USA
General militaria.
Send S.A.S.E. for list.

HAVLICEK
18603 Rte 173E
Harvard, IL 60033, USA
General militaria.
Send SASE for list.

JIM DETERS
1300 Loch Lane
Lake Forest, IL 60045, USA
German militaria.
List $1.00.

LARRY RINKER
P.O. Box 8406
Bartlett, IL 60103, USA
U.S. WWII militaria.
Send S.A.S.E. for list.

STEVE PEDERGNANA, JR.
P.O. Box 1062
Oak Park, IL 60304, USA
U.S. military surplus.
S6; 12; 1500; $30.00
Catalogue available. 1 or 2 times per year.
CONTACT: Steve Pedergnana, Jr.
SALES DEPT: S. J. Pedergnana, Jr.
PURCHASING DEPT: S. J. Pedergnana, Jr.
FOR SALE: Genuine United States armed forces surplus clothing, equipment, patches, survival gear, etc. Send $1.00 for current catalog. Pedergnana, POB 1062, Oak Park, IL 60304.

ERNEST E. NELKEN, JR.
203 Dorsetshire Drive
Steger, IL 60475, USA
312-756-3557
Imperial German militaria.
Catalogue available.

PASS IN REVIEW
P.O. Box 153
Hinsdale, IL 60521, USA
General militaria

FOX MILITARY EQUIPMENT CO.
P.O. Box 452
Hinsdale, IL 60522, USA
General militaria.
Catalogue available.

MILITARIA & VARIA
P.O. Box 292
Riverside, IL 60546, USA
20th century general militaria.
List $3.00. Quarterly.

TASK FORCE MILITARIA
2341 W. Belmont
Chicago, IL 60618, USA
General militaria.
Send SASE for List.

GENERAL MILITARIA

HUTCHINSON HOUSE
P.O. Box 41021
Chicago, IL 60641, USA
General militaria.
Catalogue available.

LOVETREE
Box 8462
Cicero, IL 60650, USA
WWII German militaria

DAVID E. J. PEPIN
P.O. Box 354
Grant Park, IL 60940, USA
Japanese militaria collector.

DARRELL R. LULLING
Rt #1, Thunderbird Estates
Oakwood, IL 61858, USA
NVA/VC & Chicom militaria.

W. W. 2 PRODUCTS, LTD.
3859 Hartford
St. Louis, MO 63116, USA
314-772-9732
General militaria & reproductions.
Catalogue $3.00.

W.W.#2 LTD.
P.O. Box 2063
St. Louis, MO 63158, USA
314-772-9732
General militaria & reproductions.
224 page catalogue $3.00.

RUTLADER ANTIQUES
8247 Wornall Road
Kansas City, MO 64114, USA
General militaria

CHIP MINX
4024 NW Claymont Drive
Kansas City, MO 64116, USA
German & Russian WWI militaria collector.

R. B. BOWDEN
P.O. Box 152
Rantoul, KS 66079, USA
General militaria.
Send 50 cents for List.

KEITH MELTON
P.O. Box 18421
Wichita, KS 67218, USA
Espionage collector

D. VOEGLER
5435 Essen Lane, Apt. #5
Baton Rouge, LA 70809, USA
General militaria

CENTURION CORPORATION
219 N. Jackson Street
El Dorado, AR 71730, USA
(501) 862-0021
General militaria mail-order.
Catalogue available.
CONTACT: Eric Richardson, President

LANCER MILITARIA
P.O. Box 886
Mount Ida, AR 71957, USA
General militaria

FREEDOM ARSENAL
P.O. Box 117
Beaver, AR 72613, USA
General militaria

BIEN HOA PRODUCTIONS
P.O. Box 56
Fayetteville, AR 72702, USA
Vietnam militaria.
Free List, 22 cent stamp.

QUINCY SALES
P.O. Box 700113
Tulsa, OK 74170, USA
(918)-743-7048
General militaria & insignia.
S4; 25;
Catalogue available. Annually.
CONTACT: Jackie Eyre
SALES DEPT: Jackie Eyre
PURCHASING DEPT: Jackie Eyre
48-page catalog of shoulder patches, wings, badges, insignia, etc. U.S. & foreign $2.00. 20-page catalog of U.S. medals, ribbons, etc. $1.00. Both catalogs $2.50

MODERN MILITARIA
P.O. Box 773245
Houston, TX 77215, USA
(713)-782-0258
Communist Bloc militaria.
S3,S4,S6,S13;
List available. Bi-annually.
CONTACT: Beryl Barnett, Jr.
Specializing in communist bloc militaria. Collect, sell and trade. Firearms, parts, uniforms, headgear, field gear. Research all aspects of COMBLOC military for writers and artists.

CATTLE BARON LEATHER & CUTLERY COMPANY
P.O. Box 100724
San Antonio, TX 78201, USA
(512)-697-8900
Bullwhips, holsters, knives, & German military bomber jackets.
S4,S5,S11,S13; 14;
Catalogue available. Bi-annually.
CONTACT: Jerry Ardolino, President
PURCHASING DEPT: Jerry Ardolino, President
Adventurers! Cattle Baron Leather & Cutlery Co.'s color catalog shows the finest expedition bullwhips, shoulder holsters, handmade knives, German bomber jackets, European pilot's watches - more. World Class products used by adventurers & military in more than 21 countries. Send $3.00.

LONE STAR MILITARIA
P.O. Box 29701
San Antonio, TX 78229, USA
(512)-647-3140
Texas militaria
S6; 1;
CONTACT: Martin Callahan
SALES DEPT: Martin Callahan
PURCHASING DEPT: Martin Callahan
Buy-Sell-Trade: Texas militaria, 1836-1945. Especially wanted: Texas military collar insignia, Texas U.C.V. U.C.V., G.A.R. memorabilia, Texas military uniforms, and Texas military photographs.

TEXAS MILITARY COLLECTIBLES

P.O. Box 433
Leander, TX 78641, USA
General militaria.
$5.00 brings 6 catalogues per year.

GLENDOWER
Box 2917
Pampa, TX 79066, USA
General militaria.
List $1.00.

RON COURSEY
1804 College
Midland, TX 79701, USA
General militaria

GARY FIEST
500 West Wall, #302
Midland, TX 79701, USA
Scottish items from all periods.
Write for details.

WIND & PROP
4219 Durango
Odessa, TX 79762, USA
U.S. militaria.
List available.

THE SUPPLY SERGEANT
9509 E. Colfax Avenue
Aurora, CO 80010, USA
General militaria.
List available.

BILL BONSER
Rt #1, Box 131
Loveland, CO 80537, USA
Cavalry collector

CECIL SANDERSON
P.O. Box 1092
Lyman, WY 82937, USA
Vietnam military collector

ANTIQUE GUN SHOPPE
E. 2640 Ponderosa Blvd.
Post Falls, ID 83854, USA
(208)-773-1320 (208)-773-0106
General militaria, museum, antique artillery.
S6,S8,S11,S12; 28;
CONTACT: Glen E. Mattox
SALES DEPT: Glen E. Mattox
We offer a free museum showing development of U.S. fighting arms - For sale - A complete line of antique arms and military items including cannons and Gatling Gun.

DR. L. G. McNEAL
85 Quarter Circle Drive
Nibley, UT 84321, USA
801-752-7842
U.S. cavalry items

RICHARD SIMMONS
P.O. Box 5680
Phoenix, AZ 85006, USA
Military mail-order auction.
S7;
Catalogue available. Quarterly.
Military mail auction catalog. Issued four times yearly covering militaria from United States, Germany and all countries and eras. Next issue $2.00. Consignee's write.

GENERAL MILITARIA

CARL SCIORTINO
P.O. Box 6424
Scottsdale, AZ 85261, USA
(602)-483-2140
General military medals & insignia.
S4; 9;
Catalogue available. Every 18 months.
CONTACT: Carl Sciortino
U.S. Civil War through WWII insignia, wings, uniforms, books, WWI patches, numbered campaign medals. Wanted: C.W. Corps badges. Carl Sciortino Militaria, Box 6424, Scottsdale, AZ 85261.

JOSEPH M. LEONARD
837 West Santa Cruz Drive
Temple, AZ 85282, USA
602-994-6871 602-966-9477
General British militaria.
List available.

JAMES A. MERRELL
P.O. Box 1045
Sun City, AZ 85372, USA
3rd Reich memorabilia

COLLECTORS CACHE
P.O. Box 17808
Tucson, AZ 85731, USA
General militaria

ALBUQUERQUE COLLECTORS WORLD
2242 Wyoming N.E.
Albuquerque, NM 87112, USA
General militaria

RECON ONE
P.O. Box 6978
Stateline, NV 89449, USA
General militaria

ROGER STEELE
Suite 282
8715 Hollywood Blvd.
Hollywood, CA 90028, USA
Nazi German militaria

DAVID VINAR
1610 North Martel Avenue, #12
Los Angeles, CA 90046, USA
(213)-851-4492
General militaria.
S4,S12; 29; 3000;
Catalogue available. Monthly.
SALES DEPT: David Vinar
PURCHASING DEPT: David Vinar

ROBERT S. McCARTER
6336 Wilshire Blvd.
Los Angeles, CA 90048, USA
WWII AAF survival items.

RAY A. DUNLAP,SR.
973 Oneonta Drive
Los Angeles, CA 90065, USA
U.S. militaria & insignia.

LORIN RELIN
5014 Cascade Court
Culver City, CA 90230, USA
Russian militaria.
List available.

MYERS MUSEUMS, UNLIMITED - A CORPORATION
16521 Akron Street
Pacific Palisades, CA 90272, USA
(213)-454-3686
General militaria.
S4,S12,S13; 20+;
Catalogue available. 3 times per year.
CONTACT: Robert D. Myers, President
SALES DEPT: R. Myers (Robert)
PURCHASING DEPT: R. Myers (Robert)
Catalog sales (subscription $5.00/year). Containing: militaria; memorabilia; collectibles; related literature; natural history; curios; relics; artifacts; oddities; wax figures; accoutrements; explosive ordnance; books and manuals.

BRITISH COLLECTIBLES,LTD.
2113 Wilshire Blvd.
Santa Monica, CA 90403, USA
British militaria.
Catalogue $3.00.

BAIRD COLLECTABLES
P.O. Box 444
Los Alamitos, CA 90720, USA
213-920-1213
Police badges & insignia.
Catalogue available.

GENE GUILAROFF
P.O. Box 4048
Sunland, CA 91040, USA
Vietnam militaria collector.

ART SOWIN
8436 Samra Drive
Canoga Park, CA 91304, USA
U.S. Cavalry collector.

THE BLACK WATCH
8825 Reseda Blvd.
Northridge, CA 91324, USA
213-701-5177
General militaria.
Catalogue available.

ANTIQUE MILITARIA
Troglodyte Hall
P.O. Box 4843
Thousand Oaks, CA 91359, USA
General militaria

WILLIAM FINKLER
13854 Cumpston Street
Van Nuys, CA 91401, USA
General militaria.
List $.50.

MILITARY MUSEUM
Box 267
Pomona, CA 91769, USA
215-697-1700
General militaria

DAVE PARHAM
320 Aldwych Road
El Cajon, CA 92020, USA
General militaria.
Free list with SASE.

W. D. GRISSOM,SR.
836 Orange, Suite 353P
Coronado, CA 92118, USA
General militaria.
S4; 3;
Catalogue available. 6 times annually.
SALES DEPT: W. D. Grissom,Sr.
W. D. Grissom is the man who introduced an unheard-of new element into medal dealing - price competition. Get his Free list and see for yourself.

S/SGT. JOHN L. ROMEO
72417 Sunnyslope Road
29 Palms, CA 92277, USA
General militaria

HODGES
P.O. Box 1527
Idyllwild, CA 92349, USA
General militaria

WEST COAST MILITARIA
P.O. Box 974
Orange, CA 92666, USA
General militaria.
S4,S12; 10;
Catalogue available. Quarterly.
CONTACT: Mark V. Kraatz
SALES DEPT: Marcia Kraatz
PURCHASING DEPT: Mark Kraatz

LITTLE JOHN'S ANTIQUE ARMS
777 S. Main Street
Orange, CA 92668, USA
General militaria & antique arms.

BRITANNIA DESIGNS
P.O. Box 477
San Juan Capistrano, CA 92693, USA
(714)-496-0477
British militaria.
S4,S12,S13; 12;
Catalogue available. Bi-annually.
CONTACT: Keith & Paul Greenhalgh
SALES DEPT: Keith
PURCHASING DEPT: Keith
British Militaria: Cap/hat badges, uniforms, webbing equipment, hats/berets, T-shirts with British cap badge designs, blazer badges, rank badges, Scottish clan badges. Pith Helmets (Kitchener/Wolselgy).

RICHARD S. DURAN
P.O. Box 63
Fillmore, CA 93015, USA
U.S.M.C. items

CENTURION MILITARY HOBBIES
3087 N. Main Street
Walnut Creek, CA 94596, USA
General militaria.

GOLDEN GATE MILITARY
3091 N. Main Street
Walnut Creek, CA 94596, USA
415-939-2259
Military mail auction.
Catalogue $6.00, $24.00 for 4 issues.
CONTACT: Mike Milliman

GENERAL MILITARIA

LOUIS DEMERS
2749 Twin Oaks Lane
San Jose, CA 95127, USA
(408)-259-0514
Japanese militaria
S4,S7; 13;
AD MANAGER: Louis Demers
Japan only mail bid sales. Monthly publication, "The Study of The Orders, Medals and Badges of Japan".

MICHAEL GAINEY
791 Shirley Blvd.
Arcata, CA 95521, USA
General militaria.
List available.

KNIGHTS OF MILITARIA,LTD.
P.O. Box 97
Phillipsville, CA 95559, USA
German militaria

THE MILITARY SHOP OF HAWAII
1921 Kalakaua Avenue
Honolulu, HI 96815, USA
General militaria.
Catalogue available.

R. STEPHEN DORSEY MILITARIA
P.O. Box 263
Eugene, OR 97440, USA
503-937-3348
U.S. general militaria.
List $1.00.
CONTACT: R. Stephen Dorsey

CASCADE GENERAL STORE
90 North Main Street
Ashland, OR 97520, USA
503-482-3040
General militaria.
Catalogue 3 for $5.00 per year.

BYGONE WARRIOR MILITARIA
P.O. Box 70211
Seattle, WA 98107, USA
(206)-283-5141 (206)-784-1700 Shop
General militaria.
S4; 6;
Catalogue available. 3 or 4 times annually.
CONTACT: Christopher C. Bruner
SALES DEPT: Christopher C. Bruner
Catalogue subscription $6.00 for four issues. Specializing in United States 1898-1945, German World War One and foreign. Uniforms, insignia, medals, field gear, ordnance, wings, paper.

EUGENE KRIDLER
103 Huckleberry Crest
Sequim, WA 98382, USA
General military insignia

WHY NOT COLLECTABLES
701 North 6th Street
Kelso, WA 98626, USA
(206)-423-7931
U.S. womens militaria.
S7,S12;
CONTACT: Ken Kaighin
Wanted Womens uniforms & related items. Wave, WAC, Marine Spars, CAP, etc. Personal papers, books, medals, awards, 1914 to 1956.

YE OLDE CURIOSITY SHOPPE
P.O. Box 115
Haberfield
Sydney, N.S.W. 2045, Australia
General militaria.
Catalogue available.

MILITARY ANTIQUES
91 Shenton Road
Swanbourne, Australia
384-1218
General militaria.
CONTACT: John Burridge

JOHANNES A. FLOCH
Fach 181
Vienna, A-1191, Austria
General militaria.

WEITHOFER
Siebensterng. 25
Wien, 1070, Austria
(02 22) 93 58 654
Austrian militaria.

W. E. MOSER
Nussdorfer Strasse 70
Wien, 1090, Austria
(02 22) 3 48 05 24
General militaria.

BALMING-MILITARIA
Scheldestraat 100
Antwerpen, 2000, Belgium
General militaria.

TWENTY-SIX
Rue Schaltin, 21-23
Spa, 4880, Belgium
General militaria.

BOB DRUMMOND
P.O. Box 2
Champlain Post Office
Lasalle, Quebec H8P 3H9, Canada
General militaria.

PETER SIMUNDSON
6929 Estorial Road
Mississauga, Ontario L5N 1N2, Canada
General militaria.

AD HOMINEM
Rt #3
Orillia, Ontario L3V 6H3, Canada
General militaria.
List $1.00.

CANAM
Fruitland, Ontario LOR 1LO, Canada
416-643-4357
Militaria & gun parts.
Catalogue $1.00.

JOHN C. DENNER CO.
RR 1, Box 122
North Lancaster, Ontario KOC 1ZO, Canada
613-525-3840
General militaria.
Catalogues $6.00 per year, $2.00 each.

MARWAY MILITARIA,INC.
418 Begg Crescent
Saskatoon, Sask. S7H 4P5, Canada
306-477-2391
General militaria.

MILARM COMPANY,LTD.
10969 - 101st Street
Edmonton, Alberta T5H 2S9, Canada
General militaria.
Catalogue $3.00.

SPECIAL FORCES MX QM
P.O. Box 690, Unit #3
Thornhill, Ontario L3T 4A5, Canada
General militaria.

THE MILITARY EXCHANGE LIMITED
95 Yonge Street
Toronto, Ontario M5C 1S8, Canada
364-9550
General militaria.
Catalogue $2.00.

MILITARY ANTIQUARIAN
P.O. Box 8096
Ottawa, Ontario K1G 3H6, Canada
General militaria.

EDWARD E. DENBY & ASSOCIATES
1206 Yonge Street
Toronto, Ontario M4T 1W1, Canada
416-921-2493
Military medals & insignia.

ANTIQUE MILITARIA
C.P. 3037
St. Roch
Quebec, P.Q. G1K 6X9, Canada
General militaria.
Catalogue $2.00.

PENMAN'S
47 Princess Street
Saint John, N.B. E2L 1K1, Canada
657-7544
General militaria.
CONTACT: A. W. Penman

H & D ARMS CO.,LTD
815-A Bank Street
Ottawa, Ontario K1S 3V7, Canada
613-235-0645
General militaria.
List available.
CONTACT: James Hendry

UNIQUE IMPORTS
Postbox 64
Virum, DK 2830, Denmark
General militaria, 3rd Reich repros.
Catalogue available.

JOHN P. WITTER
Satomaki, Salpa 23
Vantaa 37, 01370, Finland
Finnish, Estonian, Latvian, Lithuanian collector.

NORDARMS
Itainen Rantakatu 48-50 G 208
Turku 81, 20810, Finland
General militaria.
CONTACT: Pertti Ketola

GENERAL MILITARIA

CARREL SPORTS
Carrel Sports
Les Deux Alpes, 38860, France
General militaria.
CONTACT: Pierre-Alain Carrel

G. CEDOU
39 Rue Sainte-Famille
Toulouse, 31200, France
General militaria.

J. MARC FOURNIER
10 Rue Claude Genoux
Albertsville, 73200, France
German WWII militaria.
List available, send SASE.

ETS J. Y. SEGALEN
4 Route de Rebigue
B.P.45
Castanet, Tolosan 31320, France
Military insignia.

LE POILU
20, rue Emile Duclaux
Paris, 75015, France
306-77-32
General militaria.
CONTACT: Pierre Besnard

MARK ANDREWS
10 Chatham Street
Ramsgate, Kent Great Britain
Elite forces militaria.
List $1.00.

NORMAN W. COLLETT
47 Queen Victoria Street
London, EC4N 4SA, Great Britain
General militaria.

JOHN EDWARDS
7 Nutleigh Grove
Hitchin, Herts SG5 2NH, Great Britain
General militaria.

CHRIS FARLOWE
159 Offord Road
Islington, London N1, Great Britain
General militaria.

DONALD HALL
P.O. Box 65
Cheshunt, Herts. EN8 0QR, Great Britain
General militaria

A. J. MARRIOTT-SMITH
4 New Road
S. Darenth, Kent DA4 9AR, Great Britain
Aviation mail order.
S4;
Send $5.00 for our quarterly list.

NICHOLAS MORIGI
14, Seacroft Road
Broadstairs, Kent CT10 1LT, Great Britain
843-602243 (24 HRS)
General militaria.
Catalogue $2.00.

T. H. POPE
49 Wordsworth Road
Loughborough, Leicester Great Britain
General militaria

JOHN RICHARDSON
15 Overslade Crescent
Coundon Green
Coventry, CV6 2AW, Great Britain
General militaria

ANDREW SPENCER
32 Rotcher Hill
Holmfirth, W. Yorkshire Great Britain
General militaria

JEREMY TENNISWOOD
28 Gordon Road
Aldershots, Hants GU11 1ND, Great Britain
0252-319791
General militaria.
S3,S6,S11,S12,S13; 20;
Catalogue available. As needed.
CONTACT: Jeremy Tenniswood
SALES DEPT: Jeremy Tenniswood
PURCHASING DEPT: Jeremy Tenniswood
International Dealer, Retail and Trade, buy, sell and exchanges. Military firearms, medals, badges, headress, uniform swords and bayonets. Seperate subject lists $1.00

A. R. FABB BROS,LTD.
Waldeck House
Waldeck Road
Maidenhead, Berkshire Great Britain
General militaria

BATTLE ORDERS SHOGUN UK
71 A Eastbourne Road
Willingdon
Eastbourne, BN20 9NR, Great Britain
011-44 3212 7254 TLX: 896691
General militaria. Want lists solicited.

KEPI-BLANC
64 Station Road
Blackpool, Great Britain
French militaria. Will trade for German Third Reich.

MILITARY HERITAGE MUSEUM
The Fort
New Haven, Sussex BN9 9DL, Great Britain
General militaria dealer.
Catalogue available.

SMALL'S ARMY STORE
17-19 Bradshaw Gate
Bolton, Lancashire BL1 1E1, Great Britain
25494
General militaria. Want lists solicited.

THE LONDON STAMP EXCHANGE,LTD.
5 Buckingham Street
Strand
London, WC2N 6BS, Great Britain
General militaria

TRADITION,LTD.
10 Mayfair Street
Mayfair
London, W1Y 7LB, Great Britain
British militaria.
List available.
CONTACT: Geoff White

WARNER'S MILITARY SPECIALISTS
2 The Apple Market
Eden Street
Kingston-upon-Thames, Surrey KT1 1JE, Great Britain 01-546 7527 (24 HRS)
General militaria.
S4;
Catalogue available. Quarterly.
CONTACT: Geoffrey George Warner

ADRIAN FORMAN
13, Shepherd Market
Mayfair, London W1, Great Britain
01-629-6599
General militaria.
12 Lists for $10.00.

DE DONDERBUS
J. Winters
Oostzeedijk 231 A
Rotterdam, 3061VW, Holland
010 4140267 01862 2338
General militaria, antique weapons. WWI & WWII items & documentation.
S3,S9; 16;
SALES DEPT: J. Winters
PURCHASING DEPT: J. Winters

RINALDO BREGA
Via Euganea-S. Biagio, 54/A
Bresseo (PD), 35033, ITALY
051 266668 049 9901056
Finders service, identification service, and everything on Abn. & elite units.
S6;
List available.
Headgear, uniforms, insignia, equipment, mainly from airborne and elite units. Italian and other countries' surplus. Wants lists of all kinds welcome. Trade welcome.

UNIVERSAL ANTIQUES LTD.
P.O. Box
Hibernian House, The Quay
Bantry, County Cork Ireland
All sorts of French, British, & Spanish WWI, WWII, & 19th century militaria.
S4; 6;
Catalogue available. Quarterly
We are certainly the biggest Irish militaria dealers. Items from the 19th century to the present. Catalogue: $1.00. Universal Antiques Ltd., P.O. Box, Bantry, County Cork (Eire).

ALLE ANTICHE ARMI
Via Bigli 24
Milano, 20121, Italy
General militaria.

TANABERA COMPANY
4-4-10 Nishisakaidanicho
Oharano, Nishikyoku
Kyoto, Japan
331-5994
General militaria.
CONTACT: S. Iwai

H. BECK, ANTIK STUBE
Churer Strasse 69
Nendelm, FL 9491, Liechenstein
General militaria.

GENERAL MILITARIA

ARES VAPIEN & ANTIKT
Antikhallarna
Vastra Hmngaten 6
Goteborg, 411 17, Sweden
General militaria. Japanese arms & armour collector.

ADRIAN BOHLEN
Dorf 55
Thierachern, CH-3634, Switzerland
General militaria.

WOLFRAM DEIBEL
Bahnhofstrasse 19/A
Penzberg, 8122, West Germany
0 88 56/22 19
General militaria.
Catalogue $5.00.

KARLHEINZ MULLER
Barfuberstrasse 4
6250 Limburg, Lahn 1 West Germany
General militaria.

CLYDE F. PERRY
Schwansee Strasse 64
8 Munchen 90, West Germany
German militaria.

BRINKMANN-VERLAG
Landauerstrasse 2
Pirmasens, 6780, West Germany
Military book publisher.
CONTACT: Bernd Stephan

MILITARIA SUDWEST
Am Hofbohl 1
Abtsteinach 3, D-6941, West Germany
06207/2500
Military medals & general militaria.
Catalogue available.
CONTACT: Gerhard Rudloff

SIEGMUND STEPHAN
Walsroder Strasse 41
Soltau, 3040, West Germany
(0 51 91) 1 62 85
Helmets, Militaria & models.

REYNOLDS OBJEKTSTANDER GMBH
Morgensternstrasse 1
Berlin 45, 1000, West Germany
(0 30) 7 72 20 87
General militaria.
Catalogue available.

COMMANDER'S BUNKER
Postfach 1103
Farchant, 8105, West Germany
01149-8824-637
General militaria.
CONTACT: Tee J. Reeve,III

MILITARY HEADGEAR

GERMAN HELMET PARTS
P.O. Box 8
80 Forest Street
Plaistow, NH 03865, USA
603-382-7845
German helmet restoration.
Write for details & estimates.

'THE PHOENIX EXCHANGE'
P.O. Box 66
163 Troutman Road
Arcola, PA 19420, USA
(215)-933-0909
Quarterly military collectors magazine.
; 2;
CONTACT: Terry Hannon, Publisher
SALES DEPT: W. R. Bendel
PURCHASING DEPT: Terry Hannon
AD MANAGER: William Bendel
'The Phoenix Exchange' functions as the military collector's & dealer's marketplace. Articles cover a wide spectrum of militaria, but attend to each subject in detail. Send $2.00 for a sample copy.

RICHARD BIRD
1308 Pickering Circle
Largo, MD 20772, USA
301-336-4766
Military headgear.
Catalogue $1.00.

BATTLEFIELD MEMORABILIA
P.O. Box 1044
Forestville, MD 20772, USA
Military helmets.
3 catalogues for $1.00.

FLOYD R. TUBBS
191 Trudy Avenue
Munroe Falls, OH 44262, USA
Helmets & headgear dealer.
List available.

RICO'S
P.O. Box 1604
Dayton, OH 45401, USA
Headgear & special forces insignia.
List available.

GARY FOX
409 Slattery Bldg.
Shreveport, LA 71101, USA
U.S. military headgear.
Write for details.

HUGH L. WADE
17314 Raymer Street
Northridge, CA 91325, USA
Helmets & headgear.

MILITARY INSIGNIA

DELTA SUPPLY
P.O. Box 936
Boylston, MA 01505, USA
(617)-732-6881
U.S. & foreign elite unit insignia.
S6,S7,S11; 7;
Send $2.00 for List. 1 - 2 times annually, with suppliments. CONTACT: Hal S. Feldman
SALES DEPT: Hal S. Feldman
PURCHASING DEPT: Hal S. Feldman
Special Forces, airborne, "Elite" forces. All types of worldwide insignia and militaria. Wholesale and retail mail sales. Elite forces mail-auction starts Spring 1987.

THE CANONNEER
P.O. Box 24013
Westside Station
Jersey City, NJ 07304, USA
U.S. military insignia.
List $1.00 plus 2 stamps.
CONTACT: G. McDonald

DENNIS M. COVELLO
219 Van Houten Avenue
Wyckoff, NJ 07481, USA
(201)-891-8524
U.S. cloth embroidered insignia.
S4,S12;
Catalogue available. As required.
CONTACT: Dennis M. Covello
U.S. patches: Great selection, prices, service. Army list($.40 with long SASE), oval/flash ($.22), Navy ($.22), USN shoulder tabs list, both old and current ($.40).

BILL BRUCE, INSIGNIA
Box R-1
APO New York, NY 09146, USA
0661-74366
General military insignia dealer.
S6,S7,S11; 3;
Catalogue available. Monthly.
CONTACT: David W. Bruce
PURCHASING DEPT: Bill Bruce
Military insignia, top quality wings, bullion, hand-made, etc. Selling collections. SASE (long) for lists. Imperial Germany, current NATO & Bundeswehr.

DIFFERENCE
66 West 38th Street
New York, NY 10018, USA
212-730-7255
Military insignia.

M. ALBERT MENDEZ
P.O. Box 1227
Rockefeller Center Station
New York, NY 10185, USA
Military insignia.

CHASIN INSIGNIA
3725 Henry Hudson Parkway
Riverdale, NY 10463, USA
General militari.
S4; 4;
Catalogue available. Updated every 6-8 months.
CONTACT: Steven B. Chasin
U.S. military insignia WWII to the present. Specializing in current issue insignia. All types.

BRIAN BENEDICT
P.O. Box 237
Carmel, NY 10512, USA
(914)-225-7474
General militaria.
S6; 13;
SALES DEPT: Brian Benedict
PURCHASING DEPT: Brian Benedict
Wholesaler of embroidered emblems & metallic insignia in large lots. Also rare militaria insignia.

EMBLEMS AND BADGES,INC.
P.O. Box 365
Monroe, NY 10950, USA
Military insignia.
Catalogue $.50.

MILITARY INSIGNIA

JOHN R. BURKHART
4980 Doyle Road
Pittsburgh, PA 15227, USA
(412)-881-2541
General military insignia.
S4,S12; 6;
Catalogue available. As warranted.
CONTACT: John R. Burkhart
SALES DEPT: John R. Burkhart
PURCHASING DEPT: John R. Burkhart
Sale of insignia, WWI to present. Specializing in patches from WWII and large selection of theatre-made Vietnam patches. Want lists solicited.

C. J. DiGIACOMO
900 Florida Avenue
Pittsburgh, PA 15228, USA
General military insignia

JIM POWELL
P.O. Box 113-P
14 North Mountain Street
Newberg, PA 17240, USA
(717)-423-5236
General military insignia.
S4; 6;
Catalogue available. Bi-annually.
CONTACT: James O. Powell
SALES DEPT: Jim Powell
PURCHASING DEPT: Jim Powell

PHOENIX MILITARIA CORPORATION
P.O. Box 66
163 Troutman Road
Arcola, PA 19420, USA
(215)-933-0909
General militaria & publisher of militaria collectors books & magazines.
S6; 19; 23,000; $60.00
Catalogue $3.00. Semi-annually.
CONTACT: Terry Hannon, President
SALES DEPT: Irene Karas
PURCHASING DEPT: Terry Hannon
Send $3.00 or 3 pieces of military insignia for our latest catalogue of U.S. & world-wide militaria. Thousands of items listed including: books, manuals, medals, insignia, equipment, uniform components, badges, etc.

'THE PHOENIX EXCHANGE'
P.O. Box 66
163 Troutman Road
Arcola, PA 19420, USA
(215)-933-0909
Quarterly military collectors magazine.
; 2;
CONTACT: Terry Hannon, Publisher
SALES DEPT: W. R. Bendel
PURCHASING DEPT: Terry Hannon
AD MANAGER: William Bendel
'The Phoenix Exchange' functions as the military collector's & dealer's marketplace. Articles cover a wide spectrum of militaria, but attend to each subject in detail. Send $2.00 for a sample copy.

DOVER ARMY-NAVY STORE, INC.
222 West Loockerman Street
Dover, DE 19901, USA
(302)-736-1959
Complete line of military goods as found in a traditional Army-Navy store.
S1,S2,S3,S6,S12; 22;

Catalogue available. Annually.
CONTACT: Jerome Zaback
SALES DEPT: Jerry or Frank Zaback
PURCHASING DEPT: jerry Zaback
We are probably the largest retail distributor of military insignia, both Police & Military on the East Coast.

TROOPER LUKE'S
P.O. Box 446
3019B Oak Green Court
Ellicott City, MD 21043, USA
(301)-465-8637
Replica Vietnam War insignia.
S6,S12,S13; 5;
Flyer available. Bi-annually.
CONTACT: Luke Flannery
SALES DEPT: Luke Flannery
PURCHASING DEPT: Luke Flannery
Specializing in quality reproductions of Vietnam era and later patches and badges. Navy, SEAL teams, Riverine and Air Force. Flyer available upon request.

PETER J. McDERMOTT
803 Natures Run
Severna Park, MD 21146, USA
Airborne insignia.
Send SASE + ($.39) for list.

JERRY CHODZINSKI
P.O. Box 6005
Baltimore, MD 21231-0005, USA
Military insignia

EAGLE ENTERPRISES
P.O. Box 391
Dumfries, VA 22026, USA
(703)-221-5763 (703)-221-9825
U.S.M.C. militaria 18
S3,S4;
CONTACT: R. T. Spooner
SALES DEPT: R. T. Spooner
PURCHASING DEPT: R. T. Spooner
Buy, Sell, Trade, U.S. Marine insignia, chevrons, etc. We buy USMC collections. Tell us what you've got and what you want for it.

HARRY PUGH
5009 N. 24th Street
Arlington, VA 22207, USA
General militaria
S4;
Foreign military parachutist-elite insignia of 80 plus nations. Also foreign aviation, repro. MAC-SOG patches and USSF blazer badges.

SHENANDOAH SALES
P.O. Box 130
Goldvein, VA 22720, USA
(703)-439-3986
General military insignia.
S6; 5;
Catalogue available.
CONTACT: Arnold W. Mathias
SALES DEPT: Arnold W. Mathias
PURCHASING DEPT: Arnold W. Mathias
Retail and wholesale mail order of patches, wings, ovals, flashes, DIs.

RICHARD L. DAVIS
3529 Abington Road
Columbia, SC 29203, USA

803-252-6586
WWII AAF wings

BILL PRICE
349 N. 22nd Court East
Bradenton, FL 33508, USA
(813)-747-8221
Cloth insignia: US military, NASA, US & foreign law enforcement & fire service.
S6; 40;
List available. As needed.
SALES DEPT: Bill Price
Insignia (cloth) - US military, NASA, US & foreign law enforcement, US & foreign fire service. Lists available. Write for details.

INSIGNIA
P.O. Box 3133
Naples, FL 33939, USA
U.S. military insignia.
Catalogue available.
CONTACT: H. J. Saunders

RALPH CURTIS - BOOKS
P.O. Box 183
Sanibel, FL 33957, USA
(813)-472-5470
Military patches, books & U.S. insignia.
S6,S12; 6;
Catalogue available. Quarterly.
CONTACT: Ralph Curtis
SALES DEPT: Ralph Curtis
PURCHASING DEPT: Ralph Curtis
U.S. patches, insignia, medals, wings, military books. Mail-order only. Specialize in U.S. Marine Corps.

RICHARD W. SMITH INSIGNIA
P.O. Box 2118
Hendersonville, TN 37077, USA
(615)-822-3353 (615)-824-6157
U.S. military patches & books.
S4; 14;
Catalogue available. Annually.
CONTACT: Richard W. Smith
SALES DEPT: Richard W. Smith
PURCHASING DEPT: Richard W. Smith
Mail order sales of US cloth patches for all branches. Books on US cloth military insignia. Lists cost $1.00. Refundable with order.

BRITISH REGALIA IMPORTS
P.O. Box 50473
Nashville, TN 37205, USA
British & Scottish military regalia.
S4; 2;
Catalogue available. Semi-annually.
CONTACT: Brian T. Sinclair-Whitely
British/Scottish military insignia and general regalia. See our classified Ad in Phoenix Exchange under Insignia for details.

J. D. HAGELBERGER
P.O. Box 20
Kunkle, OH 43531, USA
Military buttons.

MILITARY INSIGNIA

INSIGNIA SHOP
615 Wellington Court, #5
Waterloo, IA 50702, USA
(319)-235-0431
General militaria.
S6,S11,S13; 41;
Catalogue available. Annually.
CONTACT: David R. Northey
SALES DEPT: David R. Northey
PURCHASING DEPT: David R. Northey
Offer all military regalia items and repair/restoration of ribbons & medals.

ARTHUR J. GRAU,JR.
3608 Douglas Ave #401
Racine, WI 53402, USA
(414)-639-9677
Researcher of private military schools.
S4;
Dealer in ROTC and private military school insignia. Also author of "Military Schools of Yesteryear", articles and stories.

QMD BOOK & INSIGNIA EXCHANGE
4540 Morningside Avenue
St. Paul, MN 55110, USA
(612)-426-9768
Foreign militaria insignia - Middle East & Peacekeeping Forces insignia.
S4; 10;
Catalogue available. Annually, w/quarterly updates. CONTACT: David V. Olson
SALES DEPT: D. V. Olson
PURCHASING DEPT: D. V. Olson
AD MANAGER: D. V. Olson
QMD Book & Insignia Exchange specializes in insignia from Middle East countries & Peacekeeping Forces insignia and the respective topical books of the area.

BOB HRITZ
P.O. Box 1055
Glendale Heights, IL 60139, USA
312-620-0156
Military aviation wings collector.

FILL WOZNIAK
2903 N. Dawson Avenue
Chicago, IL 60618, USA
Military insignia

BUSHPILOT WINGS
P.O. Box 20378
Dallas, TX 75220, USA
(214)-381-2311
Insignia, clothing, knives, books, signal devices.
S4,S5,S6; 4;
Catalogue available. Annually.
CONTACT: Ron Whipple
SALES DEPT: Ron Whipple
PURCHASING DEPT: Ron Whipple
Mail order Bushpilot Wings, insignia, caps, T-shirts, survival gear, books and aviation oriented novelties.

AMCRAFT
Rt 1, Box 124A
Crockett, TX 75835, USA
409-544-3724
Military insignia.
Collectors brochure $2.00.

ROYCE E. 'BO' SCOTT
1013 Willowbriar Lane
Deer Park, TX 77536, USA
Military insignia

JMC MILITARY INSIGNIA
8000 Village Oak,G-39
San Antonio, TX 78233, USA
Military insignia

LEASURES TREASURES
2801 W. Colorado Avenue
Colorado Springs, CO 80904, USA
303-635-8539
Military insignia.
List available.

THE INSIGNIA EXCHANGE
1046 N. Wedgewood Drive
Mesa, AZ 85203, USA
(602)-834-7604
Worldwide military insignia.
S4; 8;
Catalogue available. Annually.
CONTACT: Donald R. Strobaugh
SALES DEPT: Donald R. Strobaugh
World-wide parachutist wings: Airborne; Special Forces; Elite; Aviation items. Sales list includes 90+ countries $1.00 in USA/Canada, $2.00 other countries. Will buy collections.

MILITARY PINMAN
873 Oneonta Drive
Los Angeles, CA 90065, USA
Military insignia

KEN NOLAN,INC.,P.M.DIV.
P.O. Box C-19555
16901 P.M. Milliken Avenue
Irvine, CA 92713, USA
(714)-863-1532
Surplus, insignia, medals & uniforms.
S4; 29; 100,000; $60.00
Catalogue $1.00. 3 or 4 times annually.
Serving individuals, military, police for 29 years. 90% of orders shipped within 24 hours. Military field clothing, boots, insignia, nameplates, etc. Send $1.00 for catalog.

ALL EMBLEM COMPANY
Dept. Military
P.O. Box 31078
Seattle, WA 98103, USA
New military insignia.
Catalogue $2.00.

MILITARY LIBRARIES

UNIVERSITY OF MASSACHUSETTS LIBRARY
Serials Section
Amherst, MA 01003, USA
Military library collection.

BOSTON PUBLIC LIBRARY
Serial Section
Boston, MA 02117, USA
Military library collection

HARVARD COLLEGE LIBRARY
Acquisition Department
Cambridge, MA 02138, USA
Military library collection

MAHAN LIBRARY
Naval War College
Newport, RI 02840, USA
Military library.

DARTMOUTH COLLEGE LIBRARY
Dartmouth College
Hanover, NH 03755, USA
Military library.

MAINE MARITIME ACADEMY
Library
Castine, ME 04420, USA
Military library.

CONNECTICUT STATE LIBRARY
231 Capital Avenue
Hartford, CT 06106, USA
Military library.
CONTACT: Attn: F. Desplanques

U.S. COAST GUARD ACADEMY LIBRARY
New London, CT 06320, USA
Military library.

YALE UNIVERSITY LIBRARY
Accessions Department
New Haven, CT 06520, USA
Military library.

PRINCETON UNIVERSITY LIBRARY
Princeton University
Princeton, NJ 08544, USA
Military library.

ALEXANDER LIBRARY
Rutgers University
New Brunswick, NJ 08901, USA
Military library.

U.S. ARMY LIBRARY
Wildflecken
APO New York, NY 09026, USA
Military library.

USAREUR LIBRARY & RESOURCE CENTER
HQ USAREUR & 7th Army
APO New York, NY 09403, USA
Military library.

NEW YORK PUBLIC LIBRARY
P.O. Box 2240
Grand Central Station
New York, NY 10017, USA
Military library.

THE NEW YORK HISTORICAL SOCIETY
170 Central Park West
New York, NY 10024, USA
(212)-873-3400
Regional historical association. Museums, library, quarterly, newsletter.
CONTACT: James B. Bell

MILITARY LIBRARIES

UNITED STATES MILITARY ACADEMY
LIBRARY
Building 757
Serials Unit #3
West Point, NY 10996, USA
Staff library not open to public.
Research questions relative to the U.S.M.A. and the museum collections are answered by the staff according to available time. Staff available by appointment for serious research.

BENET LABORATORIES
Technical Library Division
Watervliet Arsenal
Watervliet, NY 12189, USA
Military library.

NEW YORK STATE LIBRARY
Albany, NY 12201, USA
Military library.

NEW YORK HISTORICAL SOCIETY
LIBRARY
Cooperstown, NY 13326, USA
Military library.

CORNELL UNIVERSITY LIBRARY
Serials Department
Ithaca, NY 14853, USA
Military library.

PENNSYLVANIA STATE UNIVERSITY
LIBRARY
Pennsylvania State University
University Park, PA 16802, USA
Military library.

PENNSYLVANIA STATE UNIVERSITY
Serials Records Section
Room 46, Box 1601
Harrisburg, PA 17120, USA
Military library.

WIDENER COLLEGE
Wolfgram Library
Serials Department
Chester, PA 19013, USA
Military library.

FREE LIBRARY OF PHILADELPHIA
Auto. Reference Collection
Logan Square
Philadelphia, PA 19103, USA
Military reference library.

CHESTER COUNTY LIBRARY
Periodical Department
400 Exton Square Parkway
Exton, PA 19341, USA
Military library.

USUHS
Attn: Library
4301 Jones Bridge Rd.,Bldg.70
Bethesda, MD 20014, USA
Military library.

THE PENTAGON ARMY LIBRARY
Attn: Acquisitions
Room 1A-518
Washington, DC 20310, USA
Military library.

U.S. ARMY CENTER OF MILITARY
HISTORY
Attn:DAMH-HSR-L,Pulaski Bldg.
20 Massachusetts Avenue, N.W.
Washington, DC 20314, USA
U.S. Army military history center.

NAVAL HISTORICAL CENTER
M/F Navy Department Library
Washington, DC 20374, USA
U.S. Navy military history center.

LIBRARY OF CONGRESS
Continuations Unit
Order Division 9018034
Washington, DC 20540, USA
Military reference library.

SMITHSONIAN INSTITUTION
MHT Library
Smithsonian Institute
Washington, DC 20560, USA
Military reference library.

UNIVERSITY OF MARYLAND
McKeldin Library
Serials Department
College Park, MD 20742, USA
Military reference library.

U.S. NAVAL ACADEMY
Library
Serials & Binding Department
Annapolis, MD 21402, USA
Military library.

VIRGINIA STATE LIBRARY
Serial Section
Richmond, VA 23215, USA
Military library.
CONTACT: Attn: Mrs. Gaines

SPECIAL SERVICE LIBRARY
Building P-9023
Fort Lee, VA 23801, USA
Military library.

G. C. MARSHALL LIBRARY
Drawer 920
Lexington, VA 24450, USA
Military reference library.

DUKE UNIVERSITY LIBRARY
Periodicals Department
Durham, NC 27706, USA
Military reference library.

TECHNICAL SERVICE LIBRARY
Building AT 2747
Fort Bragg, NC 28307, USA
Military library.

THE DANIEL LIBRARY
The Citadel
Charleston, SC 29409, USA
Military library.
CONTACT: c/o Col. J. M. Hillard

CONRAD TECHNICAL LIBRARY
USASC & FG Bldg. 29807
Fort Gordon, GA 30905, USA
Military reference library.

JOHN C. PACE LIBRARY
University of West Florida
Pensacola, FL 32504, USA
Military reference library.

U.S. AIR FORCE HISTORICAL RESEARCH
CENTER
Maxwell AFB, AL 36112, USA
Military research library

MITCHELL MEMORIAL LIBRARY
Acquisitions Department
Mississippi State University
Mississippi State, MS 39762, USA
Military reference library.

U.S. ARMY ARMOR SCHOOL LIBRARY
Building 2369
Fort Knox, KY 40121, USA
U.S. Army museum & library.

FORT CAMPBELL POST LIBRARY
Fort Campbell, KY 42223, USA
Military library.

INDIANA STATE LIBRARY
140 N. Senate Avenue
Indianapolis, IN 46204, USA
Military reference library.

ACADEMIC LIBRARY
USASC Post Headquarters
Ft. Benjamin Harrison
Indianapolis, IN 46216, USA
Military library.

ALLEN COUNTY PUBLIC LIBRARY
Periodical Section
P.O. Box 2270
Fort Wayne, IN 46801, USA
Military reference library.

UNIVERSITY MICROFILMS,INC.
Serials Processing Department
300 North Zeeb Road
Ann Arbor, MI 48106, USA
Military reference library.

UNIVERSITY OF WISCONSIN MILWAUKEE
LIBRARY
Serials Department
P.O. Box 604
Milwaukee, WI 53201, USA
Military reference library.

MINNESOTA HISTORICAL SOCIETY
Library-Serials Unit #771
690 Cedar Street
St. Paul, MN 55101, USA
Regional historical society

VETERANS ADMINSTRATION LIBRARY
#133
VA Hospital
Downey, IL 60064, USA
Military library.

THE NEWBERRY LIBRARY
Serials Division
60 West Walton Street
Chicago, IL 60610, USA
Military library.

MILITARY LIBRARIES

MILNER LIBRARY
Serials Department
Illinois State University
Normal, IL 61761, USA
Military reference library.

ILLINOIS STATE LIBRARY
Serials Section
Centennial Building
Springfield, IL 62756, USA
Military reference library.

SOUTHERN ILLINOIS UNIVERSITY
Morris Library
Periodicals Department
Carbondale, IL 62901, USA
Military reference library.

PUBLIC LIBRARY #11
1301 Olive Street
St. Louis, MO 63103, USA
Military reference library.

UNIVERSITY OF KANSAS LIBRARY
Periodical Department
University of Kansas
Lawrence, KS 66044, USA
Military reference library.

AFO LIBRARY
Office of the Adjutant General
Jackson Barracks
New Orleans, LA 70146, USA
Military library.
CONTACT: Mrs. Oalmann

LOUISIANA STATE UNIVERSITY LIBRARY
Louisiana State University
Serials Department
Baton Rouge, LA 70803, USA
Military reference library.

MORRIS SWETT LIBRARY
U.S. Army Field Arty. School
Snow Hall (W-2671)
Fort Sill, OK 73503, USA
Military library.

LIBRARY CENTER
Building 1850
Fort Hood, TX 76544, USA
Military library.

CONFEDERATE RESEARCH CENTER
Hill College
P.O. Box 619
Hillsboro, TX 76645, USA
(817)-582-2555
Museum of Texan Confederate forces. Reference material available.
Extensive Civil War museum alonga with research center.

TEXAS A & M UNIVERSITY LIBRARY
Texas A & M University
Serials Department
College Station, TX 77843, USA
Military library.

TEXAS UNIVERSITY LIBRARY
University of Texas at Austin
Serials Department
Austin, TX 78712, USA

Military reference library.

U.S. ARMY AIR DEFENSE SCHOOL LIBRARY
P.O. Box 5040
Fort Bliss, TX 79916, USA
Military library.

SERGEANTS MAJOR ACADEMY
Learning Resource Center
Building 11203
Fort Bliss, TX 79918, USA
Military library.

U.S.A.F. ACADEMY LIBRARY
Academy Library (DFSELS)
U.S.A.F. Academy, CO 80840, USA
Military library.

UNIVERSITY OF WYOMING
University of Wyoming Library
Box 3334, Periodicals Dept.
Laramie, WY 82071, USA
Military reference library.

UNIVERSITY OF UTAH LIBRARYS
Serials Department
Salt Lake City, UT 84112, USA
Military reference library.

ARIZONA STATE UNIVERSITY
Library
Periodicals Department
Tempe, AZ 85287, USA
Military reference library.

UNIVERSITY OF ARIZONA
The Library
Tucson, AZ 85721, USA
Military reference library.

WESTERN NEW MEXICO UNIVERSITY
Miller Library
Serials Department
Silver City, NM 88061, USA
Military reference library.

NEW MEXICO MILITARY INSTITUTE
The Library
Roswell, NM 88201, USA
Military library.

SAN FRANCISCO PUBLIC LIBRARY
Periodical Department
Civic Center
San Francisco, CA 94102, USA
Military reference library.

UNIVERSITY OF CALIFORNIA LIBRARY
University of California
Davis, CA 95616, USA
Military reference library.

CALIFORNIA STATE LIBRARY
10th & N Streets
Sacramento, CA 95809, USA
Military reference library.

SEATTLE PUBLIC LIBRARY #577
Serials Division
1000 Fourth Avenue
Seattle, WA 98104, USA
Military reference library.

LIBRARY OFFICE
Building 2109
Fort Lewis, WA 98433, USA
Military reference library.

MILITARY PERIODICALS

'JOURNAL OF ELECTRONIC DEFENSE'
685 Canton Street
Norwood, MA 02062, USA
(617)-769-9750
Journal of the Association of Old Crows.
S10,S13; 8;
CONTACT: Hal Gershanoff, Pub. & Editor
SALES DEPT: Celeste Donohue
PURCHASING DEPT: Joyce Van Wagener
AD MANAGER: Celeste Donohue
Offical publication of the Assn. of Old Crows (The Electronic Defense Assn.), JED focuses on EW, C3, C3I, C3CM, planning/procurement, avionics, and military computing.

'THE ARTILLERYMAN QUARTERLY'
P.O. Box 129
4 Water Street
Arlington, MA 02174, USA
617-643-7900
Black powder artilleryman's magazine. 4 issues $12.00 annually, sample $3.00.
CONTACT: Cutter & Locke,Inc.

'THE CIVIL WAR BOOK EXCHANGE & COLLECTOR'S NEWSPAPER'
P.O. Box C
4 Water Street
Arlington, MA 02174, USA
(617)-646-2010
Civil War collectors magazine.
S12,S13;
CONTACT: Cutter & Locke,Inc.

'NAVAL WAR COLLEGE REVIEW'
U.S. Naval War College
Newport, RI 02840, USA
Contemporary military publication.

'MAN-AT-ARMS'
222 W. Exchange Street
Providence, RI 02903, USA
1-800-341-1522
Military history publication. 1 year (six issues): $18.00.

'NATIONAL SECURITY AFFAIRS'
Brown University
Providence, RI 02912, USA
Contemporary military publication.

WEAPONS & WARFARE PRESS
218 Beech Street
Bennington, VT 05201, USA
(802)-447-0313
Military book publisher.
S6; 19; 4,000; $50.00
Catalogue available. Monthly.
CONTACT: Ray Merriam
Publish magazine, hundreds of books on every aspect of military history. Sell thousands of current titles of other publishers (retail only). See our advertisement.

MILITARY PERIODICALS

'VIET-NAM NEWSLETTER'
P.O. Box 122
Collinsville, CT 06022, USA
Contemporary military publication.

'BRADLEY AIR MUSEUM NEWSLETTER'
10 Christine Drive
East Heartland, CT 06108, USA
Museum newsletter.
S13;

'THE HOWLING GALE'
Box A-37
Chase Hall, USGA
New London, CT 06320, USA
U.S. Coast Guard publication.

'GUNG-HO'
Charlton Building
Derby, CT 06418, USA
Contemporary military publication.

'THE TORCH'
369 S. Leonard
Waterbury, CT 06708, USA
Italian-American war veterans magazine. Quarterly.

'ARMY AVIATION MAGAZINE'
One Crestwood Road
Westport, CT 06880, USA
Contemporary military publication.

'WORLD INTELLIGENCE REPORT'
P.O. Box 1265
Merchantville, NJ 08109, USA
Military intelligence newsletter. 24 issues at $32.00 per year.
CONTACT: R. C. Smith

'WEST POINT MUSEUM BULLETIN'
U.S. Military Academy
West Point, NY 10096, USA
Contemporary military publication.

'NEW BREED'
P.O. Box 428
Nanuot, NY 10954, USA
Contemporary para-military magazine.

'JING BAO JOURNAL'
14th Air Force Ass'n, Inc.
P.O. Box 285
Selden, NY 11784, USA
Journal of 14th Air Force Association. Flying Tigers 14th Air Force Assn., Inc. Association of war veterans who served in China 1942-1945.

'ARMS COLLECTORS JOURNAL'
P.O. Box 385
16 River Road
Mechanicville, NY 12118, USA
518-664-9743
Monthly arms collectors magazine. $12.00 per year.
CONTACT: David Petronis, Publisher

WORLD WAR I AEROPLANES
15 Crescent Road
Poughkeepsie, NY 12601, USA
914-473-3679 Aero 818-243-6820 Skyways
Provides service info on WWI aeroplanes.
S6,S10; 25; 2200; 100.00
CONTACT: Leonard E. Opdycke
SALES DEPT: L. E. Opdycke
AD MANAGER: Richard Alden
Provides service (information, names, projects, materials, books) to modellers, builders & restorers of aeroplanes (1900-1919, 1920-1940) through two journals, WW I Aero & Skyways.

'ARMS COLLECTING'
P.O. Box 70
Alexandria Bay, NY 13607, USA
Firearms collectors magazine.
CONTACT: Museum Restoration Service

'THE COURIER'
P.O. Box 1863
Williamsville, NY 14221, USA
(716)-634-8324
Civil War collector magazine. Bi-monthly, advertising accepted.
; 2; ; 5000+; ; $65.00
CONTACT: Fred Hoffman
SALES DEPT: Fred Hoffman
PURCHASING DEPT: Fred Hoffman
AD MANAGER: Fred Hoffman
CIVIL WAR - If you are a Civil War buff, buy/sell CW memorabilia, or provide a service for buffs, "The Courier" will help you! Send for your FREE sample copy! The Courier, PO Box 1863, Williamsville, NY 14221.

'CIVIL WAR TIMES'
P.O. Box 8200
2145 Kone Road
Harrisburg, PA 17105, USA
717-657-9555
Civil War historical interest magazine.
CONTACT: Doug Moul

'ANTIQUES & AUCTION NEWS'
P.O. Box 500
Mount Joy, PA 17552, USA
(717)-653-9797
Antique trade paper

'MILITARY IMAGES'
RD #2, Box 124A
East Stroudsburg, PA 18301, USA
(717)-476-1388
U.S. military history magazine
S4; 9; 3,000; $75.00
CONTACT: Harry Roach
AD MANAGER: Lawrence Bixley
MI covers U.S. military history 1839-1900. Emphasis on photography. Concentration on Civil War. Biographies, regimentals, analyses of uniforms, equipment, insignia. Book reviews.

PHOENIX MILITARIA CORPORATION
P.O. Box 66
163 Troutman Road
Arcola, PA 19420, USA
(215)-933-0909
General militaria & publisher of militaria collectors books & magazines.
S6; 19; 23,000; $60.00
Catalogue $3.00. Semi-annually.
CONTACT: Terry Hannon, President
SALES DEPT: Irene Karas
PURCHASING DEPT: Terry Hannon
Send $3.00 or 3 pieces of military insignia for our latest catalogue of U.S. & world-wide militaria. Thousands of items listed including: books, manuals, medals, insignia, equipment, uniform components, badges, etc.

'THE PHOENIX EXCHANGE'
P.O. Box 66
163 Troutman Road
Arcola, PA 19420, USA
(215)-933-0909
Quarterly military collectors magazine.
; 2;
CONTACT: Terry Hannon, Publisher
SALES DEPT: W. R. Bendel
PURCHASING DEPT: Terry Hannon
AD MANAGER: William Bendel
'The Phoenix Exchange' functions as the military collector's & dealer's marketplace. Articles cover a wide spectrum of militaria, but attend to each subject in detail. Send $2.00 for a sample copy.

'NATIONAL GUARDSMAN'
One Massachusetts Ave., N.W.
Washington, DC 20001, USA
Journal of the National Guard Association.

'COVERT ACTION BULLETIN'
P.O. Box 50272
Washington, DC 20004, USA
Contemporary military publication

'DEFENCE WEEK'
915 15th St., N.W., Suite 400
Washington, DC 20005, USA
Contemporary military publication.

'AMERICAN LEGION MAGAZINE'
1608 K. Street N.W.
Washington, DC 20006, USA
Journal of the American Legion.

'THE AMERICAN RIFLEMAN'
1600 Rhode Island Ave., N.W.
Washington, DC 20006, USA
Journal of the National Rifle Association.

'JEWISH VETERAN'
1712 New Hampshire Avenue
Washington, DC 20009, USA
Military veterans magazine.

'DEFENCE & FOREIGN AFFAIRS'
1777 T Street, N.W.
Washington, DC 20009, USA
Contemporary military magazine.

'NEWSLETTER'
GPO 30051
Washington, DC 20014, USA
International military archives journal.

'SERGEANTS'
4235 28th Ave., Suite 707
Marlow Heights, DC 20031, USA
Air Force sergeants newsletter.

'NAVAL AFFAIRS'
1303 New Hampshire Ave., N.W.
Washington, DC 20036, USA
Naval magazine.

MILITARY PERIODICALS

'ARMED FORCES JOURNAL'
1414 22nd Street, N.W.
Washington, DC 20037, USA
Contemporary military magazine.

'ARMY RESERVE MAGAZINE'
Pentagon DAAR IO
Washington, DC 20310, USA
Army reservist publication.

'ARMY MUSEUM NEWSLETTER'
Centre of Military History
Washington, DC 20314, USA
Military museum newsletter.
CONTACT: Chief of Military History

'MILITARY LIBRARIAN BULLETIN'
National War College Library
4th and P Streets, S.W.
Washington, DC 20315, USA
Military librarian newsletter.

'NAVY CHAPLAINS BULLETIN'
Bureau of Naval Personnel
Washington, DC 20370, USA
Navy chaplains newsletter.

'NAVAL AVIATION NEWS'
Potomac Annex, Building 6
23 & E. Streets, N.W.
Washington, DC 20372, USA
Naval aviation magazine.

'U.S. NAVY MEDICINE'
Department of the Navy
Bureau of Medicine
Washington, DC 20372, USA
Naval medicine journal.

'HOTLINE HQMC'
Headquarters Marine Corps
Washington, DC 20380, USA
Marine Corps HQ newsletter.

'RETIRED MARINES'
Headquarters Marine Corps
Washington, DC 20380, USA
Magazine for retired Marines.

'ALL HANDS'
Bureau of Naval Personnel
U.S.G.P.OFF.
Washington, DC 20402, USA
Naval magazine.

'ARMS CONTROL ABSTRACTS'
Library of Congress
Washington, DC 20560, USA
Military journal.

'MILITARY MEDICINE'
P.O. Box 104
Kensington, MD 20795, USA
Military medical journal.

'U.S. NAVAL PROCEEDINGS'
U.S. Naval Institute
Annapolis, MD 21402, USA
(301)-268-6110
Journal of U.S. Naval Institute.
S6,S10;
Catalogue available. Bi-annually.
CONTACT: Capt. James A. Barber, USN(Ret)

SALES DEPT: Ms. Jean Tullier
PURCHASING DEPT: Ms. Jean Tullier
AD MANAGER: Mr. James Burke
The U.S. Naval Institute is the leading organization for naval and maritime professionals. The Naval Institute has 95,000 members worldwide who enjoy monthly issues of the Institute's highly respected magazine, 'PROCEEDINGS'.

'WORLD WAR II MAGAZINE'
105 Loudoun Street, S.W.
Leesburg, VA 22075, USA
703-771-9400
Bi-monthly military history magazine.

'MILITARY HISTORY MAGAZINE'
105 Loudoun Street, S.W.
P.O. Box 8
Leesburg, VA 22075, USA
703-771-9400
Bi-monthly military history magazine.
CONTACT: Adam Landis, Publisher

'MARINE CORPS GAZETTE'
P.O. Box 1775
Quantico, VA 22134, USA
Marine Corps media publication.

'LEATHERNECK'
P.O. Box 1775
Quantico, VA 22134, USA
U.S.M.C enlisted men's magazine.

'ARMY'
2425 Wilson Blvd.
Arlington, VA 22201, USA
Military service publication.

'NAVAL ABSTRACTS'
Center for Naval Analysis
2000 N. Beauregard Street
Alexandria, VA 22311, USA
Naval magazine.

'SOLDIERS'
Bldg. 2, Door 11
Cameron Station
Alexandria, VA 22314, USA
(202)-274-6671 (202)-274-6691
U.S. Army publication.
S13; 40;
CONTACT: HQDA/OSA/OCPA/CID

'NORTH SOUTH TRADER'
725 Caroline Street
Fredericksburg, VA 22401, USA
Civil War collectors & re-enactors magazine. 6 issues $18.00 annually.

'DIE RITTERKREUZTRAGER'
4125 Silbury Road
Richmond, VA 23234, USA
(804)-271-1864 (804)-275-9390
Third Reich collectors newsletter.
S13; 1;
CONTACT: Earl E. Cousins
SALES DEPT: Earl E. Cousins
PURCHASING DEPT: Bonita K. Cousins
AD MANAGER: Earl E. Cousins

'THE GUN GAZETTE'
P.O. Box 2685

Warren Robins, GA 31099, USA
912-922-3307
Firearms collectors magazine. 12 issues at $10.00 per year.

'INFANTRY'
ATSH-SE
Ft. Benning, GA 31905, USA
Army Infantry School magazine.

'8TH AF NEWS'
P.O. Box 3556
Hollywood, FL 33023, USA
Military veterans newsletter.

'WILDCAT'
15158 N.E. 6th Avenue
Miami, FL 33162, USA
(305)-944-2416
Gun & militaria publication. 12 issues $5.00.
S4,S11,S12;
CONTACT: Richard Cecilio
AD MANAGER: Joanne Cecilio
Monthly publication! Reaching all GUN and MILITARIA enthusiasts. Free 40 word ad to subscribers. "The best $5.00 value for collectors."

'MILITARY POLICE JOURNAL'
USAPS
Ft. McClellan, AL 36205, USA
Journal of the military police.

'THE ARMY FLIER'
Fifth Avenue, Bldg 121
Fort Rucker, AL 36362, USA
Army aviation magazine.

'U.S. ARMY AVIATION DIGEST'
P.O. Box 699
Fort Rucker, AL 36362, USA
Army aviation journal.
S13;

'ARMOR'
P.O. Box O
Ft. Knox, KY 40121, USA
Army armor magazine.

'CAMP CHASE GAZETTE'
3984 Cinn.-Zanesville Rd., N.E.
Lancaster, OH 43130, USA
(614)-653-5818
Civil War re-enactors magazine. 10 issues for $15.00 annually.
S4,S5,S12,S13; 14;
CONTACT: William P. Keitz
SALES DEPT: Pam Keitz
PURCHASING DEPT: Pam Keitz
AD MANAGER: Pam Keitz
The only Civil War periodical that gives a current & concise list of reenactments by date & state with info on who to contact. Reprints of CW stationary, envelopes, military forms & books.

'WARSHIP INTERNATIONAL'
1729 Lois Court
Toledo, OH 43613, USA
Contemporary military magazine. Sample copy $3.75.

MILITARY PERIODICALS

'GUN & KNIFE SHOW CALENDAR'
Rt. 2, Box 25
Sharpsville, IN 46068, USA
317-963-5282
Publisher. $9.00 per year.
CONTACT: Linda Piel, Secretary

'INDIANA COMBAT VETERAN'
1402 North Shadeland Avenue
Indianapolis, IN 46219, USA
Veterans magazine.

'BANZAI'
P.O. Box 141
Gowen, MI 49326, USA
Japanese militaria newsletter. Sample $3.00.

'EX-CBI ROUNDUP'
P.O. Box 102
Laurens, IA 50554, USA
WWII China-Burma-India veterans magazine.

'THE ANTIQUE TRADER WEEKLY'
P.O. Box 1050
Dubuque, IA 52001, USA
319-588-2073
General antique newspaper publisher. Published weekly.
CONTACT: E. A. Babka, Publisher

'SOUND-OFF'
P.O. Box 1848
Milwaukee, WI 53201, USA
Military magazine.

'COVERT NEWS'
Scanning Unlimited
9230 W. National Ave., #6
Milwaukee, WI 53227, USA
Covert newsletter.

'POLICE COLLECTORS NEWS'
RR #1, Box 14
Baldwin, WI 54002, USA
Police collectors newsletter.

'THE GUN REPORT'
P.O. Box 111
Aledo, IL 61231, USA
Firearms magazine

'THE REGIMENTAL OBSERVER'
6 Lambert Lane
Springfield, IL 62704, USA
(217)-546-2423
Civil War newsletter
S8,S9,S12; 7; 500;
CONTACT: John & Fran L. Satterlee
SALES DEPT: John L. Satterlee
PURCHASING DEPT: Frances H. Satterlee
AD MANAGER: Fran H. Satterlee

'LE MERCENAIRE'
P.O. Box 507
Fredericktown, MO 63645, USA
(314)-783-6965
Monthly newsletter on Intelligence. Sample $2.00.
S4,S9; 10;
CONTACT: George Ellis
SALES DEPT: G. Ellis
PURCHASING DEPT: G. Ellis
AD MANAGER: D. Ellis
Le Mercenaire is a monthly intelligence newsletter which concisely and effectively covers worldwide strategic news including terrorism, communist subversion and covert operations.

'VFW' MAGAZINE
Broadway & 34th Street
Kansas City, MO 64111, USA
Journal of the Veterans of Foreign Wars.

'SHOW ME LEGIONNAIRE'
P.O. Box 179
Jefferson City, MO 65102, USA
Military veterans magazine

'GUN LIST'
P.O. Box 7387
Columbia, MO 65205, USA
(314)-442-7788
Tabloid listing guns for sale. Monthly, advertising accepted.
S13; 3;
SALES DEPT: Sue Golden
AD MANAGER: Sue Golden
GUNLIST is a monthly publication with over 5,000 guns forsale from dealers, collectors and individuals nationwide.

'THE MILITARY REVIEW'
Funston Hall
Fort Leavenworth, KS 66027, USA
Army active service publication.

'SHOTGUN NEWS'
P.O. Box 669
Hastings, NE 68901, USA
Firearms advertising magazine.

'CARTRIDGE' NEWSLETTER
1912 Sandra Avenue
Metairie, LA 70003, USA
Cartridge collectors newsletter.

'ARKANSAS LEGIONAIRE'
1415 W. Seventh Street
Little Rock, AR 72201, USA
Military veterans magazine.

'SCABBARD & BLADE JOURNAL'
205 Thatcher Hall
Oklahoma State University
Stillwater, OK 74074, USA
Military magazine.

'BATTLEFLAG'
1832 Highland Drive
Carrollton, TX 75006, USA
Military magazine.

'DER GAULEITER'
P.O. Box 721288
Houston, TX 77272, USA
Monthly German militaria magazine.

'THE BACKWOODSMAN MAGAZINE'
Rt. 8, Box 579
Livingston, TX 77351, USA
(409)-566-4600
19th century backwoods survivalist magazine. 6 issues $11.00 annually.
S1,S2,S3,S4,S5,S6; 6;
CONTACT: Charlie Richie, Publisher
AD MANAGER: Lynne Richie
A magazine offering articles on woodslore, trapping, primitive survival, old-time military campaigns, 19th century how-to, muzzleloading, and Indian equipment.

'AIRMAN MAGAZINE'
Bldg 1509
Kelly AFB, TX 78241, USA
(512)-925-7757 (512)-925-7758
U.S.A.F. enlisted man's magazine.
CONTACT: HQ AFSINC/IIC

'DISPATCH MAGAZINE'
P.O. Box CAF
Harlingen, TX 78551, USA
Journal of The Confederate Air Force.

'TEXAS LEGION NEWS'
P.O. Box 789
Austin, TX 78767, USA
Veterans magazine.

'SURVIVE'
5735 Arapahoe Avenue
Boulder, CO 80303, USA
Survivalist magazine

'SOLDIER OF FORTUNE'
P.O. Box 693
Boulder, CO 80306, USA
Adventurers magazine.

'THE TALON'
Box 6066, U.S.A.F. Academy
Denver, CO 80840, USA
Air Force Academy magazine.

'MILITARY INTELLIGENCE'
P.O. Box 569
Ft. Huachuca, AZ 85613, USA
Active military service publication.

'OLYMPIC COLLECTORS NEWSLETTER'
Box 41630
Tucson, AZ 85717, USA
Collectors magazine.
CONTACT: W. A. Wilson

'RIFLE MAGAZINE'
6471 Airpark Drive
Prescott, AZ 86301, USA
(602)-445-7810
Firearms magazine.
S6; 20;
Catalogue available. Annually.
CONTACT: Dave Wolfe
SALES DEPT: Mark Harris
PURCHASING DEPT: Mark Harris
AD MANAGER: Sana Kosco

'HANDLOADER MAGAZINE'
6471 Airpark Drive
Prescott, AZ 86301, USA
(602)-445-7810
Technical firearms magazine.
S4; 20;
Catalogue available. Annually.
CONTACT: Dave Wolfe
SALES DEPT: Mark Harris
PURCHASING DEPT: Mark Harris
AD MANAGER: Sana Kosco

MILITARY PERIODICALS

'USAF FIGHTER WEAPONS REVIEW'
57 Fighter Weapons Wing
Nellis A.F.B., NV 89191, USA
Active military service publication.

'DISABLED AMERICAN VETERAN NEWS'
1816 S. Figeroa Street
Los Angeles, CA 90015, USA
Veterans magazine.

'BULLETIN, VETS OF A.E.F. SIBERIA'
316 N. Crescent Heights Blvd.
Los Angeles, CA 90048, USA
Veterans magazine.

'BULLETIN'
10224 Lareina Avenue
Downey, CA 90241, USA
Veterans magazine.
CONTACT: National Order of Trench Rats

'AIR COMBAT'
7950 Deering Avenue
Canoga Park, CA 91304, USA
Vintage military aviation magazine.

'AIR CLASSICS'
7950 Deering Avenue
Canoga Park, CA 91304, USA
Vintage aviation magazine.

'MILITARY MODELLER'
7950 Deering Avenue
Canoga Park, CA 91304, USA
Military modeling magazine.

'THE KEPI'
P.O. Box 699
Baldwin Park, CA 91706, USA
U.S. Civil War collector magazine. $18.00 annually for 6 issues.

'AFV-G2'
P.O. Box 820
La Puente, CA 91747, USA
Armor enthusiast magazine.

'AMERICAN HANDGUNNER'
Suite 200
591 Camine De La Reina
San Diego, CA 92108, USA
Firearms magazine.

'ABA NEWSLETTER'
P.O. Box 11247
San Diego, CA 92111, USA
Military magazine.

'THE PERISCOPE'
Commander, Submarine Force
U.S. Pacific Fleet, FSPO
San Diego, CA 92120, USA
Active Naval service publication.

'AAHS JOURNAL'
3614 Pendleton Avenue
Santa Ana, CA 92704, USA
Journal of American Aviation Historical Society.

'SEABEE COVERALL'
Naval Construction Battalion
Port Hueneme, CA 93043, USA
Active Naval service publication.

'BAYONET'
1st Brigade, 7th Inf. Div.
Fort Ord, CA 93941, USA
Active military service publication.

'390TH NEWSLETTER'
2060 Colorado Avenue
Turlock, CA 95380, USA
Military magazine.

'MILITARY'
2122 28th Street
Sacramento, CA 95818, USA
(916)-457-8990
Monthy recent military history magazine.
S13; 2;
CONTACT: Armond Noble
AD MANAGER: Helen Noble
MILITARY, a monthly magazine devoted to WWII, Korea, Viet-nam and today. Latest news, articles written by actual troopers. FREE sample copy.

'GRUNT'
P.O. Box 8717
Tamuning, GU 96911, USA
Military magazine.

'MINIE NEWS'
14 Sunlight Crescent
East Brighton, Victoria 3187, Australia
03-615-1088 03-592-4968 after 5
American Civil War Round Table of Australia. Monthly newsletter.
S10;
CONTACT: Barry Crompton

'MUFTI'
Anzac House
4 Collins Street
Melbourne, Victoria 3000, Australia
63-4571
Quarterly paper of The Returned Services League. Advertising accepted.
S10;
CONTACT: Miss E. Page, Sec. to Editor

'AUSTRALIAN GUNNER'
35 Willis Street
Hampton, Victoria 3188, Australia
Military magazine.

'CAPS & FLINTS'
110 Pellatt Street
Beaumaris, Victoria 3191, Australia
Military magazine.

'DEFENCE FORCE JOURNAL'
(Army Office)
Russell Offices
Canberra, ACT 2600, Australia
Military magazine.

'DISPATCH'
67 Beech Street
Coogee, NSW 2034, Australia
Military magazine.

'DOUBLE DIAMOND'
P.O. Box 81
Ashburton, Victoria 3147, Australia
Military magazine.

'DUTY FIRST'
Royal Australian Regiment
Victoria Barracks
Paddington, NSW 2021, Australia
Military magazine.

'GALLIPOLI GAZETTE'
12 Loftus Street
Sydney, NSW 2000, Australia
Military magazine.

'GREY & SCARLET'
35 Willis Street
Hampton, Victoria 3188, Australia
Military magazine.

'MINATURE MANEUVRES'
50 Clissold Parade
Campsie, NSW 2194, Australia
Military magazine.

'NULLI SECUNDUS'
24 Hall Street
Belmore, NSW 2192, Australia
Monthly magazine of the 2/2nd Australian Infantry Bn.

'ORDNANCE'
35 Willis Street
Hampton, Victoria 3188, Australia
Military magazine.

'PACIFIC DEFENCE REPORTER'
3 Queen's Street
Chippendale, NSW 2008, Australia
Military magazine.

'RAAF ACADEMY JOURNAL'
35 Willis Street
Hampton, Victoria 3188, Australia
Military magazine.

'ROTA'
25/441 Alfred Street
North Sydney, NSW 2060, Australia
Military magazine.

'SABRETACHE'
P.O. Box 67
Lyneham, ACT 2602, Australia
Journal of the Military Historical Society of Australia.

'WAR & SOCIETY'
RMS Duntroon, ACT 2600, Australia
Military magazine.

'NATO'S SIXTEEN NATIONS'
Intern'l Press Center, Box 61
1, Blvd. Charlemagne
Brussels, 1040, Belgium
(02) 230-2517
International Defense Industry journal. 6-8 issues per year. Ads accepted.

'SPORT INTERNATIONAL'
119 Ave. Franklin Roosevelt
Bruxelles, B-1050, Belgium
Military & shooting magazine.

MILITARY PERIODICALS

'TANK NEWS'
Square Van Bever 33
Brussels, 1180, Belgium
Journal of the Tank Museum. Quarterly, published in French.
S4;
CONTACT: Guy Van De Casteele, Sec.

'ARMY MOTORS'
P.O. Box 7279
Edmonton, Alberta T5E 6C8, CANADA
Quarterly journal of the Military Vehicle Collectors Club.

'ADSUM'
Edifice 513, Chambre 155
Forces Canadiennes, Valcartier
Courcelette, Quebec G0A 1R0, Canada
Active military personnel magazine.

'BAGOTVILLE PHARE BEACON'
Canadian Forces, Bagotville
Alouette, Quebec G0V 1A0, Canada
Active military personnel magazine. Advertising accepted.

'BORDEN CITIZEN'
Canadian Forces Base Borden
Borden, Ontario L0M 1C0, Canada
Active military personnel magazine. Accepts advertising, Photographic cap.

'CHATAIR'
P.O. Box 200
Canadian Forces, Chatham
Curtis Park, N.B. E0C 2E0, Canada
Active military personnel magazine. Advertising accepted.

'CONTACT'
Canadian Forces Base Trenton
Astra, Ontario K0K 1B0, Canada
Active military personnel magazine. Advertising accepted.

'COURIER'
Box 2350
Canadian Forces Base Cold Lake
Medley, Alberta T0A 2M0, Canada
Active military personnel magazine. Advertising accepted.

'DER KANADIER'
CFPO 5000
Canadian Forces Europe
Belleville, Ontario K0K 3R0, Canada
Active military personnel magazine. Advertising accepted.

'EAGLE'
Canadian Forces Station Masset
Masset, B.C. V0T 1M0, Canada
Active military personnel magazine. Advertising accepted.

'ENSIGN'
P.O. Box 280
Canadian Forces, Cornwallis
Cornwallis, Nova Scotia B0S 1H0, Canada
Active military personnel magazine. Advertising accepted.

'GAZETTE'
Room 114, Bldg. H-10
Canadian Forces, Gagetown
Oromocto, N.B. E0G 2P0, Canada
Active military personnel magazine. Advertising accepted.

'GULF WINGS'
Canadian Forces, Summerside
Slemon Park, P.E.I. C0B 2A0, Canada
Active military personnel magazine. Advertising accepted.

'L'ALAIN'
Forces Canadiennes, St. Jean
Richelain, Quebec J0J 1R0, Canada
Active military personnel magazine. Advertising accepted.

'LE RAMPART'
College Militaire
Saint-Jean, Quebec J0J 1R0, Canada
Active military personnel magazine. Advertising accepted.

'LE/THE PARAPET'
Forces Canadiennes, Montreal
St. Hubert, Quebec J3Y 5T4, Canada
Active military personnel magazine. Advertising accepted.

'LEGION MAGAZINE'
Canvet Publications, Ltd.
359 Kent Street, Suite 504
Ottawa, Ontario K2P 0R7, Canada
613-235-8741
Monthly Canadian veterans magazine. Advertising accepted.

'LOOKOUT'
Canadian Forces Base Esquimalt
Victoria, B.C. V0S 1B0, Canada
Active military personnel magazine. Accepts advertising.

'MONCTON PROVIDER'
P.O. Box 190
Canadian Forces Base, Moncton
Moncton, N.B. E1C 8L2, Canada
Active military personnel magazine. Advertising accepted.

'NATIONAL NEWSLETTER'
223 Sayward Building
1207 Douglas Street
Victoria, B.C. V8W 2E7, Canada
384-1331
Newsletter of F.M.U.S.I of Canada.
CONTACT: D. C. Smith, Sec.-Treasurer

'PETAWAWA BASE POST'
Canadian Forces Base Petawawa
Petawawa, Ontario K8H 2X3, Canada
Active military personnel magazine. Advertising accepted.

'SEALANDAIR'
Canadian Forces Base Edmonton
Lancaster Park, Alberta T0A 2H0, Canada
Active military personnel magazine. Advertising accepted.

'STAG'
Canadian Forces Base Shilo
Shilo, Manitoba R0K 2A0, Canada
Active military personnel magazine. Accepts advertising.

'THE 'BRIDGE'
Canadian Forces, Falconbridge
Ridgeview, Ontario P0M 2T0, Canada
Active military personnel magazine. Advertising accepted.

'THE ARGUS'
P.O. Box 99
Canadian Forces Base Greenwood
Greenwood, Nova Scotia B0P 1N0, Canada
Active military personnel magazine. Advertising accepted.

'THE DISPATCH'
Canadian Forces Base Toronto
Downsview, Ontario M3K 1Y6, Canada
Active military personnel magazine. Advertising accepted.

'THE FALCON'
Canadian Forces Base Ottawa
Ottawa, Ontario K1A OK5, Canada
Active military personnel magazine. Accepts advertising.

'THE GUNRUNNER'
Box 565
Lethbridge, Alberta T1J 3Z4, Canada
403-327-3030
Firearms collectors magazine. Monthly. Advertising accepted.

'THE MOUNTAINEER'
MPO 612
Canadian Forces, Chilliwack
Vedder Crossing, B.C. V0X 2E0, Canada
Active military personnel magazine. Advertising accepted.

'THE PLAINSMAN'
P.O. Box 240
Canadian Forces Base Moose Jaw
Bushell Park, Sask. S0H 0N0, Canada
Active military personnel magazine. Advertising accepted.

'THE ROUNDUP'
Currie Barracks, Bldg. D-2
Canadian Forces Base Calgary
Calgary, Alberta T3E 1T8, Canada
Active military personnel magazine. Advertising accepted.

'THE SHIELD'
Canadian Forces Base North Bay
Hornell Heights, Ontario P0H 1P0, Canada
Active military personnel magazine. Advertising accepted.

'THE WARRIOR'
P.O. Box 190
Canadian Forces, Shearwater
Shearwater, Nova Scotia B0J 3A0, Canada
Active military personnel magazine. Advertising accepted.

MILITARY PERIODICALS

'TOTEM TIMES'
Canadian Forces Base Comox
Lazo, B.C. V0R 2K0, Canada
Active military personnel magazine. Advertising accepted.

'TOWER TIMES'
Canadian Forces Base Kingston
Kingston, Ontario K7L 2Z2, Canada
Active military personnel magazine. Advertising accepted.

'TRIDENT'
P.O. Box 3308
Halifax South, Nova Scotia B3J 3JI, Canada
Active military personnel magazine. Advertising accepted.

'VOXAIR'
Canadian Forces Base Winnipeg
Westwin, Manitoba R2R OTO, Canada
204-889-3963
Active military personnel magazine. Bi-weekly, accepts advertising.
CONTACT: Carl L. Fitzpatrick, Editor

'CAHS JOURNAL'
P.O. Box 224
Willowdale, Ontario M2N 5S8, Canada
Military magazine.

'CANADIAN GUNNER'
CFB Shilo, Manitoba R0K 2A0, Canada
Military magazine.

'CANADIAN MILITARY ENGINEER'
National Defence HQ
DGMEO
Ottawa, Ontario K1A 0K2, Canada
Military magazine.

'CANADIAN MILITARY JOURNAL'
3450 Durocher St., Suite 8
Montreal, Quebec H2X 2E1, Canada
Military magazine.

'C.S.M.M.I. JOURNAL'
P.O. Box 38
West Hill, Ontario M1E 4R4, Canada
Quarterly journal of the Canadian Society of Military Medals & Insignia.

'DESERT RATS NEWSLETTER'
58 Langton Road
London, Ontario M5N 2MI, Canada
Military magazine.

'FLIGHTLINES'
P.O. Box 35
Mount Hope, Ontario L0R IW0, Canada
Canadian warplane quarterly magazine.

'GUN TALK'
P.O. Box 1334
Regina, Saskatchewan S4P 3B8, Canada
Firearms magazine.

'MEDALS & INSIGNIA'
P.O. Box 1263
Guelph, Ontario N1H 6H6, Canada
Quarterly journal of the Canadian Society of Military Medals & Insignia.

'NEWSLETTER'
Imperial Veterans in Canada
280 St. Mary Avenue
Winnipeg, Manitoba R3C 0M6, Canada
Newsletter of the Imperial Veterans in Canada.

'O.M.M.C. JOURNAL'
330 Sussex Drive
Ottawa, Ontario K1A 0M8, Canada
Journal of the O.M.M.C.

'PERSONNEL NEWSLETTER'
Office of Personnel
National Defence Headquarters
Ottawa, Ontario Canada
Interoffice newsletter.

'R.C.M.I. YEARBOOK'
426 University Avenue
Toronto, Ontario M5G 1S9, Canada
Annual of the R.C.M.I.

'SENTINEL'
Room 1524
100 Metcalf Street
Ottawa, Ontario K1A 0K2, Canada
Military magazine.

'THE DOSSIER'
11 Gilmore Drive
Brampton, Ontario L6V 3K3, Canada
Military magazine.

'WARBIRDS'
P.O. Box 429
Lacombe, Alberta Canada
Magazine of The Western Warbirds Ass'n.

'WARGAMER'
8635 Gilley Avenue
Burnaby, BC V5J A21, Canada
Military magazine.

'CANADIAN DEFENCE QUARTERLY'
Defence Publications Ltd.
310 Dupont Street
Toronto, Ontario M5R 1V9, Canada
(416) 364-9344
Quarterly journal of the Canadian defence industry. Ads accepted.

'AN COSANTOIR'
Parkgate
Dublin, 8, Eire
771881
The Defence Forces monthly magazine. Advertising accepted.
CONTACT: Advertising Coordinator

'THE IRISH SWORD'
Newman House
86 St. Stephen's Green
Dublin, 2, Eire
Military magazine.
CONTACT: University College of Dublin

'HOMMES LIBRES'
36 Rue du Commerce
Paris, France
Military magazine.

'KEPI BLANC'
B.P. 78
Aubagne, 13673, France
Magazine of the French Foreign Legion.
CONTACT: La Legion Etranger

'NEWSLETTER'
C/O University of Paul Valery
Montpelier, Cedex F-34032, France
Newsletter of the International Commission of Military History.

'REVUE HISTORIQUE DES ARMES'
Dept. de l'Info. Militaire
6 Rue St. Charles
Paris, 75015, France
Journal of the Department of Military Information.

'1st THE QUEEN'S DRAGOON GUARDS'
Combined Service Publications
P.O. Box 4
Farnborough, Hampshire GU14 7LR, Great Britain STD (0252) 515891
Journal of 1st Queen's Dragoon Guards. Annual, accepts advertising.

'4th/7th ROYAL DRAGOON GUARDS'
Combined Service Publications
P.O. Box 4
Farnborough, Hampshire GU14 7LR, Great Britain STD (0252) 515891
Journal of 4th/7th Royal Dragoon Guards. Annual, accepts advertising.

'5th ROYAL INNISKILLING DRAGOON GUARDS'
Combined Service Publications
P.O. Box 4
Farnborough, Hampshire GU14 7LR, Great Britain STD (0252) 515891
Annual journal of the 5th Royal Inniskilling Dragoon Guards.

'A.A.C. JOURNAL'
Combined Service Publications
P.O. Box 4
Farnborough, Hampshire GU14 7LR, Great Britain STD (0252) 515891
Annual journal of the Army Air Corps. Advertising accepted.

'ACORN'
Combined Service Publications
P.O. Box 4
Farnborough, Hampshire GU14 7LR, Great Britain STD (0252) 515891
Annual journal of The Life Guards. Advertising accepted.

'AFTER THE BATTLE'
3 New Plaistow Road
Stratford
London, E15 3JA, Great Britain
01-534-8833
Military photo-history magazine. No advertising accepted.
CONTACT: W. G. Ramsey, Editor

MILITARY PERIODICALS

'ARMED FORCES'
Ian Allan, Ltd.
Coombelands House, Addlestone
Weybridge, KT15 1HY, Great Britain
0932-58511
Monthly British Armed Forces & NATO magazine. Circ. 30M. Ads accepted.
CONTACT: Patricia Bramley, Ad manager

'ARMS & ARMOUR SOCIETY JOURNAL'
162 Marsh Lane
Stanmore, Middlesex HA7 4RU, Great Britain
Journal of The Arms & Armour Society. Bi-annual, Circ. 800, Ads accepted.
CONTACT: M. J. Sarche, Ad manager

'ARMY MEDICAL SERVICE MAGAZINE'
Combined Service Publications
P.O. Box 4
Farnborough, Hampshire GU14 7LR, Great Britain STD (0252) 515891
Magazine of the Army Medical Service. Tri-annual, advertising accepted.

'ARMY MUSEUM'
National Army Museum
Royal Hospital Road
London, SW3 4HT, Great Britain
01-730-0717 ext. 49
National Army Museum annual magazine. Annual, No advertising accepted.
S4;
CONTACT: Miss E. Talbot Rice, Info Off.

'ARMY QUARTERLY'
1 West Street
Tavistock, Devonshire PL19 8DS, Great Britain
0822-3577/2785
Army Quarterly & Defence Journal. Advertising accepted. Circ. 20,000.
CONTACT: P. Tinson

'BACK BADGE'
Combined Service Publications
P.O. Box 4
Farnborough, Hampshire GU14 7LR, Great Britain STD (0252) 515891
Bi-annual journal of The Gloucestershire Regiment. Advertising accepted.

'BLUE & ROYAL'
Combined Service Publications
P.O. Box 4
Farnborough, Hampshire GU14 7LR, Great Britain STD (0252) 515891
Annual journal of The Blues & Royals. Advertising accepted.

'BRITANNIA MAGAZINE'
Combined Service Publications
P.O. Box 4
Farnborough, Hampshire GU14 7LR, Great Britain STD (0252) 515891
The Royal Naval College, Dartmouth. Tri-annual, advertising accepted.

'BRITISH ARMY REVIEW'
1st Avenue House, Room 205
40/48 High Holborn
London, WC1V 6HE, Great Britain
01-430-7478
Official publication of the British Army. No advertising accepted.

CONTACT: H. B. C. Watkins, Editor

'CADET JOURNAL & GAZETTE'
Combined Service Publications
P.O. Box 4
Farnborough, Hampshire GU14 7LR, Great Britain STD (0252) 515891
Bi-monthly magazine of Army Cadet Force Association. Advertising accepted.

'CASTLE NEWSPAPER'
Combined Service Publications
P.O. Box 4
Farnborough, Hampshire GU14 7LR, Great Britain STD (0252) 515891
Royal Anglian Regiment newspaper. Tri-annual, accepts advertising.

'DELHI SPEARMAN'
Combined Service Publications
P.O. Box 4
Farnborough, Hampshire GU14 7LR, Great Britain
Journal of 9th/12th The Royal Lancers. Annual, accepts advertising.

'DUKE OF EDINBURGH'S ROYAL REGIMENT JOURNAL'
Combined Service Publications
P.O. Box 4
Farnborough, Hampshire GU14 7LR, Great Britain STD (0252) 515891
Annual journal of the Duke of Edinburgh Royal Regiment. Advertising accepted.

'FALLING LEAF'
21, Metchley Lane, Harborne
Birmingham, B17 OHT, Great Britain
021-426-2915
Quarterly journal of The Psywar Society. Advertising accepted.
CONTACT: Rod Oakland, Sec.

'FIRM & FORESTER'
Combined Service Publications
P.O. Box 4
Farnborough, Hampshire GU14 7LR, Great Britain STD (0252) 515891
Bi-annual magazine of Worcestershire & Sherwood Foresters. Ads accepted.

'FUSILIER'
Combined Service Publications
P.O. Box 4
Farnborough, Hampshire GU14 7LR, Great Britain STD (0252) 515891
Journal of Royal Regiment of Fusiliers. Bi-annual, accepts advertising.

'GATE'
Combined Service Publications
P.O. Box 4
Farnborough, Hampshire GU14 7LR, Great Britain STD (0252) 515891
Journal of R.C.S. Apprentices College. Tri-annual, accepts advertising.

'GREEN HOWARDS' GAZETTE'
Combined Service Publications
P.O. Box 4
Farnborough, Hampshire GU14 7LR, Great Britain STD (0252) 515891
Quarterly magazine of the Green Howards.

Advertising accepted.

'GUNNER'
Combined Service Publications
P.O. Box 4
Farnborough, Hampshire GU14 7LR, Great Britain STD (0252) 515891
Monthly magazine of the Royal Artillery. Advertising accepted.

'GUNS REVIEW'
Ravenhill Publishing Co., Ltd.
Standard House, Bonhill St.
London, EC2A 4DA, Great Britain
01-628-4741
Monthly magazine of guns & shooting. Advertising accepted. Circ. 20,000.
CONTACT: Alan Burfoot, Ad manager

'HAWK'
Combined Service Publications
P.O. Box 4
Farnborough, Hampshire GU14 7LR, Great Britain STD (0252) 515891
Journal of the 14th/20th King's Hussars. Annual, advertising accepted.

'HONOURABLE ARTILLERY COMPANY'
Combined Service Publications
P.O. Box 4
Farnborough, Hampshire GU14 7LR, Great Britain STD (0252) 515891
Journal of Honourable Artillery Company. Bi-annual, advertising accepted.

'IRISH RANGERS NEWSPAPER'
Combined Service Publications
P.O. Box 4
Farnborough, Hampshire GU14 7LR, Great Britain STD (0252) 515891
Newspaper of Royal Irish Rangers. Quarterly, advertising accepted.

'IRON DUKE'
Combined Service Publications
P.O. Box 4
Farnborough, Hampshire GU14 7LR, Great Britain STD (0252) 515891
Magazine of Duke of Wellington's Regiment. Tri-annual, Ads accepted.

'LION & THE DRAGON'
Combined Service Publications
P.O. Box 4
Farnborough, Hampshire GU14 7LR, Great Britain STD (0252) 515891
The King's Own Royal Border Regiment. Annual, advertising accepted.

'LONDON SCOTTISH REGIMENT GAZETTE'
Combined Service Publications
P.O. Box 4
Farnborough, Hampshire GU14 7LR, Great Britain STD (0252) 515891
Gazette of The London Scottish Regiment. Quarterly, advertising accepted.

MILITARY PERIODICALS

'M.P.S.C. JOURNAL'
Combined Service Publications
P.O. Box 4
Farnborough, Hampshire GU14 7LR, Great Britain STD (0252) 515891
Journal of Military Provost Staff Corps. Bi-annual, advertising accepted.

'MARS & MINERVA'
Combined Service Publications
P.O. Box 4
Farnborough, Hampshire GU14 7LR, Great Britain STD (0252) 515891
Journal of Special Air Service Regiment. Bi-annual, advertising accepted.

'MEN OF HARLECH'
Combined Service Publications
P.O. Box 4
Farnborough, Hampshire GU14 7LR, Great Britain STD (0252) 515891
Journal of The Royal Regiment of Wales. Bi-annual, advertising accepted.

'MILITARY ILLUSTRATED'
169 Seven Sisters Road
Petersfield, Hants. GU32 2JF, Great Britain
0730 63976
Military collecting & history magazine. Bi-monthly, advertising accepted.
CONTACT: Martin Windrow, Editor

'MIND BODY & SPIRIT'
Combined Service Publications
P.O. Box 4
Farnborough, Hampshire GU14 7LR, Great Britain STD (0252) 515891
Journal of Army Physical Training Corps. Annual, advertising accepted.

'NAVY NEWS'
H.M.S. Nelson
Portsmouth, PO1 3HH, Great Britain
0705 826040 ex 24226 0705 822351 ex 24226
Monthly newspaper of The Royal Navy. Advertising accepted. Circ. 10,000.
CONTACT: Miss M. K. Brown, Manager

'OAK TREE'
Combined Service Publications
P.O. Box 4
Farnborough, Hampshire GU14 7LR, Great Britain STD (0252) 515891
The Cheshire Regiment magazine. Bi-annual, advertising accepted.

'OWL PIE'
Combined Service Publications
P.O. Box 4
Farnborough, Hampshire GU14 7LR, Great Britain STD (0252) 515891
Magazine of the Staff College Camberley. Annual, advertising accepted.

'P.W.O.'
Combined Service Publications
P.O. Box 4
Farnborough, Hampshire GU14 7LR, Great Britain STD (0252) 515891
Annual journal of the Royal Hussars. Advertising accepted.

'PEGASUS'
Combined Service Publications
P.O. Box 4
Farnborough, Hampshire GU14 7LR, Great Britain STD (0252) 515891
Journal of the Parachute Regiment. Tri-annual, accepts advertising.

'PROVOST PARADE'
Combined Service Publications
P.O. Box 4
Farnborough, Hampshire GU14 7LR, Great Britain STD (0252) 515891
Journal of the Royal Air Force Police. Bi-annual, advertising accepted.

'Q.M.O.'
Combined Service Publications
P.O. Box 4
Farnborough, Hampshire GU14 7LR, Great Britain STD (0252) 515891
Journal of 13th/18th Royal Hussars. Annual, advertising accepted.

'QUEEN'S OWN HIGHLANDER'
Combined Service Publications
P.O. Box 4
Farnborough, Hampshire GU14 7LR, Great Britain STD (0252) 515891
Journal of the Seaforth & Camerons. Bi-annual, advertising accepted.

'QUEEN'S OWN HUSSARS'
Combined Service Publications
P.O. Box 4
Farnborough, Hampshire GU14 7LR, Great Britain STD (0252) 515891
Annual journal of Queen's Own Hussars. Advertising accepted.

'QUEEN'S REGIMENT JOURNAL'
Combined Service Publications
P.O. Box 4
Farnborough, Hampshire GU14 7LR, Great Britain STD (0252) 515891
Journal of the Queen's Regiment. Bi-annual, advertising accepted.

'R.A.Ch.D.'
Combined Service Publications
P.O. Box 4
Farnborough, Hampshire GU14 7LR, Great Britain STD (0252) 515891
Journal of Royal Army Chaplains' Department. Bi-annual, Ads accepted.

'R.A.M.C. JOURNAL'
Combined Service Publications
P.O. Box 4
Farnborough, Hampshire GU14 7LR, Great Britain STD (0252) 515891
Journal of the Royal Army Medical Corps. Tri-annual, advertising accepted.

'R.A.O.C. GAZETTE'
Combined Service Publications
P.O. Box 4
Farnborough, Hampshire GU14 7LR, Great Britain STD (0252) 515891
Monthly magazine of the R.A.O.C. Advertising accepted.

'R.A.P.C. JOURNAL'
Combined Service Publications
P.O. Box 4
Farnborough, Hampshire GU14 7LR, Great Britain STD (0252) 515891
Journal of the Royal Army Pay Corps. Tri-annual, advertising accepted.

'R.E. JOURNAL'
Combined Service Publications
P.O. Box 4
Farnborough, Hampshire GU14 7LR, Great Britain STD (0252) 515891
Journal of the Royal Engineers. Quarterly, advertising accepted.

'R.E.M.E. JOURNAL'
Combined Service Publications
P.O. Box 4
Farnborough, Hampshire GU14 7LR, Great Britain STD (0252) 515891
Annual journal of the R.E.M.E. Advertising accepted.

'R.M.P. JOURNAL'
Combined Service Publications
P.O. Box 4
Farnborough, Hampshire GU14 7LR, Great Britain STD (0252) 515891
Journal of the Royal Military Police. Quarterly, advertising accepted.

'ROYAL ARTILLERY JOURNAL'
Combined Service Publications
P.O. Box 4
Farnborough, Hampshire GU14 7LR, Great Britain STD (0252) 515891
Journal of the Royal Artillery. Bi-annual, advertising accepted.

'ROYAL HAMPSHIRE REGIMENT JOURNAL'
Combined Service Publications
P.O. Box 4
Farnborough, Hampshire GU14 7LR, Great Britain STD (0252) 515891
Journal of the Royal Hampshire Regiment. Bi-annual, advertising accepted.

'ROYAL HIGHLAND FUSILIERS'
Combined Service Publications
P.O. Box 4
Farnborough, Hampshire GU14 7LR, Great Britain STD (0252) 515891
Journal of Royal Highland Fusiliers. Bi-annual, advertising accepted.

'ROYAL PIONEER'
Combined Service Publications
P.O. Box 4
Farnborough, Hampshire GU14 7LR, Great Britain STD (0252) 515891
Magazine of the Royal Pioneer Corps. Tri-annual, advertising accepted.

'ROYAL SIGNALS JOURNAL'
Combined Service Publications
P.O. Box 4
Farnborough, Hampshire GU14 7LR, Great Britain STD (0252) 515891
Journal of the Royal Corps of Signals. Tri-annual, advertising accepted.

MILITARY PERIODICALS

'ROYAL WELCH FUSILIERS JOURNAL'
Combined Service Publications
P.O. Box 4
Farnborough, Hampshire GU14 7LR, Great Britain STD (0252) 515891
Journal of the Royal Welch Fusiliers. Annual, advertising accepted.

'SILVER BUGLE'
Combined Service Publications
P.O. Box 4
Farnborough, Hampshire GU14 7LR, Great Britain STD (0252) 515891
Journal of The Light Infantry. Bi-annual, advertising accepted.

'STAFFORD KNOT'
Combined Service Publications
P.O. Box 4
Farnborough, Hampshire GU14 7LR, Great Britain STD (0252) 515891
Journal of The Staffordshire Regiment. Bi-annual, advertising accepted.

'SUSTAINER'
Combined Service Publications
P.O. Box 4
Farnborough, Hampshire GU14 7LR, Great Britain STD (0252) 515891
Army Catering Corps Journal. Bi-annual, advertising accepted.

'TAMIYA MODEL MAGAZINE INT'L'
71-B Maple Road
Surbiton, Surrey KT6 4AG, Great Britain
01-399-9956/9954
Magazine for model enthusiasts. Quarterly, advertising accepted.
CONTACT: Chris Ellis, Editor

'THE CRAFTSMAN'
Combined Service Publications
P.O. Box 4
Farnborough, Hampshire GU14 7LR, Great Britain STD (0252) 515891
Monthly magazine of the R.E.M.E. Advertising accepted.

'THE GUARDS MAGAZINE'
Room 11
Horse Guards, Whitehall
London, SW1A 2AX, Great Britain
01-930-4466 ex 2499
Journal of the Household Division. Quarterly, advertising accepted.
CONTACT: The Treasurer

'THE KINGSMAN'
Combined Service Publications
P.O. Box 4
Farnborough, Hampshire GU14 7LR, Great Britain STD (0252) 515891
King's Regiment, Manchester & Liverpool annual magazine. Advertising accepted.

'THE LANCASHIRE LAD'
Combined Service Publications
P.O. Box 4
Farnborough, Hampshire GU14 7LR, Great Britain STD (0252) 515891
Journal of Queen's Lancashire Regiment. Bi-annual, advertising accepted.

'THE LEGION'
48 Pall Mall
London, SW1Y 5JY, Great Britain
01-930-8533 01-930-8131
Magazine of the Royal British Legion. Bi-monthly, Ads accepted. Circ. 26,000.
CONTACT: Rob C. D. Truefitt, Ad mgr.

'THE NAVAL REVIEW'
Cornhill House, The Hangers
Bishop's Waltham
Southampton, SO3 1EF, Great Britain
04893-2656
Magazine of a British naval society. Quarterly, advertising accepted.
CONTACT: Rear Admiral J. R. Hill, Editor

'THE PENNANT'
15 Buckingham Gate
London, SW1E 6NS, Great Britain
01-834-0853 01-828-2508
Journal of The Officers' Pensions Society. Bi-annual, Ads accepted.
CONTACT: L. W. A. Gingell, Editor

'THE SAPPER'
Geerings of Ashford Ltd.
Cobbs Wood House, Chart Road
Ashford, Kent TN23 1EP, Great Britain
0233-33366
Journal of the Corps of Royal Engineers. Bi-monthly, Ads accepted. Circ. 7,500.
CONTACT: Debbie Pope, Ad manager

'THE WHITE LANCER & VEDETTE'
Combined Service Publications
P.O. Box 4
Farnborough, Hampshire GU14 7LR, Great Britain STD (0252) 515891
Annual journal of the 17th/21st Lancers. Advertising accepted.

'THIN RED LINE'
Combined Service Publications
P.O. Box 4
Farnborough, Hampshire GU14 7LR, Great Britain STD (0252) 515891
Argyll & Sutherland Highlanders bi-annual magazine. Ads accepted.

'THISTLE'
Combined Service Publications
P.O. Box 4
Farnborough, Hampshire GU14 7LR, Great Britain STD (0252) 515891
Bi-annual journal of the Royal Scots. Advertising accepted.

'TORCH'
Combined Service Publications
P.O. Box 4
Farnborough, Hampshire GU14 7LR, Great Britain STD (0252) 515891
Royal Army Educational Corps bi-annual journal. Advertising accepted.

'WHITE ROSE'
Combined Service Publications
P.O. Box 4
Farnborough, Hampshire GU14 7LR, Great Britain STD (0252) 515891
Prince of Wales's Own Regiment of Yorkshire tri-annual magazine. Ads, yes.

'WIRE'
Combined Service Publications
P.O. Box 4
Farnborough, Hampshire GU14 7LR, Great Britain STD (0252) 515891
Magazine of the Royal Corps of Signals. Bi-monthly, advertising accepted.

'WISH STREAM'
Combined Service Publications
P.O. Box 4
Farnborough, Hampshire GU14 7LR, Great Britain STD (0252) 515891
Royal Military Academy Sandhurst bi-annual magazine. Ads accepted.

'15th/19th THE KING'S ROYAL HUSSARS'
Combined Service Publications
P.O. Box 4
Farnborough, Hampshire GU14 7LR, Great Britain STD (0252) 515891
15th/19th The King's Royal Hussars annual journal. Advertising accepted.

'ACCESSIONS LIST'
Imperial War Museum
Lambeth Road
London, SE1, Great Britain
Military magazine.
CONTACT: Department of Printed Books

'AERO MART'
Marlborough Road
Ipswich, Great Britain
Military magazine.

'AIR PICTORIAL'
65 Victoria Street
Windsor, Berkshire Great Britain
Military aviation magazine.

'AIRFIX'
4 Surbiton Hall Close
Kingston, Surrey Great Britain
Military modelers magazine.

'AIRSHIP ASSOCIATION'
4 Caledon Road
Beaconsfield, Buckingham Great Britain
Military magazine.
CONTACT: D. C. P. Willmer

'ARMOURY HOUSE'
London, EC1 Y2BQ, Great Britain
Military magazine.

'ARMY AIR CORPS JOURNAL'
Combined Service Publications
273 Farnborough Road
Farnborough, Hants GU14 7LY, Great Britain
Annual journal of the Army Air Corps. Advertising accepted.

'ARMY ORDERS'
HMSO, P.O. Box 569
London, SE1, Great Britain
Military magazine.

'ARMY VEHICLE DATA SHEETS'
Sussex House, Parkside
Wimbledon
London, SW1 95NB, Great Britain
Military magazine.

MILITARY PERIODICALS

'AVIATION ARCHAEOLOGIST'
154 Hardhorn Road
Poulton-le-Fylde, Lancasters. Great Britain
Military aviation magazine.

'AVIATION ARTIST NEWSLETTER'
The Guild of Aviation Artists
11 Great Spilmans
London, SW22 SZL, Great Britain
Journal of The Guild of Aviation Artists.

'BLACKHORN'
P.O. Box 4
Farnborough, Hampshire Great Britain
Military magazine.

'BORDERERS CHRONICLE'
RHQ, The Barracks
Berwick-on-Tweed, Great Britain
Military magazine.

'BRITISH MODEL SOLDIERS SOCIETY BULLETIN'
16 Charlton Road, Kenton
Harrow, Middlesex Great Britain
Journal of the British Model Soldiers Society.

'BULLETIN'
The Duke of Yorks Headquarters
Chelsea
London, SW2, Great Britain
Journal of Military & Historical Society.
CONTACT: Military & Historical Society

'CATERPILLAR'
7 Welbourne Walk
Hull, N.Humberside HU4 7S7, Great Britain
Military magazine.

'CLAYMORE'
New Haig House
Logi Green Road
Edinburgh, EH7 4HR, Great Britain
Military magazine.

'CONTROL COLUMN'
127 Hawton Road
Newark, Notts NG2 44QG, Great Britain
Military magazine.

'COVANANTER'
P.O. Box 4
Farnborough, Hampshire Great Britain
Military magazine.

'CROSS & COCKADE JOURNAL'
31 Holly Road, Cove
Farnborough, Hampshire GU1 4OEA, Great Britain
Journal of the British Society of WWI Aero Historians.

'DEFENCE MATERIAL'
30 Fleet Street
London, EC4 YLAH, Great Britain
Military magazine.

'DEVONSHIRE & DORSET REGIMENTAL JOURNAL'
Combined Service Publications
P.O. Box 4
Farnborough, Hampshire GU14 7LR, Great Britain STD (0252) 515891
Journal of Devonshire & Dorset Regiment.
Bi-annual, advertising accepted.

'DIEHARDS NEWSLETTER'
Lynsore Bungalow
Upper Hardes
Canterbury, Kent CT4 6EE, Great Britain
Military newsletter.

'FLYPAST'
1 Wothorpe Road
Stamford, Lincs. PE9 2JR, Great Britain
Military aviation magazine.

'FORTRESS STUDY GROUP BULLETIN'
24 Walters Road
Hoo
Rochester, Kent ME3 9JR, Great Britain
Military magazine.

'THE GLOBE AND LAUREL'
Royal Marines Eastney
Southsea, Hampshire PO4 9PX, Great Britain
Portsmouth 822351
The Journal of The Royal Marines. Bi-monthly, accepts advertising.

'HISTORICAL BREECHLOADING'
C/O Imperial War Museum
Lambeth Road
London, SE6, Great Britain
Firearms magazine.
CONTACT: Small Arms Journal

'INTERCOM'
4 Kingsman Street
Woolwich
London, SE1 8SPT, Great Britain
Military magazine.

'JOURNAL'
RAF Museum
Hendon, NW9 5LL, Great Britain
Journal of the RAF Museum.

'JOURNAL OF SOCIETY FOR ARMY HISTORICAL RESEARCH'
Old War Office Building
London, WC1, Great Britain
Military magazine.

'DISPATCH'
11 Hamilton Avenue
Pollokshields
Glasgow, G41 4JG, Great Britain
041-427-4638
Scottish Military Coll. Soc. Journal. Advertising & inserts accepted.

'LIONESS'
Queen Elizabeth Park
Guildford, Surrey Great Britain
Journal of the WRAC Association.
CONTACT: WRAC Association

'MILITARY AIRCRAFT DATA SHEETS'
Sussex House, Parkside
Wimbledon
London, SW1 95NB, Great Britain
Military aviation magazine.

'MILITARY BALANCE'
23 Tavistock Street
London, WC2E 7NQ, Great Britain
01-379 7676
Annual military assessment. Internat'l Institute of Strategic Studies.

'MILITARY MODELLING'
P.O. Box 35
Hemel Hempstead, Herts. HP1 IEE, Great Britain
Military modelling magazine.

'NAVAL FORCES'
Monch (UK),Ltd.
84 Alexandra Road
Farnborough, Hants GU14 6DD, Great Britain
(02 52) 51 79 74
International forum for maritime power. Bi-monthly, advertising accepted.

'NAVY'
Broadway House
Broadway
Wimbledon, London W19, Great Britain
Naval magazine.

'READY'
75 High
Aldershot, Hampshire GU11 1BY, Great Britain
Military magazine.

'RED HACKLE'
P.O. Box 4
Farnborough, Hampshire Great Britain
Military magazine.

'ROYAL AIR FORCE NEWS'
94 High
Holborne
London, WC1 V6LL, Great Britain
RAF journal.
CONTACT: Ministry of Defence

'ROYAL ARMY PAY CORPS JOURNAL'
Deepcut
Camberly, Surrey Great Britain
Journal of the Royal Army Pay Corps.

'ROYAL ENGINEERS JOURNAL'
Chatham, Great Britain
Journal of the Royal Engineers.

'ROYAL MILITARY POLICE JOURNAL'
RHQ, Roussellon Barracks
Chichester, West Sussex Great Britain
Journal of the Royal Military Police.

'RUSI JOURNAL'
Institute For Defence Studies
Whitehall
London, SW1A 2ET, Great Britain
Military magazine.

'SCOTS GUARDS MAGAZINE'
RHQ,Bloomsbury Court
High Holborne
London, WC1 A2BW, Great Britain
Magazine of the Scots Guard.

'STAND TO!'
Guilton Mill, Poulton Lane
Ash
North Canterbury, Kent CT3 2HN, Great Britain
Military magazine.

MILITARY PERIODICALS

'TANK'
Geerings of Ashford Ltd.
Cobbs Wood House, Chart Road
Ashford, Kent TN23 1BR, Great Britain
(0233) 33366
Service journal of the Royal Tank Regiment.
S13;
CONTACT: Debbie Pope, Ad. Mgr.
AD MANAGER: Debbie Pope

'TANKETTE'
15 Berwick Ave. Heaton Mersey
Stockport, Cheshire SK4 3AA, Great Britain
Journal of Miniature A.F.V. Association.
Bi-monthly, inserts accepted.
CONTACT: G. E. G. Williams, Sec.

'THE RBL JOURNAL'
Pall Mall
London, SW1 Y5JY, Great Britain
Military magazine.

'THE ROSE & LAUREL'
Templer Barracks
Ashford, Kent TN23 3HH, Great Britain
Ashford 25251 Ex 208
Journal of the Intelligence Corps. Annual, advertising accepted.
CONTACT: The Corps Secretary

'THE TURRET'
5 Tamar Close
Durrington
Worthing, Sussex BN1 33JZ, Great Britain
Military magazine.

'THE WAGGONER'
RHQ RCT, Buller Barracks
Aldershot, Hants GU11 2BX, Great Britain
Journal of The Royal Corps of Transport.
Quarterly, advertising accepted.
CONTACT: Maj.(retd) C. W. P. Coan, Ed.

'TIGER AND ROSE'
P.O. Box 4
Farnborough, Hampshire Great Britain
Military magazine.

'WAR MONTHLY'
Standard House
Bonhill Street
London, EC2 A4DA, Great Britain
Military magazine.

'WAR PICTURE LIBRARY'
King's Reach Tower
Stamford Street
London, SE1 9LS, Great Britain
Military pictorial magazine.

'WARGAMES'
215A Upper Grosvenor Road
Tunbridge Wells, Kent Great Britain
Military magazine.

'WAVE'
The Naval Club
38 Hill Street
Mayfair, London W1X 8DP, Great Britain
01-493 7672-3-4
Year Book of The Naval Club London. Annual, accepts advertising.
CONTACT: The General Secretary

'WEAPONS & WAREFARE'
169 Wardour Street
London, W1, Great Britain
Military magazine.

'WORLD II'
20 Bedfordbury
London, WC2 N4BT, Great Britain
Military magazine.

'PTISI'
Technical Press S.A.
6, Gorgiou Street
Athens, 11636, Greece
Greek aerospace & defence technology bi-monthly journal. Ads accepted. Greek.

'MILITARY DIGEST'
Army Headquarters
DHQ P.O.
New Delhi, India
Military magazine.

'USI JOURNAL'
Kashmir House
Rajaju
New Delhi, India
Military magazine.

'VIKRANT'
One Todarmal Road
Bengali Market
New Delhi, 110001, India
Military magazine.

'DEFENCE TODAY'
29/2, Via Tagliamento
Rome, 00198, Italy
(06) 8444956
International Defense industry journal. Monthly, advertising accepted.
CONTACT: Paolo F. Bancale, Editor

'AVIAZIONE'
29/2, Via Tagliamento
Rome, 00198, Italy
(06) 8444956
International aviation industry monthly journal.
Advertising accepted.

'RIVISTA ITALIANA DIFESA'
Via Martiri della
Liberazione 79/3
Chiavari, 16043, Italy
(01 85) 30 86 06
International defense industry magazine. 11 issues yearly. Ads accepted. Italian.
CONTACT: Mr. Giovanni Lazzari

'CONTRAIL'
6-27 Honura-Cho
Minamaku
Hiroshima, 734, Japan
Military aviation magazine.

'ARAB DEFENCE'
Dar Assayad S.A.L.
P.O. Box 1038
Beirut, Lebanon
National defense industry journal.

'ASIAN DEFENCE JOURNAL'
6th Floor, Bangunan Bakti

91 Jin Campbell
Kuala Lampur, Malaysia
Military magazine.

'JUND OMAN'
Department of Defence
P.O. Box 14
Muscat, Oman
Military magazine.

'PAKISTAN ARMY JOURNAL'
General Staff Brance
GHQ Rawalpindi, Pakistan
Journal of the Pakistan Army.

'PAKISTAN MILITARY DIGEST'
General Staff Branch
GHQ Rawalpindii, Pakistan
Military magazine.

'VIATA MILITARA/MILITARY LIFE'
Str. Cobalcescu Nr.28
Bucharest, Romania
Military magazine.

'PIONEER'
Manpower Division
Tanglin
Singapore, 10, Singapore
Military magazine.

'JUDEAN'
Box 7309
Johannesburg, South Africa
Military magazine.

'MILITARIA'
Military Information Bureau
SADF Private Bag X289
Pretoria, 0001, South Africa
Military magazine.

'MILITARY HISTORY JOURNAL'
P.O. Box 52090
Saxonwold, Transvaal 2132, South Africa
646-5513
Journal of the South African National Museum of Military History.
S8;
CONTACT: Col. Geo. R. Duxbury, Director

'SPRINGBOK'
Duncan House
Devilliers Street
Johannesburg, 2001, South Africa
Military magazine.

'AIR FORCES OF THE WORLD'
86 Avenue Louis Casai
Case Postale 162AI, Cointrin
Geneva, Switzerland
Military aviation magazine.

'AIRCRAFT ARMAMENT'
86 Avenue Louis Casai
Case Postale 162, Cointrin
Geneva, Switzerland
Military magazine
CONTACT: 1000 Fr. Interavia S.A.

MILITARY PERIODICALS

'INTERAVIA'
P.O. Box 162
86 Louis Cassai
Geneva, Switzerland
Military magazine.

'SOVIET MILITARY REVIEW'
2, Marschal Biryuzov Street
Moscow, 123298, USSR
198-55-30 198-55-52
Soviet military journal. English.
CONTACT: N. Velikanov, Editor-in-Chief
Subscriptions may be obtained for this magazine and other Soviet periodicals, along with advertising information from: Total Circulation Services Inc., 111 Eigth Avenue, Room 1009, New York, NY 10011.

'DEUTSCHES WAFFEN-JOURNAL'
Postfach 100340
Schmollerstrasse 31
Schwabisch Hall, 7170, West Germany
(0791) 30 61
Firearms & military collectors magazine. Monthly, advertising accepted.

'DIE BUNDESWEHR'
Postfach 20 06 68
Sudstrasse 123
Bonn 2, 5300, West Germany
(02 28) 38 23
Monthly magazine of the West German Armed Forces. Advertising accepted.

'DIE FREIWILLIGE'
Postfach 3023
Osnabrueck, 4500, West Germany
Military magazine.

'DIE OASE'
Postfach 70 02 09
Alte Bahnhofstrasse 148a
Bochum, 4630, West Germany
(0234) 28 72 54
Afrika Korps veterans magazine. German. Bi-monthly, advertising accepted.

'DORNIER-POST'
Dornier AG
Postfach 2160
Munich, 2160, West Germany
Military aviation magazine.

'LUFTWAFFE'
P.O. Box 140 261
Heilsbachstrasse, 26
Bonn 1, 5300, West Germany
(02 28) 64 83
Monthly journal of the German Air Force Staff. Advertising accepted. German.
CONTACT: Monch Publishing Group

'LUFTWAFFEN REVIEW'
Postrach 7025
Bremen 34, 2800, West Germany
Military magazine.

'MILITARY TECHNOLOGY'
P.O. Box 140 261
Heilsbachstrasse, 26
Bonn 1, 5300, West Germany
(02 28) 64 83
International Defense Industry journal. 13 issues annually, accepts advertising.
CONTACT: Monch Publishing Group

'WAFFEN UND KOSTUMKUNDE'
Vohburger Str.1.
Munich 21, 8000, West Germany
Military magazine.

'WEHR TECHNIK'
P.O. Box 140 261
Heilsbachstrasse, 26
Bonn 1, 5300, West Germany
(02 28) 64 83
Monthly journal of German Association for Technology. German. Ads accepted.
CONTACT: Monch Publishing Group

'TECNOLOGIA MILITAR'
P.O. Box 140 261
Heilsbachstrasse, 26
Bonn 1, 5300, West Germany
(02 28) 64 83
International defense industry journal. Advertising, 11 issues yearly, Spanish.
CONTACT: Monch Publishing Group

'MS & T'
P.O. Box 140 261
Heilsbachstrasse, 26
Bonn 1, 5300, West Germany
(02 28) 64 83
Military simulation & training magazine. Quarterly, advertising accepted.
CONTACT: Monch Publishing Group

'HEER'
P.O. Box 140 261
Heilsbachstrasse, 26
Bonn 1, 5300, West Germany
(02 28) 64 83
Official monthly magazine of the German Army. Advertising accepted. German.
CONTACT: Monch Publishing Group

'MARINE'
P.O. Box 140 261
Heilsbachstrasse, 26
Bonn 1, 5300, West Germany
(02 28) 64 83
Monthly journal of the Naval Staff in the MoD. Advertising accepted. German.
CONTACT: Monch Publishing Group

'MARINE-RUNDSCHAU'
P.O. Box 140 261
Heilsbachstrasse, 26
Bonn 1, 5300, West Germany
(02 28) 64 83
International naval industry journal. Bi-monthly, Ads accepted. German.
CONTACT: Monch Publishing Group

'BGS'
P.O. Box 140 261
Heilsbachstrasse, 26
Bonn 1, 5300, West Germany
(02 28) 64 83
Journal of the German Border Patrol. Bi-monthly, Ads accepted. German.
CONTACT: Monch Publishing Group

'LOYAL'
P.O. Box 140 261
Heilsbachstrasse, 26
Bonn 1, 5300, West Germany
(02 28) 64 83
Bi-monthly journal of the Reserves. Advertising accepted. German.
CONTACT: Monch Publishing Group

'ASSEGAI'
P.O. Box 1943
Harare, Zimbabwe
Military magazine.

MANUFACTURERS

GLOBAL EMBROIDERY CORPORATION
5821 Adams Street
West New York, NJ 07093, USA
201-868-8300
Insignia manufacturer.

BEST EMBLEM & INSIGNIA CO.
636 Broadway
New York, NY 10012, USA
(212)-677-4332 (800)-BESTEMB
Insignia and uniform accessories.

CORNING OPTICS
605 Third Avenue
New York, NY 10016, USA
Sunglass manufacturer.
CONTACT: c/o Visual Horizons

THE FRANKLIN MINT
Information Research Services
Franklin Center, PA 19091, USA
Manufacturer of medals & banknotes.

SPORTSMEN'S ACCESSORY MFG. CO.
P.O. Box 18091
615 Reed Street
Philadelphia, PA 19147, USA
215-336-6464
Manufacturers of shooters supplies.
CONTACT: E. Tomaso

KARL'S CUSTOM WHEELS
152 Skimino Road
Williamsburg, VA 23185, USA
804-565-1997
Artillery carriage wheel manufacturer.
CONTACT: Karl Gayer

PEACE RIVER MANUFACTURING CO.
P.O. Box 932
Lake Wales, FL 33859, USA
(813)-676-4421
Modern battle dress.
S6;
Catalogue $5.00.
CONTACT: James R. Strange, Sr.

SNAG-PRUFE FASTENERS
3717 Stanton Blvd.
Louisville, KY 40220, USA
Military insignia manufacturer.
CONTACT: Joe Brownstein

MANUFACTURERS

LOREN WILLIAMS
P.O. Box 2754
Pikeville, KY 41501, USA
Repro WWII German camo uniforms.
S6,S12; 3;
Catalogue available. As needed.
CONTACT: Loren Williams
SALES DEPT: Loren Williams
PURCHASING DEPT: Loren Williams
We deal primarily in authentic reproductions of German WWII military equipment for reenactment use. Camouflage items and entire period type uniforms are our specialty.

BLUE G INC.
P.O. Box 28054
Columbus, OH 43228, USA
(614)-274-7182
Reproduction U.S. DIs & crests.
S6; 2;
CONTACT: Bill Higgins
SALES DEPT: Bob Dale
PURCHASING DEPT: Bill Higgins
Blue G specializes in reproduction of DIs/Crests, min. quantity 50 (mixed or matched); will furnish list of stock items plus quotes on new items.

ALUMA-CASE COMPANY
11205 U.S. 24
Grand Rapids, OH 43522, USA
(419)-832-6655
Display case manufacturer.
CONTACT: Lee Quate

HAMMER
P.O. Box 1393
Columbus, IN 47201, USA
German marches on cassettes.
S6,S11,S12; 6;
Catalogue available.
CONTACT: Karl Hammer
Manufacturer/Wholesaler: WW II German march cassettes, flags, pins, and posters. Custom capacity retailer: Videos and books. We buy pre-1945, 78 rpm German march records.

AARDVARK ENTERPRISES
P.O. Box 1046
Glendale Heights, IL 60139, USA
(312)-653-0133
Military vehicles & Jeep parts.
S6,S12; 6;
CONTACT: Dan Rhame
SALES DEPT: Dan Rhame
PURCHASING DEPT: Dan Rhame
Manufacturer of replica M60 MGs. Inf., A/B, patrolboat, & lightweight configurations; also limited quantities of early style models. Models are full-size, fit original GI mounts & incorporate aluminum and original GI parts, , but cannot fire and are completely legal. Prices from $500. - $1,000.

MILITARY EMBLEMS & BADGES
P.O. Box 904102
Tulsa, OK 74105, USA
(918)-743-7048
Manufacturer of patches & D.I.s.
S6; 5;
Write for Free List.
CONTACT: Pam Dilley
SALES DEPT: Pam Dilley
PURCHASING DEPT: Pam Dilley
Manufacturers of embroidery patches, emblems, also enamelcrests & badges. Best prices in U.S.A. Send for free price list.

BIANCHI INTERNATIONAL
100 Calle Cortez
Temecola, CA 92390, USA
(714)-676-5621
Holsters & accessories.
S3;
Catalogue available. Annually.
CONTACT: John E. Bianchi
SALES DEPT: Tom Wisniewski
PURCHASING DEPT: Dick Allen
Manufacturer and distributor of holsters, accessories, belts, pouches, backpacks, and other miscellaneous carrying devices. We do major custom order products such as radio cases.

MAISON TEWFIK BICHAY
40, Talaat Harb Street
Cairo, EGYPT
751458
Manufacturer of Decorations, Medals & Badges.
S1;
CONTACT: Tewfik Guindy, Gen. Mgr.

DORSET (METAL MODEL) SOLDIERS LTD.
Latimer House, Castle Street
Mere, Wilts BA12 6JE, Great Britain
(0747) 860954
Manufacturer of lead model soldiers.
S6; 10;
Catalogue available. As needed.
CONTACT: Giles Brown
SALES DEPT: Giles Brown
We make traditional lead soldiers, civilians & spare parts for collectors worldwide. Cost of our colour catalogue only $2.00 cash.

CELTIC CORPORATION
P.O. Kotli Behram
Sailkot-3, Pakistan
Embroidered insignia manufacturer.
Write for details.

CITY TRADING COMPANY
Saghir Chowk, Noshera Road
Gujranwala, Pakistan
(0432)-5263
Embroidered insignia manufacturer.
Write for details.
CONTACT: Jamil Ahmad

HANDWORK COMPANY
Sialkot-1, Pakistan
Embroidered insignia manufacturer.
Write for details.

MEDALS/ORDERS/DECORATIONS

D. LEROY CALVERT
451 E. 14th Street, Apt. 6F
East Greenwich, RI 02818, USA
Medals & military books.
List available.

FAT CAT TOKENS
P.O. Box 1147
North Windham, ME 04062, USA
(207)-892-9681
Tokens, medals, and exonumia.
S4,S13; 11;
Catalogue available. Monthly.
CONTACT: H. N. Elston
SALES DEPT: H. N. Elston
PURCHASING DEPT: H. N. Elston
Free token and medal list published monthly (20 pages) - Sent free on request - fixed price list.

JEFFREY R. JACOB
P8-3 Panther Valley
Hackettstown, NJ 07840, USA
(201)-852-2392
Consignments of Orders & Decorations.
S4; 9;
Catalogue available. 2 or 3 per year.
CONTACT: Jeffrey R. Jacob
I accept & sell on consignment, all world orders & medals(also books) - Lists two or three per year. Also do research for collectors.

MARK H. PIEKLIK
43-3B Mount Pleasant Village
Morris Plains, NJ 07950, USA
(201)-540-0846
New Jersey National Guard miniature medals.
S6; 2;
SALES DEPT: Mark H. Pieklik
New Jersey National Guard miniature medals. Miniature medals mounted. Federal miniatures available. New Jersey Medal of Honor, Medal of Merit, Recruiting Medal.

THE ORDERS AND MEDALS SOCIETY OF AMERICA
209 W. Pittsburg Avenue
Wildwood Crest, NJ 08260, USA
Military & civilian medals & decorations collectors society. Convention, Journal.
S10; 37;
CONTACT: John E. Lelle, Secretary
AD MANAGER: Contact the Secretary
Over one thousand members world wide interested in the study and collecting of Orders, decorations, Medals, civil and military, and allied material and historical data.

DOUG ELY
1031 Old Farm Road
Point Pleasant, NJ 08742, USA
(201)-899-6542
General militaria.
Catalogue $3.00.

COL. H. H. ISAACSON
7 East 85th Street
New York, NY 10028, USA
Military medal dealer.

PETER HLINKA HISTORICAL AMERICANA
P.O. Box 310
New York, NY 10028, USA
(212)-409-6407
Military medals, U.S., British, foreign & U.S. insignia, reference books.
S6; 24;
Catalogue available. Bi-annually.
CONTACT: Peter Hlinka
SALES DEPT: Peter Hlinka
PURCHASING DEPT: Peter Hlinka

MEDALS/ORDERS/DECORATIONS

MILITARY & NAVAL ANTIQUITIES
268 Gilmore Street, Dept. PM
Mineola, NY 11501, USA
(516)-248-4673
Orders, medals, decorations of the world, insignia.
S3,S4,S6,S12; 35;
Catalogue available. Quarterly.
CONTACT: Thomas S. Halpin,Jr.
SALES DEPT: Thomas Halpin
PURCHASING DEPT: Thomas Halpin
Dealer in world-wide orders, decorations and medals, specializing in British. Catalogs published quarterly at $2.00. All items guarranteed genuine unless otherwise described.

SYDNEY B. VERNON
P.O. Box 1387
Baldwin, NY 11510, USA
(516)-536-5287 (516)-763-2087
Orders, medals & decorations.
S4,S12; 19;
Catalogue available. 11 times per year.
CONTACT: Sydney B. Vernon
SALES DEPT: Sydney B. Vernon
PURCHASING DEPT: Sydney B. Vernon
Sell Orders, Medals & Decorations of the world since 1967. Eleven catalogues yearly. Subscription N. America $6.00, elsewhere $15.00 in U.S. funds.

WAR RELICO
P.O. Box 46
New Stanton, PA 15672, USA
German WWII medals.
List $1.00.

JOHN S. LAIDACKER
750 E. Second Street
Bloomsburg, PA 17815, USA
717-784-5406
British medals.
Catalogue available.

PHOENIX MILITARIA CORPORATION
P.O. Box 66
163 Troutman Road
Arcola, PA 19420, USA
(215)-933-0909
General militaria & publisher of militaria collectors books & magazines.
S6; 19; 23,000; $60.00
Catalogue $3.00. Semi-annually.
CONTACT: Terry Hannon,President
SALES DEPT: Irene Karas
PURCHASING DEPT: Terry Hannon
Send $3.00 or pieces of military insignia for our latest catalogue of U.S. & world-wide militaria. Thousands of items listed including: books, manuals, medals, insignia,equipment, uniform components, headgear, badges, etc.

'THE PHOENIX EXCHANGE'
P.O. Box 66
163 Troutman Road
Arcola, PA 19420, USA
(215)-933-0909
Quarterly military collectors magazine.
; 2;
CONTACT: Terry Hannon,Publisher
SALES DEPT: W. R. Bendel
PURCHASING DEPT: Terry Hannon
AD MANAGER: William Bendel
'The Phoenix Exchange' functions as the military collector's & dealer's marketplace. Articles cover a wide spectrum of militaria, but attend to each subject in detail. Send $2.00 for a sample copy.

C & D GALE
2404 Berwyn Road
Wilmington, DE 19810, USA
302-478-0872
Colonials, tokens, medals.
Postal auction catalogue available.

PAUL SINOR
P.O. Box 12230
Arlington, VA 22209, USA
Medals & general militaria.
Catalogue $1.00.

OLD VETERANS MEDALS
P.O. Box 920
Williamsburg, VA 23187, USA
(804)-229-5544 (804)-643-0965
Medals, Orders, and Decorations.
S1,S4,S12;
List available. Bi-monthly.
CONTACT: Guy Dewolf
SALES DEPT: Guy DeWolf
PURCHASING DEPT: Guy DeWolf
Large selection of medals, orders, decorations of the world. Unusual selection of military autographs. Free list published every 2 months, with occasional special lists in between.

HOLBROOK ARMS (MILITARIA)
7530 S.W. 36th Street
Miami, FL 33155, USA
(305)-667-0505
General militaria
S4,S8,S12,S13; 27;
CONTACT: Rolfe R. Holbrook
SALES DEPT: Rolfe R. Holbrook
PURCHASING DEPT: Rolfe R. Holbrook
Militaria; collectibles, all categories; specializing in medals, orders, decorations worldwide. Antique guns, swords, accoutrements. Museum over twenty years. Attend most major gun shows.

MEDALS OF AMERICA
8893 Pendelton Pike
Indianapolis, IN 46226, USA
(317)-898-5523
Military medals, ribbons & badges.
S4,S12,S13; 15;
Catalogue available. Annually.
CONTACT: Linda B. Foster
Provides all full size U.S. & foreign medals, miniature medals, ribbons, badges in sets or mounted in walnut finished display cases. Personal and complete service.

JOHN BERNDSEN
909 Noah Street
St. Louis, MO 63135, USA
Military medals.
List available.

JEFFREY B. FLOYD
1500 W. Wm. Cannon #261
Austin, TX 78745, USA
(512)-444-6536
Military decorations of the world.
S4; 4;
Catalogue available. Quarterly.
CONTACT: Jeffrey B. Floyd
SALES DEPT: Jeffrey B. Floyd
Buy, sell, and trade military medals and decorations of the world. Especially Imperial Germany, Britain, and the United States.

DONALD W. NIXON
P.O. Box 64190
Los Angeles, CA 90064, USA
Military medals.
Catalogue available.

RICHARD J. SELF
1425 Clark Avenue, #C211
Long Beach, CA 90815, USA
Order, medals and decorations.
List available.

KARL STEPHENS,INC.
P.O. Box 458
Temple City, CA 91780, USA
213-445-8154
Coins & medals.
List available.

GULDSTUEN OF CALIFORNIA
P.O. Box 8552
Chula Vista, CA 92012, USA
(619)-420-3745
Foreign orders, medals and decorations.
List available.
CONTACT: Tim Eriksen

W. D. Grissom,Sr.
836 Orange, Suite 353P
Coronado, CA 92118, USA
General militaria.
S4; 3;
Catalogue available. 6 times annually.
SALES DEPT: W. D. Grissom,Sr.
W. D. Grissom is the man who introduced an unheard-of new element into medal dealing - price competition. Get his Free list and see for yourself.

KEN NOLAN,INC.,P.M.DIV.
P.O. Box C-19555
16901 P.M. Milliken Avenue
Irvine, CA 92713, USA
(714)-863-1532
Surplus, insignia, medals & uniforms.
S4; 29; 100,000; $60.00
Catalogue $1.00. 3 or 4 times annually.
Serving individuals, military, police for 29 years. 90% of orders shipped within 24 hours. Military field clothing, boots, insignia, nameplates, etc. Send $1.00 for catalog.

WILLIGES
P.O. Box 445
Wheatland, CA 95692, USA
(916)-633-2732
Tokens & medals.
S6; 23; ; $40.00
Catalogue available. Monthly.
CONTACT: William A. Williges

HAWKES LTD.
P.O. Box 3706
Federal Way, WA 98003, USA
Military medals & awards.
List available.

MEDALS/ORDERS/DECORATIONS

HAWKES, LTD.
P.O. Box 3706
Federal Way, WA 98063, USA
(206)-941-2510 PST
Orders, medals & decorations.
List available. Quarterly.

MEDALS OF THE WORLD
P.O. Box 2382
Olympia, WA 98507, USA
(206)-456-4982 (206)-786-5530
Orders, medals & decorations. German language translations of medals, etc.
S1,S6,S7,S12; 2;
Catalogue available. 1 or 2 annually.
CONTACT: Charles H. Graef
SALES DEPT: Charles H. Graef
PURCHASING DEPT: Charles H. Graef
Medals of the World. Buy - Sell - Trade.

FENRAE MEDALS
P.O. Box 117
Curtin, ACT 2605, Australia
(062) 48 0175
Medals & swords.
List available.
CONTACT: Bob & Bev Towns

SPINK & SON (AUSTRALIA)PTY,LTD
53 Martin Place
Sydney, N.S.W. 2000, Australia
Sydney 275571 Telex: 27283
Orders & decorations.
Catalogues available.

BONUS EVENTUS
Aartshertoginnestraat, 27
Oostende, 8400, Belgium
Orders, medals and decorations.
List available.

EUGENE G. URSUAL
P.O. Box 8096
Ottawa, Ontario K1G 3H6, Canada
1-613-521-9691
Military medals dealer & bookseller.
Catalogue $10.00 for 5 issues.

THE BRIGADE
952 Sherburn Street
Winnipeg, Manitoba R3E 2M6, Canada
204-775-5052
Canadian medals & insignia.
List available.
CONTACT: Jim Ferens

RIDEAU
473 DesLauriers Street
Ville Saint-Laurent, Quebec H4N 1W2, Canada
Orders, Decorations & Medals.

CHRISTIAN VASSAUX
13 Rue Edouard Lucy
Blanc-Mesnil, 93150, France
Military medals.

MARK CARTER
P.O. Box 470
Slough, SL3 6RR, Great Britain
0753-34777
Medals, orders & decorations. British & foreign. Medal research service.
S6,S11,S12,S13; 6;
Catalogue available. 3 or 4 times per year.
CONTACT: Mark Carter
SALES DEPT: Mark Carter
PURCHASING DEPT: Mark Carter
Efficient and friendly medal dealer specializing in British orders, decorations and medals. Wants lists welcomed. Research service for subscribers. Competitive prices whether buying or selling.

BRIAN CLARK - MEDALS
16 Lothian Road
Middlesbrough
Cleveland, TS4 2HR, Great Britain
0642-240827
Medals, Orders & Decorations.
S4; 17;
Catalogue available. 9 per year.
CONTACT: Brian Clark
SALES DEPT: Brian Clark
British and foreign medals, orders, decorations, ribbons, military and civil related books and accessories.

RAYMOND D. HOLDICH MEDALS & MILITARIA
16 Tudor Way
Hawkwell
Hockley, Essex SS5 4EY, Great Britain
0702-201415
Medal research, repair, & mounting.
S3,S4,S12; 10;
Catalogue available. Bi-annually.
CONTACT: R. D. Holdich
SALES DEPT: L. A. Holdich
PURCHASING DEPT: Ray Holdich
We are dealers in medals, orders, & decorations & Third Reich items. We supply miniature medals, mount medals for wear, medal research & repairs undertaken.

FRED S. WALLAND
17 Gyllyngdune Gardens
Seven Kings, Essex IG3 9HH, Great Britain
01-590-4389
British medal dealer.
Catalogue $10.00 for 12 issues.

C. J. & A. J. DIXON, LTD.
23 Prospect Street
Bridlington, E. Yorkshire Great Britain
0262 676877 0262 603348
Worldwide medals & decorations, ribbons, repairs & renovation, mounting.
S1,S4,; 20; 1300;
Catalogue available. 2 to 4 per year.
CONTACT: C. J. Dixon (Chris)
SALES DEPT: C. J. Dixon
PURCHASING DEPT: C. J. Dixon
One of the U.K.'s largest mail-order medal specialists. Current catalogue over 120 pages. Subscription only U.K. 3 Pounds, rest of world 6 Pounds (Airmail) for 3 issues.

CAPITAL MEDALS
33, Victoria Road
South Woodford, London E18 1LJ, Great Britain
(01) 530 6229
General militaria.
S3,S6,S12; 12;
Catalogue available. Bi-monthly.
CONTACT: Malcolm R. Gordon
SALES DEPT: Malcolm or Jeannie Gordon
PURCHASING DEPT: Malcolm Gordon
Dealers in medals, militaria, books, research, through shop, mail-order catalogue, medal & militaria fairs.

CHARLES A. LUSTED, LTD.
96 A Calverley Road
Tunbridge Wells, Kent TN1 2UN, Great Britain
011 44 892 25731
Orders, medals, decorations & militaria, Coins, antique jewellery.
S3,S6; 40;
Catalogue available. Quarterly.
SALES DEPT: Mrs. A. M. Morton
PURCHASING DEPT: Mrs. A. M. Morton
Specialist dealers in foreign orders, decorations and medals; also British campaign and gallantry medals, long service, life saving, etc.

DYAS COINS & MEDALS
588 Warwick Road
Tyseley
Birmingham, B11 CHT, Great Britain
021 707 2808 021 706 0471
Medals & decorations & WWI medal research only.
S1,S3,S6; 21;
Catalogue available. Quarterly.
CONTACT: Malcolm John Dyas
SALES DEPT: Malcolm John Dyas
PURCHASING DEPT: Malcolm John Dyas
I issue a medal catalogue 4 times a year. I sell at my shop edged weapons, oil paintings, gold & silver items. My main lines are medals and decorations. Also coins, bank notes, military badges.

LIVERPOOL COIN & MEDAL CO.
68 Lime Street
Liverpool, L1 1JN, Great Britain
051 708 8441 051 708 5305
Medals, coins, etc.
S3,S6; 11;
Catalogue available. Monthly.
CONTACT: L. Ross
SALES DEPT: L. Ross
One of the largest dealers in British military medals. Detailed catalogue published monthly. Send self-addressed envelope for sample list.

ROBERTS MEDALS
P.O. Box 1
Brimpton
Reading, Woolhampton RG7 4NH, Great Britain
Military medals.
Catalogue available.

SPINK & SON LTD.
King Street
St. James, London SW1Y 6QS, Great Britain
01-930 7888 Telex: 916711
Orders & decorations.
Catalogues available.

MEDALS/ORDERS/DECORATIONS

TOAD HALL MEDALS
Toad Hall, Court Road
Newton Ferrers
North Plymouth, South Devon PL8 1OH, Great Britain Plymouth 872672
Medals & general militaria.
S4; 12;
Catalogue available. Bi-monthly.
CONTACT: M. Hitchings & C. Maxey
SALES DEPT: Malcolm Hitchings
PURCHASING DEPT: Malcolm Hitchings
Toad Hall Medals, well established and specializing in export and offer a personal service 7 days a week from 7 AM to 9 PM.

WEINER von WEIN MILITARY ANTIQUES
2 Market Street
The Lanes
Brighton, Sussex BN1 1HH, Great Britain
(0273)-729948
General militaria.
S4; 25; 500; #50
Catalogue available. Bi-monthly.
CONTACT: George Weiner
SALES DEPT: George Weiner
PURCHASING DEPT: Pieter Christian Reinhardt
We buy, sell, exchange authentic Arms, Armour, World-wide militaria. Specializing in the Third Reich & Kaiserreich periods, mainly nicer grade quality desirables as per our catalogue.

HENK LOOTS
P.O. Box/Bus 3173
Pretoria, 0001, South Africa
012-28-3958 012-55-3494
Military medals.

MEDAL SALES
P.O. Box 7159
Roggebaai, 8012, South Africa
General militaria.
S4,S7,S11;
Write for details.
Purchase and disposal of collectors surplus medals and medallions in fields of British Empire gallantry and lifesaving, African and Polar exploration, Palestine, Israel and Judaica.

KEOGH COINS (PTY) LTD.
P.O. Box 4694
Durban, 4000, South Africa
Military medals & South African coins.

RANDBURG
P.O. Box 2434
Randburg, 2125, South Africa
011-789-2233 011-787-8915
Coins, Gold & Silver, Medals, Edged Weapons, some Badges.
S3,S6; 7;
Catalogue available. 1 or 2 per year.
CONTACT: Cliff Van Rensburg
SALES DEPT: Cliff Van Rensburg
PURCHASING DEPT: Cliff Van Rensburg

SPINK & SON NUMISMATICS LTD.
Lowenstrasse 65
Zurich, 8001, Switzerland
Zurich 221 18 85 Telex: 812109
Orders & decorations.
Catalogues available.

ORDEN & MILITARIA
Bauernvogtkoppel 1
Hamburg 65, 2000, West Germany
040/601 4217
German militaria dealer.
S6,S12; 6;
Catalogue available. Quarterly.
CONTACT: Detlev Niemann
SALES DEPT: Detlev Niemann
PURCHASING DEPT: Detlev Niemann
Buy and sell medals, orders, decorations, paper items, all kinds of militaria 1800-1945. Mailing list free 4 times a year. 2 year guarantee for original. If it turns out to be a reproduction, I will take it back.

HANS FISCHER
6056 Heusenstamm
Mittelweg 22, West Germany
Orders, medals and decorations of German Forces of WWII.
List available.

MILITARY ART & PHOTOS

BRIAN ZINSMEISTER
2115 Stearns Hill Road
Waltham, MA 02154, USA
WWII poster collector.

AEROPRINT
P.O. Box 154
South Shore Road
Spofford, NH 03462, USA
(603)-363-4713
Aviation & nautical art.
S4,S5; 15;
Catalogue available. Annually.
CONTACT: R. L. Westervelt
SALES DEPT: Robert Westervelt

EMERICK J. HANZL
17 Lockwood Drive
Clifton, NJ 07013, USA
Photographs.

NEIKRUG PHOTOGRAPHICA LTD.
224 East 68th Street
New York, NY 10021, USA
(212)-288-7741 (212)-288-7742
Vintage photography
S1,S4,S7; 20;
CONTACT: Marjorie Neikrug
SALES DEPT: Marjorie Neikrug
PURCHASING DEPT: Marjorie Neikrug
Civil War photographs, daguereotypes, stereo and cabinet cards, other vintage photographs, contemporary photographs, photo gallery, appraisals of fine art, photography and personal property.

MEEHAB FINE ARTS LIMITED
P.O. Box 477
New York, NY 10028, USA
WWI & WWII poster.
Catalogue $3.00.

JANET LEHR, INC.
P.O. Box 617
New York, NY 10028, USA
Military photographs.

RICHARD T. ROSENTHAL
1374 East 17th Street
Brooklyn, NY 11230, USA
(718)-645-7873
Antique photographs.
S4;
Photographs of all kinds bought and sold: Daguereotypes, Ambrotypes, tintypes, CDV's, cabinet cards, and paper photographs. Send want lists and quotes. All correspondence answered.

BIGGIN HILL PUBLICATIONS
P.O. Box 786
Freeport, NY 11520, USA
WWI and WWII posters.
Sample Photos $1.00.

DAVID FICKEN AUTO PARTS
P.O. Box 1
Babylon, NY 11702, USA
(516) 587-3332
Military vehicle parts & war posters.
S6; 20;
CONTACT: David Ficken
SALES DEPT: Dave Ficken
AD MANAGER: David Ficken
We supply original new "old stock" vacuum windshield wiper motors. Military, automobile, and truck. Also have original WWI and WWII posters.

MISCELLANEOUS MAN
P.O. Box 1914
New Freedom, PA 17349, USA
WWI & WWII posters.
Catalogue $3.00.

SOLDIER PRINTS
P.O. Box 422
Bethlehem, PA 18016, USA
Military prints

WW II VIDEO
P.O. Box 7202-B
Arlington, VA 22207, USA
Military oriented video tapes.
SASE for complete listing.

BAXTER IMPORTS
P.O. Box 792
Daytona Beach, FL 32017, USA
8th A.F. photographs of B-17's in combat, crew photos & nose art.
S4; 5;
Catalogue available.
CONTACT: H. D. Baxter
We supply 8"x10" B & W photos of aircraft (mostly B-17's) in flight & on ground, strike photos, nose art, crew photos, etc. Price list $.50.

FRANK D. GUARINO
P.O. Box 89
Debary, FL 32713, USA
Civil & Indian War images.
Catalogue available.

NAM VET PHOTO SERVICE
4518 Selma Street
Sarasota, FL 33582, USA
Vietnam photographs.

MILITARY ART & PHOTOGRAPHS

E. PETTIT
1470 Wagar, Apt. 210
Cleveland, OH 44116, USA
Military photos

BURKART DOUGLAS O.
826
P.O. Box 728
Hammond, IN 46320, USA
Battle action photographs.
42 page catalogue $2.00.

HAMMER
P.O. Box 1393
Columbus, IN 47201, USA
German marches on cassettes.
S6,S11,S12; 6;
Catalogue available.
CONTACT: Karl Hammer
Manufacturer/Wholesaler: WW II German march cassettes, flags, pins, and posters. Custom capacity retailer: Videos and books. We buy pre-1945, 78 rpm German march records.

ERNEST DAY
21147 Mahon
Southfield, MI 48075, USA
313-352-3082
Artist & historian.
Custom art commissions.

THE REICH ART
P.O. Box 285
122 1/2 East Main Street
Flushing, MI 48433, USA
(313)-659-8999
Military art prints.
S6; 6;
Illustrated catalogue available. Annually.
CONTACT: James G. Thompson
SALES DEPT: Sybille Palentyn
PURCHASING DEPT: Sybille Palentyn
Presently we are providing military, historical, or ethnic art prints, as well as commissioned art, books and replicas.

AVIATION ART MUSEUM & MINNESOTA AIR & SPACE MUSEUM
P.O. Box 16224
St. Paul, MN 55116, USA
(612)-291-7925 Off. (612)-780-2435 Hang.
Military & aviation art museum.
S3,S6,S8,S10,S12; 15;
Catalogue available. Annually.
CONTACT: Dennis Eggert
SALES DEPT: Dennis Eggert
PURCHASING DEPT: Dennis Eggert
The museums collect, restore & display several thousand pieces of aviation memorabilia from all facets of aviation, including 9 aircraft. Japanese "Kate", L-13, L-5, BT-13, BT-15, TBM, SNJ, H-34 & UC-78.

DENSA AVIATION ART
536 M Morse Avenue
Schaumburg, IL 60193, USA
(312)-893-3460
Military, ship, & aviation art prints.
S1,S4; 5;
Catalogue available.
CONTACT: Jack W. Treubig
SALES DEPT: Jack Treubig
PURCHASING DEPT: Jack Treubig

Retail sales of aviation, ship and military art prints. Limited and unlimited editions.

KEN KHUANS
155 Harbor #4812
Chicago, IL 60601, USA
WWI & II posters.
Write for details.

AERO-ART UNLIMITED
5751 E. Scarlett
Tucson, AZ 85711, USA
Aviation art dealer

MILITARY CURRENCY

SUBWAY STAMP SHOP,INC.
111 Nassau Street
New York, NY 10038, USA
212-227-8637 800-221-9960
Stamp & coin supplies.
Catalogue available.

P. WILLIAMS
7 Riddle Road
Camp Hill, PA 17011, USA
Military tokens.
Write for details.

RAY MORTIMEYER
P.O. Box 545
Cuba, MO 65453, USA
Merchant trade tokens

J.H.C. COINS & COLLECTABLES
3416 Overland Road
Boise, ID 83705, USA
Coins & collectables

ESSIE KASHANI
P.O. Box 8374
Rowland Heights, CA 91748, USA
Military currency.
2 listS & 2 auction catalogues per year.

MILITARY EQUIPMENT

BERNIE DeANGELIS
Rt. #1, Box 136
Mickleton, NJ 08056, USA
609-478-6506
Cavalry equipment.

KLEPPER AMERICA
35 Union Square West
New York, NY 10003, USA
Collapsible Boats.

WORLD MILITARY OUTFITTERS
P.O. Box 187, Salina Station
Syracuse, NY 13208, USA
Military outfitters

VALOR SPORTS
800 Streets Run Road
Pittsburgh, PA 15236, USA
Military outfitters

MATTHEWS POLICE SUPPLY COMPANY
P.O. Box 1754
Matthews, NC 28105, USA

Military & Police equipment

DUNN'S,INC.
P.O. Box 449
Hwy. 57E
Grand Junction, TN 38039, USA
800-223-8667
Hunting & camping supplies.
Catalogue available.

MARK DAVIS HORTON
2202 N. Madison Street
Corinth, MS 38834, USA
(601)-287-6043 (601)-287-0313
Pre-1945 U.S. military equipment.
S6,S12; 5;
Catalogue available. Bi-annually.
SALES DEPT: Mark Horton
PURCHASING DEPT: Mark Horton
U.S. Army equipment 1833-1942. Buy-Sell-Trade. Emphasis on horse equipment: McClellan saddles, bridles, bits, etc. Also belts and accoutrements. Appraisals, repairs, and restoration.

U.S. CAVALRY,INC.
1375 N. Wilson Road
Radcliff, KY 40160, USA
(502)-351-1164 (502)-351-7000
Military clothing & equipment, insignia, edged weapons, martial arts supplies.
S1,S6; 14; 100,000; $60.00
Catalogue available. Three times per year.
CONTACT: Randy Acton
SALES DEPT: Pearl Brantley
PURCHASING DEPT: Dan Gaddis
U.S. Cavalry operates a retail store and a mail-order catalog center. It specializes in current day hard-to-find military and adventure equipment.

MASECO
P.O. Box 18826
Cleveland Heights, OH 44118, USA
Military & outdoors supplies.
CONTACT: Michael Edelstein

STEVE STUCKMAN
5034 Bethel Road
Bucyrus, OH 44820, USA
U.S. military horse equipment.
Write for details.

S.W.A.T. EQUIPMENT
Route 1, Box 539
Three Oaks, MI 49128, USA
S.W.A.T. equipment.

JON L. SHRUM'S CAPE-OUTFITTERS
Rt. 2, Box 437-C
Cape Girardeau, MO 63701, USA
Military & outdoors outfitters.

SILENT PARTNER,INC.
230 Lafayette Street
Gretna, LA 70053, USA
Bullet-proof vests

HANDICRAFTS UNLIMITED
3000 E. Colfax, Suite 157
Denver, CO 80206, USA
Survival gear

MILITARY EQUIPMENT

BY-GONE MILITARIA
P.O. Box 5361
Mesa, AZ 85201, USA
(602)-844-8737
General militaria, appraisals.
S6,S12; 28;
CONTACT: Ken Clawson
SALES DEPT: Ken Clawson, Owner
Buy, sell, trade, German, U.S.A., Japanese, Chinese Communist, North Korean, all nations & periods. I specialize and collect WWII U.S. Airborne & Russian items.

ALLIED
Drawer 5070
Glendale, AZ 85312, USA
Survival supplies

THE SURVIVAL STORE
3250 Pollux
Las Vegas, NV 89102, USA
Posters & survival supplies.

GITANA INDUSTRIES
P.O. Box 363
Malibu, CA 90265, USA
(213)-457-6400
German army Jerricans.
S6;
Catalogue sheet available.
CONTACT: D. W. Richardson
German army Jerricans. The original jerrican and still the best. New manufacture. 5 year guarantee. Genuine NATO issue.

F. C. KAISER
P.O. Box 864
Danville, CA 94526, USA
Military paint

ALCAN,INC.
P.O. Box 2187
Bellingham, WA 98227, USA
(206)-733-6214
Security & military products.
; 39;
Catalogue available. Bi-annually.
PURCHASING DEPT: Jeff Davis
Alcan Wholesalers,Inc. offers their 56-page mail-order catalog, which features unique specialty, military, and security products. Free upon request.

MILITARY VEHICLES/PARTS

DIRECT SUPPORT
P.O. Box 317
Pittsfield, MA 01202, USA
(413)-443-4841
Military vehicle parts & accessories.
S4; 3;
Catalogue available. Annually.
CONTACT: Bob Kettler
SALES DEPT: Bob Kettler
PURCHASING DEPT: Bob Kettler
Specializing in New, NOS, and used parts for M-Series 2 1/2 ton, 5 ton, lot 6x6 military trucks. Also offer restorations on trucks and tracked armored vehicles.

THE SUPPLY DEPOT
P.O. Box 1108
Greenfield, MA 01302, USA
Military vehicle canvas dealer.
Send SASE for List.

RICHARDS AUTO SALES
200 W. Main Street
AYER, MA 01432, USA
Surplus military vehicle dealer.

R. A. MOIR
P.O. Box 303
Sterling Junction, MA 01565, USA
617-422-6352 5-9 PM
Military vehicles.

ALBERT J. McADOO
108 Central Street
Acton, MA 01720, USA
Military manuals, vehicles & parts.
Send Wants with SASE.

ROBERTS MOTOR PARTS,INC.
17 Prospect Street
West Newburg, MA 01985, USA
617 363-5407
Military vehicle parts, mail-order.

JOEL GOPAN
P.O. Box 1241
Bangor, ME 04401, USA
Military vehicle parts.

ROVERS NORTH
Route 128
Westford, VT 05494, USA
802 879-0032
Military vehicles & parts.

PORTRAYAL PRESS
P.O. Box 1913-63
Bloomfield, NJ 07003, USA
(201)-743-1851
Military vehicle books.
S4,S5;
Catalogue $2.00. Bi-annually.
CONTACT: Dennis R. Spence
Gigantic assortment of books, manuals and other publications on military vehicles . . . jeeps, military Dodges, half-tracks and related references on WW2 uniforms, insignia, history, etc.

BEACHWOOD CANVAS WORKS
P.O. Box 137
Island Heights, NJ 08732, USA
201-240-3724
Military vehicle canvas manufacturer.
Catalogue $1.00.

SARAFAN AUTO SUPPLY
P.O. Box 293
23 North Madison Avenue
Spring Valley, NY 10977, USA
914 356-1080
Military vehicle & parts.

GEORGE HILTON
P.O. Box 414
Tuxedo, NY 10987, USA
914-351-5352
Military vehicle & parts.

BARNETT & SMALL,INC.
151 East Industry Court
Deer Park, NY 11729, USA
516 242-2100
Military vehicle parts dealer.

ENGINE REBUILDERS,INC.
Route 22
Pawling, NY 12564, USA
914 855-5051
Military vehicles & parts.
CONTACT: A. J. Maxwell

WALTER MILLER AUTO LITERATURE
6710 Brooklawn Parkway
Syracuse, NY 13211, USA
German military vehicle collector.
S13;
CONTACT: Walter Miller
Auto literature wanted. Also Nazi items.

DURYEA MOTORS
4875 South Lake Road
Brockport, NY 14420, USA
(716)-637-8134 (716)-637-9572
Military vehicle & Jeep parts.
; 32;
CONTACT: Erwin Duryea
SALES DEPT: Erwin Duryea
PURCHASING DEPT: Erwin Duryea
Parts new & used for most military vehicles - Specialize Jeeps - Buy and sell any type military vehicle. Duryea Ford, Brockport, NY 14420. 716-637-8134.

SAFEPAC,INC.
P.O. Box 11008
Rochester, NY 14611, USA
716-436-4250
Military vehicles.

WILLIAM J. RICCA
P.O. Box 25
New Tripoli, PA 18066, USA
Military vehicle parts & accessories.

FRANK'S SURPLUS,INC.
P.O. Box 555
Bryn Athyn, PA 19009, USA
215-947-5616
Military vehicle.
List available.

CLARK SURPLUS TRUCK & EQUIPMENT
P.O. Box 10168
Charleston, WV 25357, USA
(304)-342-3176
Military vehicles & parts.

A. B. LINN
Rt. 1, Box 485
Salisbury, NC 28144, USA
704-637-9076
Military vehicle parts

MILITARY VEHICLES/PARTS

BUZ BOWLING COMPANY
P.O. Box 2282
Charlotte, NC 28211, USA
(704)-542-1545
Military & civilian trucks & Jeeps.
S11; 5;
CONTACT: C. B. Bowling,III
SALES DEPT: Buz Bowling
PURCHASING DEPT: Buz Bowling
Military vehicles for sale. Sold as is, restored and built to your specifications. Can also locate and rebuild specialized vehicles. Delivery arranged, photos available.

STRICKLAND'S SURPLUS,INC.
5927 Carolina Beach Road
Wilmington, NC 28403, USA
919 791-8003
Military vehicles & Jeep parts

WHITE OWL PARTS COMPANY,INC.
3201 W. Vernon Avenue
Kinston, NC 28501, USA
919-522-2586
Military vehicles & Jeep parts

SOUTHEASTERN EQUIPMENT CO.,INC
P.O. Box 5438
2950 Old Savannah Road
Augusta, GA 30906, USA
404-798-4361 TELEX NO. 545460
Surplus military vehicles.
CONTACT: Dolores Branham, Ad Manager

NORTHWEST TRUCK PARTS,INC.
P.O. Box 7245
1900 N. Orange Blossom Trail
Orlando, FL 32804, USA
(305)-849-0641
Military trucks, jeeps & parts.
S3,S6,S7; 32;
CONTACT: Dwane Steward, President
SALES DEPT: Stephen Steward
PURCHASING DEPT: Stephen Steward
Sale of reconditioned military vehicles. New, rebuilt, used parts. Wholesale to retail.

CREATIVE WORKSHOP
118 N.W. Park Street
Dania, FL 33004, USA
305 920-3303
Military vehicles & parts.

SAM WERNER
P.O. Box 327
Tracy City, TN 37387, USA
615 592-3601
Military vehicles & parts.

SOUTHERN PARTS CORPORATION
1268 N. Seventh Street
Memphis, TN 38107, USA
901 527-5601
Military vehicles & parts.

G. I. TRUCKS
P.O. Box 1
Dry Ridge, KY 41035, USA
606 428-2400
3/4-ton M-series military vehicles.

SAM WINER MOTORS,INC.
P.O. Box 6258
3417 W. Waterloo Road
Akron, OH 44312, USA
216 628-4881
Military vehicles & parts.

NELSON'S SURPLUS JEEPS & PARTS
1024 East Park Avenue
Columbiana, OH 44408, USA
216-482-5191
Military vehicles & parts.

OHIO AUTOMOTIVE PARTS
4012 Spring Grove Avenue
Cincinnati, OH 45223, USA
Military vehicles & parts.

ARMY JEEP PARTS,INC.
80 East Water Street
Chillicothe, OH 45601, USA
614-775-5302
Military vehicles & Jeep parts.
Write for details & estimates.
CONTACT: David W. Uhrig,President

MICHAEL J. SINKOVIC - CANVAS
1775 South 900 East
Zionsville, IN 46077, USA
Canvas for military vehicles.
CONTACT: Michael J. Sinkovic

JEEPACRES
Rt. 1, Box 237
New Carlisle, IN 46552, USA
219 654-8649
Military vehicles & Jeep parts.

VINTAGE POWER WAGONS
302 South 7th Street
Fairfield, IA 52556, USA
515-472-4665
Military vehicles.

WALLY'S
966 Sheridan Road
Kenosha, WI 53140, USA
414-552-8330
Military vehicle parts

WILLYS MINNEAPOLIS
P.O. Box 803
Lakeville, MN 55044, USA
612 469-3115 TELEX 29-0418
Military vehicles & parts.
CONTACT: Louis Larson, General Manager

TWIN CITY AUTO & MILITARY PARTS
839 Edgerton
St. Paul, MN 55101, USA
612 774-3339
Military vehicles & Jeep parts.
CONTACT: Jeff Eisenberg

AARDVARK ENTERPRISES
P.O. Box 1046
Glendale Heights, IL 60139, USA
(312)-653-0133
Military vehicles & Jeep parts.
S6,S12; 6;
CONTACT: Dan Rhame
SALES DEPT: Dan Rhame
PURCHASING DEPT: Dan Rhame
Manufacturer of replica M60 MGs. Inf., A/B, patrolboat, & lightweight configurations; also limited quantities of early style models. Models are full-size, fit original GI mounts & incorporate aluminum and original GI parts, , but cannot fire and are completely legal. Prices from $500. - $1,000.

IRVING MILLER & CO.
1330 South Canal Street
Chicago, IL 60607, USA
312 226-3930
Military vehicles & Jeep parts.

BERG'S TRUCK PARTS
1608 S. Wabash Avenue
Chicago, IL 60616, USA
312 427-6930
Military vehicles & Jeep parts.

VAN DE VELDE LTD.
2934 S. Laramie Avenue
Cicero, IL 60650, USA
Military vehicles & Jeep parts.

THE STEWART CORPORATION
124 South Church Street
Rockford, IL 61101, USA
815 962-5541
Military vehicles & Jeep parts.

ALBERT F. WILLIAMS
RR #2, Valley View Farm
Carmi, IL 62821, USA
618 382-8262
Military vehicles & Jeep parts.

PRECISION BILT PARTS,INC.
1819 Troost
Kansas City, MO 64108, USA
816 471-1552(OFFICE) 913 381-0659(HOME)
Military vehicles & Jeep parts.

MID-AMERICA PARTS & EQUIPMENT
1212 E. 19th Street
Kansas City, MO 64108, USA
800 821-5493
Military vehicles & Jeep parts.

FIRST OZARK ARMORED COMPANY
Rt #3, Box 184
Eureka Springs, AR 72632, USA
Military vehicles.
Write for details.

A & P TRUCK SALVAGE COMPANY
1320-28 South Robinson
Oklahoma City, OK 73109, USA
405 232-6353
Military vehicles & Jeep parts.

STEVE DAVIS
3909 Cashion
Oklahoma City, OK 73112, USA
405-942-0233
Military vehicles

MILITARY VEHICLES/PARTS

RAPCO
P.O. Box 921
Keller, TX 76248, USA
(817)-379-6875
Replica military vehicle & Jeep parts.
S6; 6;
Catalogue available. Annually.
CONTACT: Mark Chapin
SALES DEPT: Mark Chapin
PURCHASING DEPT: Mark Chapin
Data plates, decals, paint, etc. for MB/GPW M38 and A1, W.C. Dodge and M-37. Send a SASE for a complete catalog.

REDBUD SURPLUS
Redbud Point Rte. 3
Roanoke, TX 76262, USA
817 430-1486
Military vehicles & Dodge parts.
CONTACT: Dave Anwyll

TERRANG AXEL'S OFFROAD CENTER
8201 Old Laporte Road
Houston, TX 77012, USA
713 926-3044
European military vehicles & Jeep parts.

HOUSTON TRUCK & EQUIPMENT,INC.
6716 Long Drive
Houston, TX 77087, USA
713 644-5443
Military vehicles & Jeep parts.

BRENT MULLINS JEEP PARTS
206 Helena Street
Bryan, TX 77801, USA
(409)-690-0203
WWII Jeep parts.
S4; 2;
Catalogue available. Annually.
CONTACT: Brent Mullins
WWII and M-series Jeep parts, new and used.

ACE ENTERPRISES
606 Champions Row
Victoria, TX 77904, USA
Military vehicles & parts.

PUEBLO AUTO PARTS
201 N. Santa Fe Avenue
Pueblo, CO 81002, USA
303 544-7435
Military vehicles & Jeep parts.

4-WHEEL DRIVE BY VAN
24000 Ventura Boulevard
Calabasas, CA 91302, USA
213 347-0386
Military vehicles & Jeep parts.

SURPLUS CITY JEEP PARTS
11794 Sheldon Street
Sun Valley, CA 91352, USA
818 767-3666 800 443-8038
Military vehicles & Jeep parts.

FOUR WHEELER SUPPLY
16612 Beach Blvd.
Huntington Beach, CA 92647, USA
714 842-6064
Military vehicles & Jeep parts.

ANTELOPE VALLEY EQUIPMENT AND TRUCK PARTS
44532 Trevor
Lancaster, CA 93534, USA
1-800-521-8518
Military vehicles & Jeep parts.

UNITED PARTS SERVICE - UNIMOG ERSATZTEILE
P.O. Box 827
Palo Alto, CA 94301, USA
(415)-364-9184
Unimogs & Unimog spare parts.
S1,S6; 5;
Catalogue available. Every 3 years.
SALES DEPT: Michael Smith
PURCHASING DEPT: Michael Smith
Select German military Unimogs. Unimog parts; new, used and exchanged. Full service facility. Publications and accessories. Your one-stop source for all Unimog needs.

UNIMOGS UNLIMITED
P.O. Box 200
Caravanserai Ranch
Azalea, OR 97410, USA
(503)-837-3636
Military Mercedes-Benz UNIMOG trucks.
S3,S6,S10,S11,S13; 10;
CONTACT: James Ince
SALES DEPT: James Ince
PURCHASING DEPT: James Ince
America's oldest importer of military Mercedes-Benz UNIMOG 4x4 trucks. All types, special versions. Mail-order, European search services. Largest inventory. Computerized original MBZ parts department.

MILITARY MUSEUMS

SPRINGFIELD ARMORY MUSEUM
State & Federal Streets
Armory Square
Springfield, MA 01105, USA
(413)-734-6477
U.S. Army military arms museum.
Museum has the largest collection of military small arms in the world.

FORT DEVENS MUSEUM
Fort Devens, MA 01433, USA
(617)-796-3732
U.S. Army military museum.
The museum depicts the history of Fort Deven and the units stationed there.

HIGGINS ARMORY MUSEUM
100 Barber Avenue
Worcester, MA 01606, USA
(617)-853-6015
Medieval arms & armour museum.
; 55; ; 500; ; $60.00
Catalogue available.
SALES DEPT: The Director
Higgins Armory Museum exhibits mainly 16th and 17th century arms and armor and related artifacts. A metal conservator conducts restoration/repairs in newly equipped laboratory.

AMERICAN ANTIQUARIAN SOCIETY
185 Salisbury Street
Worcester, MA 01609, USA
Museum & Library
CONTACT: Marcus A. McCorison

SCOTTISH RITE MASONIC MUSEUM
Attn: Librarian
P.O. Box 519
Lexington, MA 02173, USA
Military museum.

HISTORICAL MUSEUM & ART GALLERY
Bennington, VT 05201, USA
Military museum.

SUBMARINE FORCE LIBRARY & MUSEUM
Sub Base New London
Groton, CT 06349, USA
(203)-449-3174
Military submarine museum & library.

WINCHESTER GUN MUSEUM
275 Winchester Avenue
New Haven, CT 06511, USA
Firearms museum

MUSEUM OF THE CITIZEN SOLDIER
P.O. Box 373
Maplewood, NJ 07040, USA
(201)-763-1313
U.S. Army National Guard Museum

U.S. ARMY COMMUNICATIONS - ELECTRONICS MUSEUM
Myer Hall
Bldg 1207, Ave. of Memories
Fort Monmouth, NJ 07703, USA
(201)-532-2445
U.S. Army military museum.
The museum's collection illustrates the development of the art of military communication, and the contributions of the U.S. Signal Corps.

PICATINNY ARSENAL AMMUNITION MUSEUM
Picatinny Arsenal
Dover, NJ 07801, USA
(201)-328-2797 (201)-328-3222
U.S. Army military museum.
The museum collection has over 9000 items, including ammunition, battlefield pick-ups, and paper items having to do with the early years of the arsenal.

FORT TICONDEROGA MUSEUM
One Wall Street
New York, NY 10005, USA
Military museum

METROPOLITAN MUSEUM OF ART
Fifth Avenue & 82nd Street
New York, NY 10028, USA
Military collection in museum.
CONTACT: Attn: The Library

CONGRESSIONAL MEDAL OF HONOR SOCIETY
Intrepid Sea-Air-Space Museum
West 46th Street & 12th Avenue
New York, NY 10036, USA
(212)-582-5355
Association of M.O.H. recipients.
Hall of Honor Museum and national archives containing displays, books, records and artifacts about the Medal of Honor.

MILITARY MUSEUMS

U.S. ARMY CHAPLAIN MUSEUM
Fort Wadsworth, NY 10305, USA
(212)-447-5100 ex590
U.S. Army military museum.
Depicts the history of the U.S. Army chaplaincy.

FORT WADSWORTH MUSEUM
Staten Island, NY 10305, USA
(212)-447-5100 ex731
U.S. Army military museum.
The museum displays weapons, uniforms, documents, paintings, posters, and other memorabilia related to the history of Fort Wadsworth.

WEST POINT MUSEUM
U.S. Military Academy
Building 601, #64
West Point, NY 10996, USA
(914)-938-2203 (914)-938-3201
Museum of the U.S. Army Military Academy.
Museum of military history of the world, with emphasis on American and European weapons, art, uniforms, flags and equipments, 1600-present.

HARBOR DEFENSE MUSEUM
Fort Hamilton
Brooklyn, NY 11252, USA
(718)-630-4349
Army Coast Defense museum.
CONTACT: Russell S. Gilmore, Director
Army museum of Coast Defense with associated civilian organization, The Fort Hamilton Historical Society. Charter memberships available. Harbor Defense Museum, Fort Hamilton, Brooklyn, NY 11252-5335.

WATERVLIET ARSENAL MUSEUM
Watervliet Arsenal
Watervliet, NY 12189, USA
(518)-273-4610
U.S. Army military museum.
The museum portrays the use of cannon throughout history.

HISTORY & ART INSTITUTE
125 Washington Avenue
Albany, NY 12210, USA
Military museum.

NEW YORK STATE MILITARY MUSEUM
State Capitol
Albany, NY 12224, USA
(518)-474-1631
U.S. Army National Guard Museum
Collection features over 500 flags carried by New York Units, mostly in the Civil War, plus much more in the way of Civil War relics.

FORT WILLIAM HENRY MUSEUM
Fort William Henry
Lake George, NY 12845, USA
Military & regional history museum.

PLATTSBURGH AIR FORCE BASE MUSEUM
380th BW (SAC)
Plattsburgh AFB, NY 12903, USA
(518)-565-6165
Military aviation museum.
CONTACT: Capt. Vicki Martin
Portrays the history of Plattsburgh Air Force Base, the 380th Bomb Wing/Group and military history of the North Country dating back to the American Revolution.

HOMEVILLE MUSEUM
49 Clinton Street
Homer, NY 13077, USA
(607)-749-3105
American civil & military museum.
; 10;
CONTACT: Kenneth M. Eaton
Private museum specializing in items from the Civil War period, World War II, Japanese, and Lackawanna Railroad.

FORT ONTARIO MUSEUM
Fort Ontario
Oswego, NY 13126, USA
Military & regional history museum.
CONTACT: Wallace Workmaster

REMINGTON FIREARMS MUSEUM
Catherine Street
Ilion, NY 13357, USA
894-9961
Museum of Remington firearms.

ASSOCIATION OF OLD FORT NIAGARA
Youngstown, NY 14174, USA
Military & regional history museum.

FORT LIGONIER MEMORIAL FOUNDATION
South Market Street
Ligonier, PA 15658, USA
Military & regional history museum.

PENNSYLVANIA MILITARY MUSEUM: 28TH DIVISION SHRINE
P.O. Box 148
Boalsburg, PA 16827, USA
(814)-566-6263
U.S. Army National Guard Museum

HESSIAN POWDER MAGAZINE
Carlisle Barracks, PA 17013, USA
(717)-245-3434
U.S. Army military museum.
The collection depicts the history of Carlisle and Carlisle Barracks from the French and Indian Wars to the present. It is also the repository of several large military collections.

OMAR N. BRADLEY MUSEUM
Carlisle Barracks, PA 17013, USA
(717)-245-3461
U.S. Army military museum.
Gen. Omar N. Bradley's collection of military items, and his personnal effects.

PENNSYLVANIA HISTORY & MUSEUM COMMISSION
Museum Library
P.O. Box 1026
Harrisburg, PA 17120, USA
Military museum.

GETTYSBURG NATIONAL PARK
P.O. Box 70
Gettysburg, PA 17325, USA
Military historic site

LOYAL LEGION WAR MUSEUM
1805 Pine Steet
Philadelphia, PA 19103, USA
215-735-8196
U.S. Civil War museum. Reference library available.

THE DANDY FIRST MUSEUM
3205 Lancaster Avenue
Philadelphia, PA 19104, USA
(215)-222-3224
U.S. Army National Guard Museum
The collection consists of uniforms from the Revolutionary War to the present.

ARMED FORCES MEDICAL MUSEUM
Institute of Pathology
6825 16th Street, N.W.
Washington, DC 20306, USA
(202)-576-2341
U.S. Military Museum.

MARINE CORPS HISTORY AND MUSEUMS DIVISION
Marine Corps Historical Center
Bldg. 58, Navy Yard
Washington, DC 20374, USA
U.S. Marine Corps military history center.
CONTACT: BG. E. H. Simmons,USMC(Ret)
SALES DEPT: Marine Corps Hist. Foundation
A staff division of Headquarters, U.S. Marine Corps which operates the Marine Corps Historical Program, including museums, publication of official histories, and maintaining research resources.

MARINE CORPS MUSEUM
Marine Corps Historical Center
Bldg. 58, Navy Yard
Washington, DC 20374, USA
(202)-433-2273
Staff Div. of HQ U.S.M.C. Operates the U.S.M.C. historical program & museum.
CONTACT: BG E. H. Simmons, USMC(Ret)
Presents chronological history of Marine Corps from 1775 in standing exhibits; changing minor and major special exhibits; includes on site art, personal papers, and music collections.

U.S. NAVY MEMORIAL MUSEUM
Building 76
Washington Naval yard
Washington, DC 20374, USA
(202)-693-2651
U.S. Navy Museum
Items related to all phases of U.S. Naval history.

TRUXTUN-DECATUR NAVAL MUSEUM
Building 220, Room 218
Washington Naval Yard
Washington, DC 20390, USA
(202)-783-2573
U.S. Navy Museum
Collection of books, documents, models, and artifacts relating to the history of the U.S. Navy.

NATIONAL ARMED FORCES HIST'L MUSEUM ADVISORY BOARD
The Smithsonian Institute
Washington, DC 20560, USA
Military museum.

MILITARY MUSEUMS

THE NATIONAL MUSEUM OF HISTORY AND TECHNOLOGY
Constitution Avenue, N.W.
Washington, DC 20560, USA
Military museum.

AIRMEN MEMORIAL MUSEUM
5211 Auth Road
Suitland, MD 20746, USA
(301)-899-8386
A.F. Enlisted Men's history museum.
CONTACT: Lawrence Tassone, Director

FORT MEADE MUSEUM
Fort Meade, MD 20755, USA
(301)-677-6966
U.S. Army military museum.
The collection consists of artifacts relating to the First U.S. Army, Fort Meade, and the local region.

U.S. ARMY ORDNANCE MUSEUM
Aberdeen Proving Ground
Aberdeen, MD 21005, USA
(301)-278-3602 (301)-278-5722
Museum of U.S. Army ordnance.
Nearly 18,000 items related to the evolution of ordnance material comprise the museum's collection.

U.S. NAVAL ACADEMY MUSEUM
Annapolis, MD 21402, USA
(301)-267-2108
Museum of U.S. Navy militaria. Research facilities.

NATIONAL MUSEUM OF ARMY CRIMINAL INVESTIGATION
NASF Bldg.
Bailey's Crossroads
Falls Church, VA 22041, USA
Military museum.
Artifacts and memorabilia relating to the history of the Army criminal investigation effort.

U.S. ARMY ENGINEER MUSEUM
Building 1000
16th Street & Belvoir
Fort Belvoir, VA 22060, USA
(703)-664-6104
U.S. Army Engineer museum.
The museum depicts the contributions of the Corps of Engineers to the nation over the past two centuries.

MANASSAS NATIONAL BATTLEFIELD
P.O. Box 1830
Manassas, VA 22110, USA
Military historical site
CONTACT: Stuart Gregory Vogt, Historian

MARINE CORPS AIR-GROUND MUSEUM
MCDEC
Quantico, VA 22134, USA
(703)-640-2606
Museum of U.S. Marine Corps aviation. Reference library available.
CONTACT: LtCol Rudy T. Schwanda, OIC
Integrates aircraft, artillery, tanks, vehicles, infantry weapons of a period into exhibits telling the story of air-ground team weapons, tactics, and organization development.

NATIONAL PARK SERVICE
5502 Port Royal Road
Springfield, VA 22151, USA
Administrator of military museums.

THE OLD GUARD MUSEUM
1st Battalion, 2nd Infantry
Fort Meyer, VA 22211, USA
(202)-692-9721
U.S. Army military museum.
The museum contains items connected with the history of the 3rd Infantry.

AMERICAN MILITARY MUSEUM
Rt. 305 (Collins Drive)
New Market, VA 22844, USA
(703)-740-8065
Military museum.
Hundreds of U.S. and enemy military relics from 1776 to Grenada. Open April 1987.

THE COLONIAL WILLIAMSBURG FOUNDATION
P.O. Drawer C
Williamsburg, VA 23185, USA
Military museum.
CONTACT: John C. Moon, Musick Master

MUSEUM OF THE CONFEDERACY
1201 E. Clay Street
Richmond, VA 23219, USA
804-649-1861
U.S. Civil War Confederate museum.

U.S. MARINE RAIDERS MUSEUM
c/o Am. Historical Foundation
1142 W. Grace Street
Richmond, VA 23220, USA
(804)-353-1812
U.S. Marine Raiders museum.
CONTACT: Rudy Rosenquist, Curator
The U.S. Marine Raider Museum displays the gear and materials used and collected during the active service of the U.S. Marine Raiders in WWII, as well as Japanese combat gear the Raiders captured and brought back.

U.S. ARMY TRANSPORTATION MUSEUM
P.O. Drawer D
Fort Eustis
Newport News, VA 23604, USA
(804)-878-3603 (804)-878-3480
Museum of U.S. army transportation.
; 1;
Museum shop catalogue available. Annually.
SALES DEPT: Transportation Museum Found.
PURCHASING DEPT: Transportation Museum Found.
The museum shop provides transportation, corps and regimental related items including, but not limited to books, models, T-shirts, caps, insignia & awards.

WAR MEMORIAL MUSEUM OF VIRGINIA
9285 Warwick Blvd.
Newport News, VA 23607, USA
(804)-247-8523 (804)-247-8522
General militaria museum. Research library.
S1;
CONTACT: John Quarstein
The War Memorial Museum of Virginia exisits to study U.S.military history since 1775, interpreting the impact of war upon American society.

FORT MONROE CASEMATE MUSEUM
P.O. Box 341
Fort Monroe, VA 23651, USA
(804)-727-3935
U.S. Army military museum.
The museum's collection depicts the history of Old Point Comfort from 1607 to the present day, and the history of the fort.

PORTSMOUTH NAVAL SHIPYARD MUSEUM
P.O. Box 248
2 High Street
Portsmouth, VA 23705, USA
(804)-393-8591
Naval shipyard museum.

U.S ARMY QUARTERMASTER MUSEUM
Bldg. P-5218
Fort Lee, VA 23801, USA
(804)-734-4203 (804)-734-5435
U.S. Army Quartermaster Corps museum.
The museum has extensive collections of just about every category of item the Quartermaster Corps has handled in the last 200 years.

APPOMATTOX COURT HOUSE NATIONAL HISTORICAL PARK
P.O. Box 218
Appomattox, VA 24522, USA
(804)-352-8987 (804)-352-8782
Military historic site.
S1,S4; 20;
SALES DEPT: Sheila Mayberry
PURCHASING DEPT: Sheila Mayberry
Appomattox Court House National Historical Park, a rural Virginia of 27 structures where Gen. R. E. Lee surrendered to Gen. U. S. Grant on April 9, 1865. Books and other printed materials on park are available through Eastern National Park and Monument Association, P.O. Box 327, A

NATIONAL PARK SERVICE
Harpers Ferry Center Library
Harpers Ferry, WV 25425, USA
Military & regional history museum and historic site.

GUILFORD COURTHOUSE NATIONAL MILITARY PARK
Greensboro, NC 27410, USA
Military historic site & museum.

CLEVELAND COUNTY HISTORICAL MUSEUM
P.O. Box 1335
Shelby, NC 28150, USA
(704)-482-8184
General military exhibits. Museum, library & newsletter.
; 10;
PURCHASING DEPT: Jim Marler
General history museum with a special military gallery. Exhibits include clothing, arms, documents, photographs and personal items. Donations of material welcome.

MILITARY MUSEUMS

82nd AIRBORNE DIVISION WAR MEMORIAL MUSEUM
Fort Bragg, NC 28307, USA
(919)-396-2328
U.S. Army military museum.
The museum collection contains weapons, uniforms, equipment, documents, artifacts, and photographs pertaining to the 82nd Airborne Division.

SPECIAL WARFARE MUSEUM
Fort Bragg, NC 28307, USA
(919)-396-4272
U.S. Army military museum.
The history of special warfare units is traced to give the visitor a look at irregular warfare, type of units, and tactics employed.

FORT JACKSON MUSEUM
Fort Jackson, SC 29207, USA
(803)-751-7419 (803)-751-7355
Military museum.
The Fort Jackson Museum exhibits interpret all periods of training in the Army from 1775 to present. The exhibits relate to local military history, the life of Andrew Jackson and Fort Jackson.

ARCHIVES & HISTORY
Box 11669
Columbia, SC 29211, USA
Military museum.

PARRIS ISLAND MUSEUM
Marine Corps Recruit Depot
Parris Island, SC 29905, USA
(803)-525-2951
Marine Corps & Port Royal history museum. Library & photo collection.
CONTACT: Dr. Stephen R. Wise, Curator

U.S. ARMY SIGNAL MUSEUM
Fort Gordon, GA 30905, USA
(404)-791-2818 (404)-780-2818
Military museum.
The collection encompasses artifacts illustrating the history of the Signal Corps, as well as exhibits relating to all the major U.S. conflicts since 1775.

ROBINS AFB MUSEUM OF AVIATION
Warner Robins ALC/XRS
Robins AFB, GA 31098, USA
(912)-926-3474
Military aviation museum.
CONTACT: Mr. Herbert E. Eschen
USAF role in aviation at Robins AFB and in the state of Georgia with emphasis on military aviation and its significance to national defense.

SAVANNAH VOLUNTEER GUARD MUSEUM
340 Bull Street
Savannah, GA 31401, USA
(912)-233-1430
U.S. Army National Guard Museum

NATIONAL GUARD TROPHY ROOM
P.O. Box 13488
1248 Eisenhower Drive
Savannah, GA 31406, USA
(912)-354-4810
U.S. Army National Guard Museum

CONFEDERATE NAVAL MUSEUM
P.O. Box 1022
Columbus, GA 31902, USA
(404)-327-9798
U.S. Civil War Confederate naval museum.
CONTACT: Robert Holcombe
Museum displays salvaged remains of Confederate gunboats Jackson and Chattahoochee, as well as other exhibits relating to Confederate naval history.

NATIONAL INFANTRY MUSEUM
Building #4 #1260
Fort Benning, GA 31905, USA
(404)-545-5413 (404)-545-2958
Infantry military museum.
Displays all types of military equipment used by the American infantryman, as well as by the infantry of many foreign nations.

FLORIDA NATIONAL GUARD AND MILITIA MUSEUM
82 Marine Street
St. Augustine, FL 32084, USA
(904)-829-2231
U.S. Army National Guard Museum

THE NAVAL AVIATION MUSEUM
U.S. Naval Air Station
Pensacola, FL 32508, USA
(904)-452-3604
U.S. Navy Museum

U.S.A.F. ARMAMENT MUSEUM
Elgin AFB, FL 32542, USA
(904)-882-4189
Military aviation armament museum
CONTACT: Mr. Russell C. Sneddon
Exhibits Air Force armament and delivery systems.

AIR FORCE SPACE MUSEUM
Eastern Space & Missile Center
Patrick AFB, FL 32925, USA
(305)-494-5933
Military aviation museum.
CONTACT: Mr. Donald W. Engel
Exhibits rockets and missiles and related space equipment of the Air Force and associated armed services. Commercial tours daily from Kennedy Space Visitor Center.

U.S ARMY MILITARY POLICE CORPS MUSEUM
U.S. Military Police School
Fort McClellan, AL 36201, USA
(205)-238-3201
Military museum.
CONTACT: Capt. Henry L. Berchak, Curator
Memorabilia and artifacts related to the history and traditions of the U.S. Army Military Police.

EDITH NOURSE ROGERS MUSEUM
U.S. W.A.C. Center & School
Fort McClellan, AL 36201, USA
(205)-238-3533
Military museum.
CONTACT: Elizabeth T. Saus, Curator
Depicts the history, tradition, and development of the U.S. Women's Army Corps.

U.S. ARMY AVIATION MUSEUM
Fort Rucker, AL 36360, USA
(205)-255-4507 (205)-255-4516
Military museum and reference library.
CONTACT: William A. Howell, Curator
The Aviation Museum collects, preserves, restores, and displays aircraft, artifacts, documents and other items which tell the story of Army aviation.

WARRANT OFFICER CANDIDATE HALL OF FAME
Fort Rucker, AL 36360, USA
(205)-255-2806 (205)-255-2815
Military museum.
CONTACT: Capt. W. E. Molden, Curator
The museum displays graphic material and artifacts concerned with the history of Army warrant officer aviators.

LINCOLN MEMORIAL UNIVERSITY
Lincoln Museum
Harrogate, TN 37752, USA
Military library & museum.
CONTACT: Edgar G. Archer, Director

MINIATURE SOLDIER MUSEUM
804 North Third Street
Bardstown, KY 40004, USA
Toy soldier museum.

PATTON MUSEUM OF CAVALRY AND ARMOUR
P.O. Box 208
Fort Knox, KY 40121, USA
(502)-624-3812
Collection of artillery, tanks, & flags.
The museum portrays the evolution, history, and tradition of U.S. cavalry and armor.

KENTUCKY MILITARY HISTORY MUSEUM
E. Main St., U.S. 60
Frankfort, KY 40601, USA
(502)-564-3265
U.S. Army National Guard Museum

KENTUCKY MILITARY HISTORY MUSEUM
P.O. Box H, Old Capitol
East Main Street
Frankfort, KY 40602, USA
U.S. militaria museum.
CONTACT: Lewis N. Hughes, Curator

DON F. PRATT MUSEUM
Heritage Hall
101st Airborne Division
Fort Campbell, KY 42223, USA
(502)-798-3215 (502)-798-4986
Military & firearms museum.
CONTACT: Charles H. Cureton, Director
The museum is centered around the history of the 101st Airborne Division (Ambl) from 1942 to the present.

NEWARK AIR FORCE STATION MUSEUM
P&P Aero Guidance Control Cen.
Newark AFS, OH 43055, USA
(614)-522-7452
Military aviation museum.
CONTACT: Mr. Robert Jones
35 major displays depicting inertial guidance repair for aircraft and missiles and USAF metrology. Open on request.

MILITARY MUSEUMS

THE CLEVELAND MUSEUM OF ART
11150 East Boulevard
Cleveland, OH 44106, USA
Military collection in museum.

U.S. AIR FORCE MUSEUM
HQ USAF Logistics Command
Air Force Museum
Wright-Patterson AFB, OH 45433, USA
(513)-255-7201
U.S.A.F. & general militaria museum.
CONTACT: Jack Hilliard
28,500 items...memorabilia depicting the history of the United States Air Force.

FORT BENJAMIN HARRISON MUSEUM
U.S.A. Administration Center
ATTN: AIZI-PTS
Ft. Benj. Harrison, IN 46216, USA
(317)-542-2989
U.S. Army military museum.
CONTACT: Curator
Museum contains artifacts from all branches of the U.S. Army from the Revolutionary War to the present, and foreign equipment from World War I.

38TH INFANTRY DIVISION MEMORIAL MUSEUM
P.O. Box 41375
Stout Field
Indianapolis, IN 46241, USA
(317)-241-3933
U.S. Army National Guard Museum

STOUT FIELD MILITARY EQUIPMENT MUSEUM
Stout Field
Indianapolis, IN 46241, USA
(317)-241-3933
U.S. Army National Guard Museum

U.S. ARMY FINANCE CORPS MUSEUM
U.S. Army
Finance and Accounting Center
Indianapolis, IN 46249, USA
(317)-542-2441 (317)-542-3021
U.S. Army military museum.
Collection includes all sorts of military and military-related paper items.

BATTLE GROUND HISTORICAL CORP.
P.O. Box 225
Battle Ground, IN 47920, USA
Military museum.
; 15;
Giftshop, theater and museum. Including displays of artifacts and weapons used in the 1800's.

SELFRIDGE MILITARY AIR MUSEUM
Det 1, MU
Selfridge ANG Base, MI 48045, USA
(313)-466-5035
Military aviation museum.
CONTACT: Robert A. Stone,(Col. Ret)
Selfridge Museum by direction depicts the history of Selfridge Field since 1917, and of the units stationed here currently or in the past. A display of photos and biography of Lt. Thomas E. Selfridge. All branches of the Department of Defense, plus the U.S. Coast Guard are represented.

THE HENRY FORD MUSEUM AND GREENFIELD VILLAGE
Dearborn, MI 48124, USA
Military collection.

MILWAUKEE PUBLIC MUSEUM
818 West Wisconsin Avenue
Milwaukee, WI 53233, USA
Military museum.

WISCONSIN VETERANS MUSEUM
Capitol 419 North
Madison, WI 53702, USA
608-266-1680
Museum of WWI & WWII Wisconsin veterans.

G.A.R. MEMORIAL HALL MUSEUM
Capitol 419 North
Madison, WI 53702, USA
608-266-1680
G.A.R. & Civil War veterans museum.

MALMSTROM AFB MUSEUM AND AIR PARK
341 CSC/CDR
Malmstrom AFB, MT 54902, USA
(406)-731-2705
Military aviation museum.
CONTACT: Mr. Gerald Hanson,(LTC Ret)
Portrays base history from 1942 when the base was originally the Great Falls Army Air Base, through the present missile era. Superlative model collection. WWII barracks room and diorama. Air park has LGM30G, KC97, B57, UH1, F101B, F84F, B25.

HISTORIC FORT SNELLING
Minnesota Historical Society
Ft. Snelling History Center
St. Paul, MN 55111, USA
(612)-726-1171
Restored 1820s stone fort.
Extensive living history programs, exhibits in fully furnished 1820's regular army post.

MINNESOTA NATIONAL GUARD HISTORICAL FOUNDATION
P.O. Box 11598
St. Paul, MN 55111, USA
(612)-725-5699
Military aviation museum.
CONTACT: Alfred Schwab,Jr.,(BG Ret)
Preserves the heritage and tradition of the Minnesota Air National Guard. Demonstrates the Minnesota military aviation story through displays of historical significance.

AVIATION ART MUSEUM & MINNESOTA AIR & SPACE MUSEUM
P.O. Box 16224
St. Paul, MN 55116, USA
(612)-291-7925 Off. (612)-780-2435 Hang.
Military & aviation art museum.
S3,S6,S12; 15;
Catalogue available. Annually.
CONTACT: Dennis Eggert
SALES DEPT: Dennis Eggert
PURCHASING DEPT: Dennis Eggert
The museums collect, restore & display several thousand pieces of aviation memorabilia from all facets of aviation, including 9 aircraft. Japanese "Kate", L-13, L-5, BT-13, BT-15, TBM, SNJ, H-34 & UC-78.

SOUTH DAKOTA AIR & SPACE MUSEUM
44 SMW/PACM
Ellsworth AFB, SD 57706, USA
(605)-399-7510
Military aviation/aerospace museum.
CONTACT: Col. Clark W. Ward
The SD Air and Space Museum is attempting to collect, preserve, and display the history of aviation and space exploration as conducted in South Dakota and to portray the history of Ellsworth AFB, the 44th Bomb Wing, 28th Bomb Wing and other USAF groups connected with base area.

OLD FORT MEADE MUSEUM AND HISTORIC RESEARCH ASSOCIATION
Military Museum
Fort Meade, SD 57741, USA
U.S. Army National Guard Museum
Exhibits Indian and military items from the Custer expedition through Vietnam.

GRAND FORKS AIR FORCE BASE MUSEUM
447 SMS/HOA
Grand Forks AFB, ND 58205, USA
(701)-594-6555
Military aviation museum.
CONTACT: 1Lt. Robert C. Walker
Portrays the history of the base and preserves significant events of the present.

MINOT HERITAGE CENTER
91 SMW (SAC)
Minot AFB, NB 58705, USA
(701)-727-3878
Military aviation museum.
CONTACT: Maj. Jay Kaseman
Portrays the history of the base and major assigned units. Open on request.

FORT SHERIDAN MUSEUM
Building 33
Fort Sheridan, IL 60037, USA
(312)-926-2137
Military museum.
The museum portrays the notable achievements of the fort and the U.S. Army.

G.A.R. MEMORIAL & VETERANS MUSEUM
23 East Downer Place
Aurora, IL 60504, USA
G.A.R. and U.S. militaria museum. G.A.R. reference material available.

THE GEORGE F. HARDING MUSEUM
86 E. Randolph Street
Chicago, IL 60601, USA
Military museum.

THE ROCK ISLAND ARSENAL MUSEUM
ATTN: SMCRI-ADW-B
Rock Island Arsenal
Rock Island, IL 61299, USA
(309)-782-5021
Military history museum of Rock Island Arsenal.
; 82;
CONTACT: Daniel T. Whiteman, Director
U.S. Army museum of the history of Rock Island Arsenal & Arsenal Island. Collection includes RIA manufactured items & 2,000 military small arms.

MILITARY MUSEUMS

MEMORIAL HALL OF FLAGS
1st Floor-North
Centennial Bldg, 2nd & Edwards
Springfield, IL 62704, USA
(217)-782-7221
U.S. Army National Guard Museum
Collection of 463 flags carried by Illinois regiments in four wars.

ADJUTANT GENERAL'S OFFICE
Old State Capitol
Fifth and Adams Streets
Springfield, IL 62706, USA
(217)-782-7221
U.S. Army National Guard Museum
The room used by the Adjutant General of Illinois from 1839 to 1876.

THE CITY ART MUSEUM OF ST. LOUIS
St. Louis, MO 63105, USA
Military collection.

KIEFNER-KANE MUSEUM
26 Sherman Road
Jefferson Barracks, MO 63125, USA
(314)-894-0363
U.S. Army National Guard Museum

LIBERTY MEMORIAL MUSEUM
100 West 26th Street
Kansas City, MO 64108, USA
816-221-1918
WWI militaria museum

FORT LEONARD WOOD MUSEUM
ATTN: ATZT-DPT-O
Fort Leonard Wood, MO 65473, USA
(314)-368-4249
U.S. Army military museum.
The museum's collection comprises weapons and uniforms from WWII, Korea, and Vietnam; items relating to the history of Fort Leonard Wood; Gen. Wood's personal collection; and a large outdoor display of guns and vehicles.

FORT LEAVENWORTH MUSEUM
Reynolds & Gibbon Avenues
Fort Leavenworth, KS 66027, USA
(913)-684-3553 (913)-684-3191
Museum of U.S. militaria 1827 to date. Reference library available.
The museum exhibits artifacts and materials interpreting the history of Fort Leavenworth and the history of the U.S. Army in the development of the Trans-Mississippi West.

U.S. CAVALRY MUSEUM
P.O. Box 2160
Fort Riley, KS 66442, USA
(913)-239-2737
U.S. Cavalry & military museum.
; 15;
Catalogue available. Bi-annually.
CONTACT: Terry Van Meter, Director
History of the American Cavalry 1775-1950. Open 7 days a week, closed Easter, Thanksgiving, Christmas, New Years. Free admission. Gift Store featuring limited edition prints, books, statues, records, and posters.

FORT ATKINSON STATE HISTORICAL PARK
P.O. Box 237
Fort Calhoun, NE 68023, USA
Military & regional history museum.
CONTACT: Steve Kemper, Superintendent

NEBRASKA GAME & PARKS COMMISSION
P.O. Box 30370
Lincoln, NE 68503, USA
State historical parks commission.
CONTACT: Ted Stutheit, Chief

FORT HARTSUFF STATE HISTORICAL PARK
Route 1
Burwell, NE 68823, USA
Military historic site & museum.
CONTACT: Roye Lindsay, Superintendent

FORT KEARNEY STATE HISTORICAL PARK
Route #4
Kearney, NE 68847, USA
Military historic site & museum.
CONTACT: Gene Hunt, Superintendent

LOUISIANA MILITARY HISTORY AND STATE WEAPONS COLLECTION
Jackson Barracks
New Orleans, LA 70146, USA
(504)-271-6262
U.S. Army National Guard Museum
Relics of the military service of Louisiana National Guard Units.

8TH AIR FORCE MUSEUM
2nd BW/PA
Barksdale AFB, LA 71110, USA
(318)-456-3065
Strategic Bombardment museum
CONTACT: Maj. Phillip E. Richards
History and development of Strategic Bombardment through the history of the 2nd Bombardment Wing, 1918 - Present (bombing theory and practice); 8th Air Force, 1942 - Present (bombing practice); Barksdale AFB, 1933 - Present (bombardment training).

FORT POLK MILITARY MUSEUM
Fort Polk, LA 71459, USA
(318)-578-4810
U.S. Army military museum.
The museum emphasizes the history of the 7th and 11th Armored Divisions. Items include U.S. and foreign weapons, ordnance, vehicles, flags, photographs.

45th INFANTRY DIVISION MUSEUM
2145 N.E. 36th Street
Oklahoma City, OK 73111, USA
(405)-424-5313
Oklahoma military history museum.
; 4;
CONTACT: Edison A. Reber, Chrm. of Bd.
PURCHASING DEPT: William L. Pigg
Museum preserves the military history of Oklahoma organizations and servicemen and displays military firearms and weapons, uniforms, insignia, original Mauldin cartoons, aircraft, vehicles, artillery, etc.

U.S. ARMY FIELD ARTILLERY & FORT SILL MUSEUM
Museum Building 437
Fort Sill, OK 73503, USA
(405)-351-5123 (405)-351-5318
U.S. Army artillery museum.
CONTACT: Herbert C. Morrow, Director
The collection is devoted to the history of the Field Artillery and the history of Fort Sill.

18TH FIELD ARTILLERY REGIMENT MUSEUM
HQ and HQ Battery
2nd Battalion, 18th F.A.
Fort Sill, OK 73503, USA
(405)-351-4391
U.S. Army military museum.
The museum preserves items pertaining to the history of the 18th Field Artillery.

J. M. DAVIS GUN MUSEUM
P.O. Box 966
333 N. Lynn Riggs Blvd.
Claremore, OK 74017, USA
(918)-341-5707
Firearms museum
CONTACT: Lee Good, Director
J. M. Davis Gun Museum has 20,000 firearms and closely associated pieces. Many are military pieces and come from all periods of American history, ending with WWII.

2ND ARMORED DIVISION MUSEUM
Fort Hood, TX 76544, USA
(512)-685-5632
U.S. Army military museum.
The museum presents items relating to the history of the 2nd Armored Division.

1ST CAVALRY DIVISION MUSEUM
Fort Hood, TX 76545, USA
(817)-685-3431
U.S. Army military museum.
The museum presents items relating to the history of the 1st Cavalry Division from 1855 to the present.

AUDIE L. MURPHY GUN MUSEUM
Hill College
P.O. Box 619
Hillsboro, TX 76645, USA
(817)-582-2555
Gun & uniform museum.
Audie Murphy exhibit, Edged weapon display, Special WWII German & Japanese exhibit.

FORT CONCHO MUSEUM
213 E. Avenue D
San Angelo, TX 76903, USA
Military & regional history museum.

U.S. ARMY MEDICAL MUSEUM
Academy of Health Sciences
Fort Sam Houston, TX 78234, USA
(512)-221-2358
U.S. Army military museum.
The museum displays items related to the history and development of the Army Medical Department. Most of the items date from the early 1800's.

MILITARY MUSEUMS

FORT SAM HOUSTON MILITARY MUSEUM
Fort Sam Houston, TX 78234, USA
(512)-221-6117
U.S. Army military museum.
The museum depicts the history of the fort and the local area. There are also collections, including one relating to the 2nd Infantry Division, and another of items donated by survivors of the Bataan Death March.

HANGAR 9, EDWARDS H. WHITE II MEMORIAL MUSEUM
6570th ABG History Office
Brooks AFB, TX 78235, USA
(512)-536-2203
Aviation/Aerospace Medicine museum.
CONTACT: Mr. Michael Martinez
Hangar 9, Edward H. White II Memorial Mluseum, houses the history of aviation/aerospace medicine and contains exhibits concerning the history of the base. Art work, , Sam and Enos the Astrochimps's space capsules and medical research and testing equipment from 1917 to present.

HISTORY AND TRADITIONS MUSEUM
AFMTC/LGHM
Lackland AFB, TX 78236, USA
(512)-671-3055
Military aviation museum.
CONTACT: Ms. Gloria M. Livingston
Exhibits objects of historical Air Force significance in aerospace evolution. Presents and shares Air Force heritage. Strives to educate and inspire military and civilian personnel, especially the trainees, and the local community.

U.S.A.F. SECURITY POLICE MUSEUM
3280 Technical Training Group
Lackland AFB, TX 78236, USA
(512)-671-2125
Military aviation museum.
CONTACT: Col. Albert Feldman
Preserves the heritage and tradition of the Air Force Security Police.

PANHANDLE-PLAINS HISTORICAL MUSEUM
P.O. Box 967
W. T. Station
Canyon, TX 79016, USA
Military & regional history museum.

DYESS AFB MUSEUM
96 BMW/CVI
Dyess AFB, TX 79607, USA
(915)-696-3001
Military aviation museum
CONTACT: LTC Richard B. Morganti
The development of Dyess AFB and the AAF. World War II service of the 96th BMG and base tenants.

U.S. ARMY AIR DEFENSE ARTILLERY MUSEUM
Fort Bliss, TX 79916, USA
(915)-568-6009 (915)-568-5412
U.S. Army military museum.
CONTACT: Sam Hoyle
The collection comprises weapons, uniforms, and other historical items dating from the inception of the Coast Artillery Corps to the present day.

FORT BLISS REPLICA MUSEUM
U.S. Army Air Defense Center
Fort Bliss, TX 79916, USA
(915)-568-2804
U.S. military history museum.
CONTACT: Office of the Curator
The museum presents the history of the U.S. Army in relation to Fort Bliss, El Paso, and the southwestern region of the United States.

U.S. ARMY MUSEUM OF THE N.C.O.
Sergeants Major Academy
Fort Bliss, TX 79916, USA
U.S. Army military museum.

UNIVERSITY OF TEXAS AT EL PASO
El Paso, TX 79968, USA
Military library & museum.
CONTACT: Attn: Gerald-Museum

LOWRY HERITAGE MUSEUM
P.O. Box 30035
Lowry AFB, CO 80230, USA
(303)-370-3028
Military aviation museum.
CONTACT: Mr. Steve C. Draper
SALES DEPT: Alice Ellis
The Lowry Heritage Museum interprets the history of Lowry AFB, its antecedents, the history of aerospace photography, the USAFA at Lowry and the history of its training schools. The museum displays 15 static aircraft, including a B-29.

THE MOUNTAIN POST MUSEUM
Ft. Carson, CO 80913, USA
(303)-579-2908 (303)-579-3256
Military museum.
CONTACT: James J. Bush
The museum displays military artifacts and memorabilia acquired primarily from veterans of units which have trained or passed through Fort Carson.

EDWARD J. PETERSON SPACE COMMAND MUSEUM
1 Space Wing/PACM
Peterson AFB, CO 80914, USA
(303)-554-4915
Military aviation museum.
CONTACT: Capt. James G. Pasierb
Tells the Peterson Field story starting with the base inception in May 1942 and depicting current events for succeeding airmen.

F. E. WARREN AIR FORCE BASE MUSEUM
90 SMW (SAC)
Warren AFB, WY 82001, USA
(307)-775-2980
Military aviation museum.
CONTACT: Mr. Jess Parson
Portrays the unique history of the longest continuously active Air Force base, the Jolly Roger fame of WWII and the strategically important history and mission of the current 90th Strategic Missile Wing.

MUSEUM OF FLIGHT
OO-ALC/XRH
Hill AFB, UT 84056, USA
(801)-777-6618
Military aviation museum.
CONTACT: Mr. James Brandell
The mission is to portray the history of Hill AFB and its role in associated maintenance, supply repair, and training functions, and the history of the Ogden Air Logistics Center and the units, past and present.

FORT DOUGLAS MUSEUM OF MILITARY HISTORY
U.S. Army Support Detachment
Salt Lake
Fort Douglas, UT 84113, USA
(801)-524-4098
U.S. Army military museum.
The museum tells the story of the U.S. Army in Utah and the inter-mountain west, with the emphasis on the Civil War and the history of Fort Douglas from 1876 to 1880.

JOHN M. BROWNING ARMORY
625 East 5300 South
Ogden, UT 84402, USA
(801)-392-7581
U.S. Army National Guard Museum
Ehibits approximately 260 original test model small arms developed by John M. Browning and his descendants.

ARIZONA MILITARY MUSEUM
5636 E. McDowell Road
Phoenix, AZ 85008, USA
Regional military history museum.

U.S. ARMY INTELLIGENCE MUSEUM
U.S.A. Intelligence Center
Fort Huachuca, AZ 85613, USA
(602)-538-3740 (602)-538-3706
Military museum.
CONTACT: Thomas Cavanaugh
The museum displays confiscated sabotage, espionage, and subversion items, photo-interpretation collections, and documents relating to the history of intelligence operations.

FORT HUACHUCA HISTORICAL MUSEUM
P.O. Box 766
Fort Huachuca, AZ 85613, USA
(602)-538-3638 (602)-538-5736
Military & regional history museum and library.
CONTACT: Jmes P. Finley, Curator
The museum's collection focuses on the military's contribution to southwestern history.

NATIONAL ATOMIC MUSEUM
Defense Nuclear Agency
Kirtland AFB, NM 87115, USA
(505)-264-4223
U.S. Military Museum.

RESCUE MEMORIAL MUSEUM
1550 Aircrew Tng & Test Wing
Kirkland AFB, NM 87117, USA
(505)-844-5140
Military aviation museum.
CONTACT: Capt. Joseph Dye
Exhibits memorabilia of historical significance in both Air Force Rescue and Weather Services.

MUSEUM OF NEW MEXICO
P.O. Box 2087
Santa Fe, NM 87503, USA
Military & regional museum.

MILITARY MUSEUMS

WHITE SANDS MISSILE RANGE MISSLE PARK
White Sands Missile Range
NM 88002, USA
(915)-678-1134
U.S. Army military museum.
About 40 missiles and rockets, which were actually tested here, are displayed.

MUSEUM OF NATURAL HISTORY
Research Library
900 Exposition Blvd.
Los Angeles, CA 90007, USA
Military collection.

FORT MAC ARTHUR MUSEUM
Fort Mac Arthur Museum
San Pedro, CA 90731, USA
(213)-643-0667
Military aviation museum.
CONTACT: LtCol. Gary L. MacKenzie
Illustrates the history of the USAF in the Los Angeles area, the history of Air Force Systems Command, and Fort Mac Arthur in the Age of Aerospace.

HENRY HUNTINGTON LIBRARY & ART MUSEUM
San Marino, CA 91108, USA
Military collection.

NAVAL TRAINING CENTER HISTORICAL MUSEUM
Naval Training Center
San Diego, CA 92133, USA
(714)-225-5495
U.S. Navy Museum

MARCH FIELD MUSEUM
22 AREFW/PACM
March AFB, CA 92518, USA
(714)-655-3725
Military aviation museum.
CONTACT: Maj. Michael Frietas
Depicts the story of March Field and present base activities and accomplishments.

CIVIL ENGINEER CORPS/SEABEE MUSEUM
Code 2232
U.S. NCBC
Port Hueneme, CA 93043, USA
(805)-982-5163 (805)-982-4621
Museum of civil engineers & seabees, library and gift shop.
S1,S4; 40;
Catalogue available. Annually.
SALES DEPT: Andrea King
PURCHASING DEPT: Andrea King
Seabee Museum - Port Hueneme, CA. A depository for thousands of momentos of the history of the Naval Construction Force Units, called the Seabees.

CALIFORNIA ARMY NATIONAL GUARD MUSEUM
P.O. Box G
Camp San Luis Obispo, CA 93406, USA
(805)-543-8287
U.S. Army National Guard Museum

NAVAL SHIPYARD MUSEUM
837 Fourth Street
Bremerton, WA 93910, USA
(206)-373-2991
U.S. Navy Museum.

U.S. ARMY MUSEUM, PRESIDIO OF MONTEREY
Monterey, CA 93940, USA
(408)-242-8547
Military museum.
CONTACT: Margaret B. Adams
The general collection pertains to the military history of the Monterey peninsula.

PRESIDIO ARMY MUSEUM
Building #2
San Francisco, CA 94129, USA
(415)-561-3319 (415)-561-4115
Military history museum & Haven reference library.
CONTACT: J. Phillip Langellier, Curator
Displays artifacts and other related material dealing with the history of the Presidio of San Francisco from the Spanish occupation in 1776 through the present.

TRAVIS AIR FORCE MUSEUM
60 MAW/HOX
Travis AFB, CA 94535, USA
(707)-438-5605
Military aviation museum.
CONTACT: Lt. Col. Louis J. Tobin
Depicts the history of Travis Air Force Base and airlift in the Pacific.

MILITARY MEDAL MUSEUM & RESEARCH CENTER
448 North San Pedro Street
San Jose, CA 95110, USA
(408)-298-1100 (408)-241-1656
General militaria medal museum. Appraisals & identifications available.
CONTACT: John Langton,Jr.

CASTLE AIR MUSEUM
HQ 93rd Bomb Wing (SAC)
Castle AFB, CA 95342, USA
(209)-726-4140
Military aviation museum.
CONTACT: Lt. Landis McGauhey
Depicts history of Strategic Air Operations with primary emphasis on the activities of Castle Air Force Base and the 93rd Bomb Wing.

McCLELLAN AVIATION MUSEUM
HQ Sacramento Air Log. Center
McClellan AFB, CA 95652, USA
(916)-643-3192
Military aviation museum.
CONTACT: Mr. Max Gobbel
History of McClellan AFB: McClellan begining, ground breakinag, construction, dedication (1839-1939). Logistics support pre-World War II and World War II (1939-1945), troubled peace - Korean conflict (1945-1954), Viet Nam (1964-1974), revolution in air logistics (1974

SILVER WINGS AVIATION MUSEUM
323 Flying Training Wing
Mather AFB, CA 95655, USA
(916)-364-2177
Military aviation museum.
CONTACT: Mr. Derrel Fleener
Exhibits artifacts relevant to flight and the history of navigation.

CALIFORNIA STATE MILITARY RESERVE
P.O. Box 214405
2829 Watt Avenue
Sacramento, CA 95821, USA
Military museum.

EDWARD F. BEALE MUSEUM
9 SRW/CCX BLD 2471
Beale AFB, CA 95903, USA
(916)-634-2038
Military aviation museum.
CONTACT: Col. Robert B. McConnell
Depicts the Beale Story dating from 1800 and will continue to depict current events of historic importance for succeeding generations.

SCHOFIELD BARRACKS POST MUSEUM
DPTS, USASCH
APO San Francisco, CA 96557, USA
(808)-655-0438
Military museum.
The museum contains material related to the history of the Barracks and its tenant units.

OREGON NATIONAL GUARD MILITARY MUSEUM & RESOURCE CENTER
Camp Withycombe
Clackamas, OR 97015, USA
(503)-656-9331
Military museum & history center.

MUSEUM OF FLIGHT
9404 East Marginal Way South
Seattle, WA 98108, USA
(206)-767-7373
Aircraft & aviation museum.
S1,S4; 3;
AD MANAGER: Cynthia Green
Independent, non-profit cultural institution dedicated to preservation of aircraft and related aviation artifacts. Educational facilities, archives, library. New "Great Gallery" addition opens in late Spring 1987.

FORT LEWIS MILITARY MUSEUM
ATTN: AFZH-PTPM
Fort Lewis, WA 98433, USA
(206)-968-4796 (206)-968-5835
U.S. Army military museum.
The museum displays artifacts relating to the history of the U.S. Army in the Northwest, with the emphasis on Fort Lewis and the units stationed there.

McCHORD AIR FORCE BASE MUSEUM
McChord AFB, WA 98438, USA
(206)-984-2935
Military aviation museum.
CONTACT: Mr. Fred Johnson
Preserves and displays aspects of McChord's history and that of the various units which have served there.

MILITARY MUSEUMS

FAIRCHILD AIR FORCE BASE MUSEUM
92 CSG/CDR
Fairchild AFB, WA 99011, USA
(509)-247-2100
Military aviation museum.
CONTACT: Capt. D. Emoki
Museum depicts military history of Spokane, Washington area, including Fort George Wright (1898-1956), Felts Field (1924-1949), Geiger AFB (1940-1965), and Fairchild AFB from 1942-Present. Unit histories of USAF and ANG units stationed in the area are also covered.

MUSEE LANGLOIS
Rue Sadi-Carnot
Caen, 14000, France
Military art museum.
CONTACT: M. B. d'Ymouville

MUSEE MUNICIPAL DES BEAUX-ARTS
63, avenue Charles-de-Gaulle
Rochefort, 17300, France
(46) 99.20.70
Museum of local history.
CONTACT: M. Georges Barret

MUSEE DEL LA MANUFACTURE D'ARMES DE TULLE
1, place Albert-Faucher
Tulle, 19012, France
(55) 20.20.09
Armory museum.
CONTACT: M. Jean-Marie Soubrennie

MUSEE ANTOINE CHAINTREUIL
Hotel de Ville
66, ave. du Marechal Tassigny
Pont-De-Vaux, Ain 01190, France
(85) 37.31.01
French military museum, post Revolution museum.
CONTACT: M. Guy Doucet

MUSEE GENERAL-ESTIENNE
Berry-au-Bac
Guignicourt, Aisne 02190, France
(23) 22.45.25
Museum of Gen. Estienne & the French armored forces in both World Wars.
CONTACT: M. Jacques Bahin

MUSEE DE LA COOPERATION FRANCO-AMERICAINE
Blerancourt
Chauny, Aisne 02300, France
(23) 52.60.16
American Revolution & both World Wars museum & library.
CONTACT: M. Rosenberg

TOUR-MUSEE DU CHATEAU
Chateau de Coucy
Coucy-Le-Chateau, Aisne 02380, France
(23) 52.70.05
Fortress museum of the 13th century.
CONTACT: M. Le Maire

SALLE HISTORIQUE 1814
Essises
Chateau-Thierry, Aisne 02400, France
(23) 82.80.45
Napoleonic museum. Depicts the Battle of Caquerets in February 1814.

CONTACT: M. Christian Trehel

MUSEE MUNICIPAL
32, rue Georges-Ermant
Laon, Aisne 02000, France
(23) 23.22.05
Medievel weapons, 13th thru 15th centuries.
CONTACT: Mme. Rollas

MUSEE DE LA CAVERNE DU DRAGON
Chemin des Dames
Oulche-la-Valle-Foulon
Beaurieux, Aisne 02160, France
(23) 22.40.43
WWI subterranean fortress museum.
CONTACT: Mm. Gerard de Francqueville

MUSEE ALEXANDRE DUMAS
24, rue Demoustier
Villers-Cotterets, Aisne 02600, France
(23) 96.23.30
Museum of General Dumas.
CONTACT: M. Marcel Leroy

MUSEE D'ART ET D'ARCHEOLOGIE
3, place du Colonel-Laussedat
Moulins, Allier 03000, France
(70) 44.22.98
Collection of 500 edged weapons & guns from the middle ages to the present.

MUSEE DEPARTEMENTAL DE GAP
6, Avenue du Marechal-Foch
Gap, Haute-Alpe 05000, France
(92) 51.01.58
Museum containing military arms from the 17th century thru the 20th.
CONTACT: M. Georges Dusserre

EXPOSITION PERMANENTE 'VAUBAN DANS LES ALPES'
Citadelle de Montdauphin
Guillestre, Haute-Alpe 05600, France
(92) 45.00.14
17th century fortress designed by Vauban and containing a museum.
CONTACT: M. Godard

MUSEE DES CHASSEURS A PIED
Rue du General-Van-den-Berg
Antibes, Alpes-Marit 06600, France
(93) 34.39.58
Museum presenting the chasseurs a pied from the 2nd Empire to the present day.
CONTACT: M. Le Commandant Darbois

MUSEE NAVAL ET NAPOLIONIEN
Batterie du Grillon
blvd. Kennedy, Cap-d'Antibes
Antibes, Alpes-Marit 06600, France
(93) 61.45.32
Naval museum of the Napoleonic period.
CONTACT: M. Vidal

MUSEE D'ART ET D'HISTOIRE DE PROVENCE
2, rue Mirabeau
Grasse, Alpes-Marit 06130, France
(93) 36.01.61
17th & 18th century military museum.

MUSEE MASSENA
65, rue de France
Nice, Alpes-Marit 06000, France
(93) 88.11.34
Museum of arms from the 15th, 16th, and 19th centuries & Napoleonic exhibits.
CONTACT: M. Claude Fournet

MUSEE NAVAL DE NICE
Tour Bellanda
place du Chateau, Alpes-Marit 06000, France
(93) 80.47.61
17th, 18th & 19th century naval museum.
CONTACT: Commander Belic

MUSEE INTERNATIONAL DE LA FORCE PUBLIQUE
Citadelle de Villefranche-Mer
Villefranche, Alpes-Marit 06230, France
(93) 55.45.12
Museum of the Army & Police from the 18th century to the present day.
CONTACT: Mlle. Madeleine Servera

MUSEE VIVAROIS CESAR FILHOL
Musee Vivarois Cesar Filhol
Annonay, Ardeche 07100, France
(75) 33.24.51
Local history museum.

MAISON DE LA DERNIERE CARTOUCHE
Ave. de la Derniere Cartouche
Bazeilles, Ardennes 08140, France
Franco-Prussian War museum.
CONTACT: M. Le Commandant Crequit

MUSEE DE LA BATAILLE DES ARDENNES
Novion-Porcien, Ardennes 08270, France
(24) 20.20.13
WWII military museum.
CONTACT: M. Jean Alexandre

MUSEE DE RETHELOIS ET DE PORCIEN
Tribunal de Rethel
Rethel, Ardennes 08300, France
Military museum depicting local history from the 17th Century to WWII.
CONTACT: M. Chanonier

MUSEE MILITAIRE
Chateau de Sedan
Sedan, Ardennes 08200, France
(24) 29.03.28
Military museum presenting arms & other items from the 17th Century to WWII.

FORT DE VILLY LA FERTE
Villy-sur-Chiers
Margut, Ardennes 08370, France
(24) 22.06.72
Maginot Line museum.
CONTACT: M. Pierre Buard

MUSEE DE L'ARIEGE
Chateau des Comtes de Foix
Foix, Ariege 09000, France
Museum of arms from the Middle Ages and the Orient.

MILITARY MUSEUMS

BRIENNE-LE-CHATEAU
Ancienne Ecole Militaire
rue de l'Ecole Militaire
Brienne-le-Chatea, Aube 10500, France
(25) 77.80.31
Museum of Napoleon's campaign of 1814.
CONTACT: M. Philippe Bera

MUSEE DE LA RESISTANCE
Rue Boursault
Mussy-Sur-Seine, Aube 10250, France
(25) 38.40.10
Museum of the WWII French Resistance.
CONTACT: M. Le Maire

MUSEE DES BEAUX-ARTS ET D'ARCHEOLOGIE
Ancienne abbaye Saint-Loup
Troyes, Aube 10000, France
(25) 43.49.49
Museum of Beaux-arts and of Archaeology.
CONTACT: Jean-Pierre Saint-Marie

MUSEE MILITAIRE DE GRUISSAN
Chex M. Roger Bosc
Gruissan, Aude 11430, France
(68) 94.00.08
Museum of elite French forces from the 19th century to the present date.
CONTACT: M. Roger Bosc

MUSEE FENAILLE
3, rue Saint-Just
Rodez, Aveyron 12000, France
(65) 68.12.08
Museum of arms from the 14th century to the 16th century.
CONTACT: M. Louis Balsan

MUSEE DE LA LEGION ETRANGERE
Quartier Vienot
Aubagne, 13400, France
(42) 03.03.20
Museum of the French Foreign Legion.
CONTACT: Ministere de la Defense

BASILIQUE NOTRE-DAME-DE- LA GARDE
Marseille, France
WWII military museum.

MUSEE DE LA MARINE
Palais de la Bourse
La Canebiere
Marseille cedex 1, 13222, France
(91) 91.91.51
Naval museum covering from the 9th century to the present day.
CONTACT: M. F. Reynaud

MUSEE GROBET-LABADIE
140, Bd de Longchamp
Marseille cedex 1, 13001, France
(91) 62.21.82
Museum of arms from the 16th and 17th centuries.
CONTACT: Mme. Marielle Latour

SALLE D'HONNEUR ET MUSEE DE L'INSTITUTION DES INVALIDES
Puyloubier
Puyloubier, 13114, France
Museum of the French Foreign Legion.

MUSEE DE L'EMPERI
Chateau de l'Emperi
Salon-de-Provence, 13300, France
(90) 56.22.36
Military museum covering the period from 1700-1918.
CONTACT: M. Raoul Brunon

SALLE D'HONNEUR DE L'ECOLE DE L'AIR
Chateau de Vincennes
Vincennes, 94300, France
Museum of French military aviation.

EXPOSITION PERMANENTE DU DEBARQUEMENT
Arromanches, Calvados 14117, France
(31) 22.34.31
Museum of the 1944 invasion of France.
CONTACT: M. Raymond Triboulet

MUSEE DES BALLONS
Chateau de Balleroy
Balleroy, Calvados 14490, France
(31) 21.60.61
Museum depicting the history of ballooning.
CONTACT: Mme. Leprieur-Decker

MUSEE BARON GERARD
Cours des Tribunaux
Bayeux, Calvados 14400, France
(31) 92.14.21
Museum of military art.
CONTACT: Mme. Durand

MUSEE DE LA REINE MATHILDE
Rue Leonard-Leforestier
Bayeux, Calvados 14400, France
(31) 92.05.48
Museum of 11th century military history.
CONTACT: Mme. Coic

MUSEE MEMORIAL DE LA BATILLE DE NORMANDIE
Boulevard Fabian-Ware
Bayeux, Calvados 14400, France
(31) 92.93.41
WWII military museum.
CONTACT: Dr. Jean Pierre Benamou
Museum of the 1944 Battle of Normandie. Including 120 mannequins, 6 tanks, 10 guns, and thousands of other military artifacts.

MAIRIE DE BEAUMONT
Beaumont-en-Auge, Calvados 14950, France
Museum of military art.

MUSEE DES TROUPES AEROPORTEES DE PEGASUS-BRIDGE
10, ave. du Commandant-Kieffer
Benouville, Calvados 14970, France
(31) 93.31.22
WWII airborne museum.
CONTACT: M. Raymond Triboulet

MUSEE DE LA MARINE
Musee du Vieux Honfleur
1, rue de la Prison
Honfleur, Calvados 14600, France
17th & 18th century naval museum.
CONTACT: M. Pierre Orange

MUSEE #4 COMMANDO
Place a.-Thomas
Ouistreham, Calvados 14150, France
(31) 97.18.63
WWII military museum depicting the Normandy invasion.
CONTACT: M. Jean Longuet

PRIEURE DE SAINT-HYMER
Hospice de Saint-Hymer
Pont-l'eveque, Calvados 14130, France
Napoleonic military museum.
CONTACT: M. De Pezeville

MUSEE DE LA LIBERATION DE NORMANDIE
Route Nationale 13
Surrain, Calvados 14710, France
(31) 22.57.56
WWII military museum.
CONTACT: M. Bonne

MUSEE DE LA BATAILLE TILLY ET ENVIRONS
Rue du 18-juin-1944
Tilly-sur-Seulles, Calvados 14250, France
(71) 80.83.11
WWII military museum.
CONTACT: M. Paul Jehenne

CASERNE DE POMPIERS
Caserne des Pompiers de Trouvi
Trouville, Calvados 14360, France
(71) 80.83.11
Firemen's museum.
CONTACT: Lieutenant Jacques Marsai

MUSEE DE LA SOCIETE ARCHEO-LOGIQUE ET HISTORIQUE
44,rue de Montmoreau
Angouleme, Charente 16000, France
14th thru 16th century military museum.

MUSEE MUNICIPAL
1, rue Friedland
Angouleme, Charente 16000, France
(45) 95.07.69
Museum of arms & armour.
CONTACT: Mme. Monique Bussac

ETABLISSEMENT DES CONSTRUCTION ET ARMES NAVALES DE RUELLE
Ruelle, Charente 16600, France
(45) 68.90.11
Naval museum.

MUSEE NAPOLEONIEN
Rue Napoleon
Ile D'aix, 17123, France
(46) 88.66.40
Napoleonic museum.
CONTACT: M. Gerard Hubert

MUSEE DE LA TOUR DE LA CHAINE
Tour de la Chaine
La Rochelle, 17000, France
(46) 41.14.68
16th century military museum.
CONTACT: M. Michel Gaillard

MILITARY MUSEUMS

MUSEE MUNICIPAL DES BEAUX-ARTS
63, avenue Charles-de-Gaulle
Rochefort, 17300, France
(46) 99.20.70
WWII military museum.
CONTACT: M. Georges Barret

MUSEE NAVAL HOTEL DE CHEUSSSES
Place de la Galissoniere
Rochefort, 17300, France
(46) 99.36.00
Naval museum.
CONTACT: M. Guigue

MUSEE DE LA SOCIETE D'ARCHEO-LOGIE DE SAINT-JEAN-D-ANGELY
Rue de Verdun
Saint-Jean-D'Ange, 17400, France
(46) 32.26.54
Military museum. Middle ages to the present day.
CONTACT: Mme. A. Lavallee-Beaudonnet

MUSEE NAVAL ET COGNACQ
Hotelde Clerjotte
avenue V-Bouthillier
17410, 17400, France
(46) 09.21.22
Naval museum. 17th century to the present day.
CONTACT: M. Guigue

MUSEE DUPUY-MESTREAU
4, Rue Monconseil
Saintes, 17100, France
(46) 93.36.71
Military museum.

MUSEE DE L'ETABLISSEMENT D'ETUDES ET DE FABRICATION ARM
6, route de Guery - B.P. 705
Bourges cedex, Cher 18015, France
(48) 50.21.01
Museum of firearms manufacture.

SALLES PEDAGOGIQUES L'ECOLE SUPERIEURE D'APPLICATION DU MA
Caserne Carnot, avenue Carnot
Bourges, Cher 18015, France
(48) 70.35.77
Military museum.

MUSEE SAINT-VIC
Cours Manuel
Saint-Amand-Montr, Cher 18200, France
(48) 96.55.20
Regional history museum.
CONTACT: M. Jean-Yves Hugoniot

MUSEE EDMOND-MICHELET
Rue Champanatier
Brive-La-Gaillard, Correze 19100, France
(55) 74.06.08
WWII French Resistance museum.
CONTACT: Mme. Edmond Michelet

MUSEE ERNEST RUPIN
1, place Albert-Faucher
Tulle, Correze 19012, France
(55) 20.10.09
Military museum displaying arms from the middle ages to the 19th century.
S9;
CONTACT: MME. CLAIRE MOSER

MUSEE DU CLOITRE ANDRE-MAZEYRIE
Place de la Cahtedrale
Tulle, Correze 19012, France
(55) 26.22.05
18th century armory museum.
CONTACT: Mme. Isabelle Dulac-Rooryck

MUSEE NAPOLEONIEN DE L'HOTEL DE VILLE
Hotel de Ville
Ajaccio, Corse 20000, France
(95) 21.48.17
Napoleonic museum.
CONTACT: Mme. Marie-Dominique Roche

MUSEE NATIONAL DE LA MAISON BONAPARTE
Rue Saint-Charles
Ajaccio, Corse 20000, France
(95) 21.43.89
Napoleonic museum.
CONTACT: M. Gerard Hubert

PALAIS FESCH
50-52, rue du cardinal-Fesch
Ajaccio, Corse 20000, France
(95) 21.48.17
Napoleonic art museum.

MUSEE D'ETHNOGRAPHIE CORSE
Palais des Gouverneur genois
Bastia, Corse 20200, France
(95) 31.09.12
Napoleonic museum.
CONTACT: Mme. Janine Serafini-Costoli

MUSEE BONAPARTE
Chateau Fort
Auxonne, Cote-D'Or 21130, France
(80) 38.10.65
Napoleonic museum.

CHATEAU DE BUSSY RABUTIN
Rue du Chateau
Bussy-Le-Grand, Cote-D'Or 21150, France
(80) 96.00.03
17th century military portrait museum.
CONTACT: M. L'Architecte des Batiments

MUSEE DU CHATEAU
Musees de Vitre
Vitre, 35500, France
(99) 75.04.54
Local history museum.
CONTACT: M. Patrice Forget

MUSEE DES EQUIPAGES MILITAIRES ET DU TRAIN
Quartier de Beaumont
rue du Plat-d'Etain
Tours Cedex, 37034, France
47.61.44.46
Museum depicting the history of the French Army baggage train, 1807-1987.
S12,S1,S4; 7;
CONTACT: Mr. Le General Villaume
SALES DEPT: A/C Caron J.P.
PURCHASING DEPT: A/C Caron J.P.

MUSEES DES BEAUX-ARTS
Place de l'Hotel-de-Ville
Agen, Lot-et-Garo 47000, France
(58) 66.35.27
Museum of Fine Arts.
CONTACT: Madame Esquirol

MUSEE DES GOUMS MIXTES MAROCAINS
Chateau de Montsoreau
Montsoreau, Maine-et-Lo 49730, France
(41) 51.70.25
Military museum depicting the history, uniforms & arms of the Goums, 1908-1956.
CONTACT: J. Andre Pasquier

MUSEE DU CHEVAL
Musee du Cheval
Chateau de Saumur
Saumur, 49400, France
(41) 51.30.46
Military museum
CONTACT: Mlle. Monique Jacob

MUSEE DU FORT DE LA POMPELLE
1914/1918
Route de Chalons-sur-Marne
Reims, Marne 51100, France
(26) 49.11.85
Military museum. WWI and 1870 - 1945.
S12; 20;
CONTACT: M. Andre Lamarche
SALES DEPT: Andre Lamarche
PURCHASING DEPT: Andre Lamarche

MUSEES DES AMIS DU VIEUX PAYS
Vieux Chateau
Dieulouard, 54380, France
83 23 78 13 83 35 16 25
Gallo-Romain museum, also militaria of WWI & WWII.
; 20;
CONTACT: Dr. Andre Mansuy

MUSEE DE L'HISTOIRE DU FER
B.P. 15
Avenue du General-de-Gaulle
Jarville, 54140, France
83.56.01.42
Museum of the history of Fire & Metallurgy of Iron.
CONTACT: Philippe Dague & Eric Necker

MUSEE D'ARGONNE
Varennes-en-Argonne, Meuse 55270, France
(29) 80.71.14
Museum of the history of the Argonne.
CONTACT: M. Bernard Guerin

MUSEE DE MUR DE L'ATLANTIQUE
Audinghen
Cap Gris-Nez, 62179, France
(21) 35-90-11 (21) 32-97-33
World War II military museum. We buy, exchange or sell 39/45 items.
; 35;
CONTACT: M. Claude Davies
SALES DEPT: D. C. Davies
PURCHASING DEPT: D. C. Davies
World War II museum located in the Batterie Todt. You will find in four of the rooms a collection of uniforms, guns and machineguns, bayonets, etc.

MILITARY MUSEUMS

MUSEE D'AVIATION DU MAS PALEGRY
Mas Palegry, route d'elne
Perpignan, 66000, France
(68) 54.08.79
Private air museum. 350 aircraft & 500 military vehicle models.
; 22;
CONTACT: M. Charles Noetinger
SALES DEPT: Charles Noetinger
PURCHASING DEPT: Charles Noetinger
Private air museum. Collection: Aircraft, models, other items. The owner is pilot and viticulturist. Construction balsawood and card models scale 1/200th on request.

MUSEE HISTORIQUE DE LYON
14, rue de Gadagne
Lyon, Rhone 69004, France
78.42.03.61
Regional history museum.
CONTACT: Mme Monique Ray
SALES DEPT: Museum Secretary
Painting, prints and artifacts concurent with the Siege of Lyon in 1793. Souvenirs of Napoleon I, and many arms.

MUSEE DE LA RESISTANCE
BP 142
3 Bis Rue Gabriel de Mortillet
Annecy, Cedex 74004, France
50.45.29 92
Museum of the French Resistance.
CONTACT: M. Alphonse Metral

MUSEE MUNICIPAL DE FECAMP
21, rue Alexandre Legros
Fecamp, 76400, France
(35) 28.31.99
Naval & edged weapons museum.
CONTACT: Marie-Helene Desjardins
Collection of French edged weaponry of the 18th & 19th Centuries.

MUSEE JEANNE D'ARC
33, place du Vieux-Marche
Rouen, 76000, France
(35) 88.02.70
Wax museum.
S2; 30;
CONTACT: M. Andre Preaux
SALES DEPT: Alain Preaux
PURCHASING DEPT: Alain Preaux
Museum commemorating the burning of Joan of Arc in 1431 in Old Market Square. Displaying models, engravings, books, reproductions, and a wax museum of 50 figures depicting the main stages of the life of Joan of Arc.

MUSEE D'HISTOIRE DE FRANCE 40
131, route de Paris
Abbeville, Somme 80100, France
42 56 19 84 45 51 82 83
Historical museum into a cultural touritic focal point.
; 6;
CONTACT: Henri de Wailly, President
SALES DEPT: Mrs. de Wailly
AD MANAGER: Mrs. de Wailly
The Historical Museum of "France 1940" is settled in Abbeville, on the river Somme, between Paris and Calais harbour. It is open every summer. The indoor and outdoor museum is strong in artillery.

MUSEE DEPART'L DES VOSGES ET MUSEE INTER'L DE L'IMAGERIE
1, Place Lagarde
Epinal, Vosges 88000, France
29.82.20.33 29.82.23.97
Museum of woodcuts & paintings of the 17th & 18th Century, also archaeology.
CONTACT: M. Bernard Huin

IMPERIAL WAR MUSEUM
Lambeth Road
London, SE1, Great Britain
Military museum, library, & film service 1850 to present.

ROYAL ARTILLERY MUSEUM
The Royal Military Academy
Woolrich, London SE18, Great Britain
Artillery museum

ROYAL FUSILIERS MUSEUM
HM Tower of London
London EC3, Great Britain
Infantry museum

BERKSHIRE & WESTMINSTER DRAGOONS MUSEUM
RHQ, 1 Elverton Street
Horseferry Road
London, SW1, Great Britain
Museum of Berks. & Westminster Dragoons.

21ST SPECIAL AIR SERVICE REGIMENT (ARTISTS) MUSEUM
17 Duke's Road
London, WC1, Great Britain
Museum of the 21st S.A.S. Regiment.

HOUSEHOLD CAVALRY MUSEUM
Combermere Barracks
Windsor, Berkshire Great Britain
Museum of the Household Cavalry.

DUKE OF EDINBURGH'S ROYAL REGT. (BERKSHIRE & WILTSHIRE)
The Wardrobe
58 The Close
Salisbury, Wiltshire SP1 2EX, Great Britain
0722 336222 Ext 2683
Museum of the Royal Berkshire & Wiltshire Regiments.
S4;

THE R.E.M.E. MUSEUM
Issac Newton Road
Arborfield
Reading, Berkshire RG2 9LN, Great Britain
0734 760421 Ext 2567
Royal Electrical Mechanical Engineers museum.
; 30;
The museum charts the history of REME since 1942 using photos, memorabilia, models, and life size tableaux and covers both the social and technical story.

THE CHESHIRE MILITARY MUSEUM
The Castle
Chester, Cheshire CH1 2DN, Great Britain
0224 27617
Museum of the Cheshire Regiment.
; 25;
SALES DEPT: Curator
Exhibits illustrating the histories of: 3rd Carabiniers; 5th Royal Inniskilling Dragoon Guards; The Cheshire Yeomanry; The 22nd (Cheshire) Regiment.

THE 3RD CARABINIERS DRAGOON GUARDS REGIMENTAL MUSEUM
The Dale, Liverpool Road
Chester, Cheshire Great Britain
Museum of 3rd Carabiniers Dragoon Guards.

DUKE OF CORNWALL'S LIGHT INFANTRY MUSEUM
The Keep
Bodmin., Cornwall PL31 1EQ, Great Britain
0208-2810
Museum of the Duke of Cornwall's Light Infantry Regiment.
; 60;
CONTACT: D.C.L.I. Trust
SALES DEPT: Major W. H. White
Traces the history of the Duke of Cornwall's Light Infantry, 1702-1959. Unique collection of paintings, documents, artifacts, weapons, and medals covering this period.

THE BORDER REGIMENT MUSEUM
The Castle
Carlisle, Cumberland Great Britain
Museum of the Border Regiment.

THE DEVONSHIRE REGIMENT MUSEUM
Wyvern Barracks
Exeter, Devon Great Britain
Museum of the Devonshire Regiment.

ROYAL ARMOURED CORPS TANK MUSEUM
Bovington Camp
Wareham, Dorset Great Britain
Museum of Royal Armoured Tank Corps.

DORSET MILITARY MUSEUM
The Keep
Dorchester, Dorset Great Britain
British military museum.

MUSEUM OF THE ROYAL CORPS OF SIGNALS
The School of Signals
Blandford Camp,Blandford Forum
Dorset Great Britain
Museum of the Royal Corps of Signals.

ESSEX REGIMENT MUSEUM
Oaklands Park,
Moulsham Street,
Chelmsford, Essex CM2 9AQ, Great Britain
0245-353066 0245-260614
Museum of the Essex Regiment Museum. Some archive material.
; 13;
CONTACT: Chelmsford Borough Council
SALES DEPT: Mrs. S. Bundock
Incorporated by the Chelmsford & Essex Museum. Portrays history of the Essex Regiment from C.1750 to C.1960 with paintings, weapons, medals, photographs, uniform and equipment.

MUSEUM OF THE GLOUCESTERSHIRE REGIMENT (GLOUCESTER CORP.)
103 Westgate Street
Gloucester, Gloucesters Great Britain
The museum of the Gloucestershire Regt.

MILITARY MUSEUMS

AIRBORNE FORCES MUSEUM
RHQ The Parachute Regt.
Browning Barracks
Aldershot, Hampshire GU11 2BU, Great Britain
0252-24431 Ext. 4619
Museum shop selling Airborne souvenirs, memorabilia, books, etc.
SALES DEPT: Maj.(Retd) G. Norton, Curator
Comprehensive display of the history of British Airborne Forces 1940-1984. Uniforms, equipment, weapons, vehicles, photographs, models, dioramas of operations, medals, captured enemy equipment, etc.

ROYAL CORPS OF TRANSPORT REGIMENTAL MUSEUM
Queen Elizabeth Barracks
Crookham, Hampshire Great Britain
Museum of the Royal Corps of Transport.

THE ROYAL GREEN JACKETS MUSEUM
Peninsula Barracks
Winchester, Hampshire Great Britain
Museum of the Royal Green Jackets.

R.A.M.C. HISTORICAL MUSEUM
Keogh Barracks
Aldershot, Hampshire Great Britain
Museum of the Royal Army Medical Corps.

THE ROYAL HAMPSHIRE REGIMENTAL MUSEUM
Searle's House, Southgate St.
Winchester, Hampshire Great Britain
Museum of the Royal Hampshire Regt.

ROYAL ARMY PAY CORPS MUSEUM
Worthy Down
Winchester, Hampshire Great Britain
Museum of the Royal Army Pay Corps.

HEREFORDSHIRE LIGHT INFANTRY (TERRITORIAL) MUSEUM
T A Centre, Harold Street
Hereford, Herefords. Great Britain
Museum of the Herefordshire L.I.

ROYAL ENGINEERS MUSEUM
Brompton Barracks
Chatham, Kent ME4 4UG, Great Britain
(0634) 44555 Ext 312
Museum of the Corps of Royal Engineers.
The Royal Engineers Museum traces the history and contribution of the corps to military and civil life in Britain and throughout the world.

THE BUFFS REGIMENTAL MUSEUM
Royal Museum
High Street
Canterbury, Kent CT1 2JE, Great Britain
Museum of the Buff's.

THE QUEEN'S REGIMENTAL MUSEUM
Howe Barracks
Canterbury, Kent Great Britain
Museum of the Queen's Regiment.

THE QUEEN'S OWN ROYAL WEST KENT REGIMENT MUSEUM
The Maidstone Museum and Art Gallery, Faith's Street
Maidstone, Kent Great Britain
Museum of the Queen's Royal West Kent Regiment.

THE LOYAL REGIMENT(NORTH LANCASHIRE) MUSEUM
Fulwood Barracks
Preston, Lancashire Great Britain
Museum of the Loyal Regiment.

THE KING'S OWN REGIMENTAL MUSEUM
City Museum
Old Town Hall, Market Square
Lancaster, Lancashire Great Britain
Museum of the King's Regiment.

EAST LANCASHIRE REGIMENTAL MUSEUM
Townley Hall
Burnley, Lancashire Great Britain
Museum of the East Lancashire Regiment.

LANCASTRIAN BRIGADE MUSEUM
Fulwood Barracks
Preston, Lancashire Great Britain
Museum of the Lancastrian Brigade.

THE SOUTH LANCASHIRE REGIMENT (PWV) MUSEUM
Peninsula Barracks
Warrington, Lancashire Great Britain
Museum of the South Lancashire Regiment.

EAST LANCASHIRE REGIMENTAL MUSEUM
Fulwood Barracks
Preston, Lancashire Great Britain
Museum of the East Lancashire Regiment.

THE KING'S REGIMENT (LIVERPOOL) MUSEUM
City of Liverpool Museum
William Brown Street
Liverpool, Lancashire 3, Great Britain
Museum of the King's (Liverpool) Regiment.

XX THE LANCASHIRE FUSILIER REGIMENTAL MUSEUM
Wellington Barracks
Bury, Lancashire Great Britain
Museum of XX the Lancashire Fusiliers.

ROYAL LEICESTERSHIRE REGIMENTAL MUSEUM
Newarke House Museum
Leicester, Leicesters Great Britain
Museum of the Royal Leicestershire Regt.

MUSEUM OF THE ROYAL LINCOLNSHIRE REGIMENT
The Keep, Sobraon Barracks
Burton Road
Lincoln, Lincolnshir Great Britain
Museum of the Royal Lincolnshire Regt.

THE ROYAL NORFOLK REGIMENT MUSEUM
Britannia Barracks
Norwich, Norfolk Great Britain
Museum of the Royal Norfolk Regiment.

ROYAL PIONEER CORPS MUSEUM
Simpson Barracks, Wooton
Northhampton, Northhampto Great Britain
Museum of the Royal Pioneer Corps.

THE MUSEUM OF THE NORTHHAMPTONSHIRE REGIMENT
Gibraltar Barracks
Barrack Road
Northhampton, Northhampto Great Britain
Museum of the Northhamptonshire Regt.

THE KING'S OWN SCOTTISH BORDERERS REGIMENTAL MUSEUM
The Barracks
Berwick-on-Tweed, Northumber. TD15 1DG, Great Britain 0289 307 426
Museum of the King's Scottish Borderers.
S4; 298;
SALES DEPT: Col. Ward
AD MANAGER: Mr. Ward

ROYAL NORTHUMBERLAND FUSILIERS REGIMENTAL MUSEUM
Abbot's Tower
Alnwick Castle
Alnwick, Northumber. NE2 4NP, Great Britain
Museum of the Royal Northumberland Fusiliers Regiment.
300 years of Regimental history, uniforms, medals, weapons, photographs, pictures, silver, etc.

SHERWOOD FORESTERS(NOTTINGHAM-SHIRE & DERBYSHIRE) MUSEUM
The Castle
Nottingham, Mottingham Great Britain
Museum of the Sherwood Foresters.

THE KING'S SHROPSHIRE & HEREFORDSHIRE LT. INF. MUSEUM
Sir John Moore Barracks
Copthorne
Shrewsbury, Shropshire Great Britain
Shropshire & Herefordshire Light Infantry museum.

1st THE QUEEN'S DRAGOON GUARDS REGIMENTAL MUSEUM
Clive House
Shrewsbury, Shropshire Great Britain
1st the Queen's Dragoon Guards museum.

THE SHROPSHIRE REGIMENTAL MUSEUM
The Castle
Shrewsbury, Shropshire Great Britain
58516
Museum of the Shropshire Yeomanry Regt.
Regimental museums of: The King's Shropshire Light Infantry; The Shropshire Yeomanry Cavalry; The Shropshire Royal Horse Artillery. Period 1755-1968. Uniforms, medals, badges, weapons, & militaria.

KING'S SHROPSHIRE LIGHT INFANTRY (TERRITORIAL) MUSEUM
The Drill Hall, Coleham
Shrewsbury, Shropshire Great Britain
Museum of the King's Shropshire L.I.

MILITARY MUSEUMS

SOMERSET MILITARY MUSEUM
The Castle
Taunton, Somerset Great Britain
(0823) 55504 (0823) 73434 Ext 663
Museum of the Somerset Light Infantry.
S4,S6;
CONTACT: Trustees
SALES DEPT: Lt Col R. G. Woodhouse
PURCHASING DEPT: Lt Col R. G. Woodhouse
Displays of 300 years of history of the county Regiments: The Somerset Light Infantry; West Somerset Yeomanry and North Somerset Yeomanry. Large display of the medals of the Regiment.

THE STAFFORDSHIRE REGIMENTAL MUSEUM
Whittington Barracks
Lichfield, Staffordshi Great Britain
Museum of the Staffordshire Regiment.

SUFFOLK REGIMENT MUSEUM
The Keep, Gibraltar Barracks
Bury St. Edmunds, Suffolk Great Britain
Museum of the Suffolk Regiment.

NATIONAL ARMY MUSEUM
Royal Hospital Road
London SW3 4HT, Great Britain
(01)-730-0717
Museum of the British Army.
; 26;
SALES DEPT: Philip Clarke
PURCHASING DEPT: Public Relations Officer
Exhibitions covering the history of the British Army from the Battle of Bosworth to the Falklands, also includes weapons, uniforms and art galleries. Admission is free.

R.A.O.C. MUSEUM
RAOC Training Centre
Deepcut
Camberley, Surrey Great Britain
Museum of the R.A.O.C.

THE QUEEN'S ROYAL SURREY REGIMENTAL MUSEUM
Surbiton Road
Kingston-Thames, Surrey Great Britain
Museum of the Queen's Royal Surrey Regt.

ROYAL MILITARY POLICE MUSEUM
Roussillon Barracks
Chichester, Sussex Great Britain
Museum of the Royal Military Police.

THE ROYAL SUSSEX REGIMENT MUSEUM
29 Little London
Chichester City Museum
Chichester, Sussex Great Britain
Museum of the Royal Sussex Regiment.

THE ROYAL WARWICKSHIRE REGIMENTAL MUSEUM
Regimental HQ, The Royal
Fusiliers, St. John's House
Warwick, Warwickshire Great Britain
Museum of the Royal Warwickshire Regt.

THE QUEEN'S OWN HUSSARS MUSEUM
The Lord Leycester Hospital
High Street
Warwick, Warwickshir Great Britain
Museum of the Queen's Own Hussars.

WARWICKSHIRE & WORCESTERSHIRE YEOMANRY MUSEUM
Drill Hall
Priory Road
Warwick, Warwicks. Great Britain
Museum of Warwickshire & Worcestershire Yeomanry.

THE WORCESTERSHIRE REGIMENT MUSEUM
RHQ WFR
Norton Barracks
Worcester, Worcesters. WR5 2PA, Great Britain
0905 354359
Museum of the Worcestershire Regiment.
The museum display forms part of the Worcester City Museum. The archives are at Norton Barracks.

GREEN HOWARDS REGIMENTAL MUSEUM
Trinity Church Square
The Market Place
Richmond, N.Yorkshire DL10 4QN, Great Britain
0748 2133
Museum of the Green Howards.
S1,S4; 15;
CONTACT: Museum Trustees
SALES DEPT: Mrs. A. Tappley
Uniforms, head-dress, battle relics, medals, campaign colours, buttons, badges from 1688 onwards. Open throughout the year except January & December. The museum shop sells regimental souvenirs.

THE WEST YORKSHIRE REGIMENT (14TH FOOT) & PWO MUSEUM
Imphal Barracks
York, Yorkshire Great Britain
Museum of West Yorkshire Regiment.

THE DUKE OF WELLINGTON'S REGIMENTAL MUSEUM
Bankfield Museum
Boothtown Road
Halifax, Yorkshire Great Britain
Museum of the Duke of Wellington's Regiment.

4TH/7TH ROYAL DRAGOON GUARDS REGIMENTAL MUSEUM
No. 3 Tower Street
York, Yorkshire TO1 1SB, Great Britain
(0904) 642036
Museum of the 4th/7th Royal Dragoon Guards Regiment.
History of the last 300 years of the 4th Royal Irish Dragoon Guards, the 7th Princess Royals Dragoon Guards, and the 4/7th Royal Dragoon Guards.

CASTLE MUSEUM
York, Yorkshire Great Britain
Military museum.

THE YORK & LANCASTER REGIMENTAL MUSEUM
Endcliffe Hall
Endcliffe Vale Road
Sheffield, Yorkshire 10, Great Britain
Museum of the York & Lancaster Regiment.

KING'S OWN YORKSHIRE LIGHT INFANTRY REGIMENTAL MUSEUM
Doncaster Museum
Chequers Road
Doncaster, S.Yorkshire Great Britain
Museum of the King's Own Yorkshire L.I.
S1,S4; 30;
SALES DEPT: Col. J. S. Cowley
Regimental museum of the King's Own Yorkshire Light Infantry.

THE EAST YORKSHIRE REGIMENT (15TH FOOT) MUSEUM
11 Butcher Row
Beverley, Yorkshire Great Britain
Museum of the East Yorkshire Regiment.

SCOTTISH UNITED SERVICES MUSEUM
Crown Square, The Castle
Edinburgh, Scotland 1, Great Britain
Museum of the Scottish United Services.

QUEEN'S OWN HIGHLANDERS (SEAFORTH & CAMERONS) MUSEUM
Fort George
Inverness-shire, Scotland Great Britain
Museum of the Queen's Own Highlanders.

THE ARGYLL & SUTHERLAND HIGHLANDERS REGIMENTAL MUSEUM
The Castle
Stirling, Scotland Great Britain
Museum of the Argyll & Sutherland Highlanders.

THE CAMERONIANS (SCOTTISH RIFLES) REGIMENTAL MUSEUM
Winston Barracks
Lanark, Scotland Great Britain
Museum of the Cameronians.

THE BLACK WATCH MUSEUM
Balhousie Castle
Perth, Scotland Great Britain
Museum of the Black Watch.

THE ROYAL SCOTS GREYS
Home HQ, The Royal Scots Grey
The Castle
Edinburgh, Scotland 1, Great Britain
Museum of the Royal Scots Greys.

THE ROYAL SCOTS REGIMENTAL MUSEUM
Regimental HQ, The Royal Scots
The Castle
Edinburgh, Scotland EH1 2YT, Great Britain
031-336 1761
Museum of the Royal Scots Regiment.
; 10;
Catalogue available. Annually.
SALES DEPT: Curator
PURCHASING DEPT: Royal Scots Regimental Shop
Royal Scots Regimental Museum, displaying medals, colours, uniforms, pictures, and militaria. Gift shop. Visitors welcome.

AYRSHIRE YEOMANRY MUSEUM
Yeomanry House
Ayr, Scotland Great Britain
Museum of the Ayrshire Yeomanry.

MILITARY MUSEUMS

GORDON HIGHLANDERS MUSEUM
Viewfield Road
Aberdeen, Scotland Great Britain
Museum of the Gordon Highlanders.

THE ROYAL HIGHLAND FUSILIERS MUSEUM
Regimental HQ
518 Sauchiehall Street
Glasgow, Scotland G2 3LW, Great Britain
041-332-0961
Museum of the Royal Highland Fusiliers.
S1; 26;
SALES DEPT: Major D. I. A. Mack
Regimental HQ of the Royal Highland Fusiliers.
Regimental museum, including small shop.

THE SCOTTISH HORSE MUSEUM
The Cross, Dunkeld
Perthshire, Scotland Great Britain
Museum of the Scottish Horse.

LOWLAND BRIGADE DEPOT MUSEUM
Glencorse Barracks
Milton Bridge, Penicuik
Midlothian, Scotland Great Britain
Museum of the Lowland Brigade Depot.

THE SOUTH WALES BORDERERS REGIMENTAL MUSEUM
The Barracks
Brecon, Wales Great Britain
Museum of the South Wales Borderers.

THE WELCH REGIMENTAL MUSEUM
The Barracks, Whitchurch Road
Cardiff, Wales Great Britain
Museum of the Welsh Regiment.

THE ROYAL WELCH FUSILIERS REGIMENTAL MUSEUM
The Queen's Tower
Caernarvon Castle
Caernarvon, Wales Great Britain
Museum of the Royal Welch Fusiliers.

SOUTH AFRICAN NATIONAL MUSEUM OF MILITARY HISTORY
P.O. Box 52090
Saxonwold, Transvaal 2132, South Africa
646-5513
Museum specializing in South African military history.
CONTACT: Col. George R. Duxbury

THE ROYAL ULSTER RIFLES REGIMENTAL MUSEUM
5 Waring Street
Belfast, N. Ireland United Kingdom
Museum of the Royal Ulster Rifles.

THE ROYAL IRISH FUSILIERS REGIMENTAL MUSEUM
Sovereign's House
The Mall
Armagh, N. Ireland United Kingdom
Museum of the Royal Irish Fusiliers.

5TH ROYAL INNISKILLING DRAGOON GUARDS REGIMENTAL MUSEUM
Carrickfergus Castle
Carrickfergus, Co. Antrim United Kingdom
Museum of the 5th Royal Dragoon Guards.

THE ROYAL INNISKILLING FUSILIERS REGIMENTAL MUSEUM
St. Lucia Barracks
Omagh, Co. Tyrone United Kingdom
Museum of the Royal Inniskilling Fusiliers.

OPTICAL EQUIPMENT

ROBERT N. CARPENTER
205 Coventry Road
Virginia Beach, VA 23462, USA
Optical equipment repaired.
Write for details & estimates.

U.S. OPTICS
P.O. Box 14206
Atlanta, GA 30324, USA
Sunglasses & eyewear

GANDER MOUNTAIN,INC.
P.O. Box 6
Wilmot, WI 53192, USA
(414)-862-2331
Outdoor sportsman supplies.
S4; 27;
Catalogue available. 8 times a year.
CONTACT: Ralph Freitag
SALES DEPT: Linda Jacobs
FREE Outdoorsman catalogs from Gander Mountain, filled with name brand, low priced gear for hunting, fishing, camping and hiking. Write Gander Mountain,Inc., P.O. Box 6, Dept. PM, Wilmot, WI 53192.

STANO COMPONENTS
P.O. Box 6274
San Bernadino, CA 92412, USA
Optical equipment.
Catalogue $4.00.

MILITARY PAPER ITEMS

PAPER AMERICANA
576P Massachusetts Avenue
Lunenburg, MA 01462, USA
(617)-582-7844
Military paper items, Colonial to WWII.
S4; 11;
Catalogue available. Quarterly.
CONTACT: Gordon Totty
SALES DEPT: Gordon Totty
PURCHASING DEPT: Gordon Totty
Military paper items: French & Indian Wars, Revolutionary War, War of 1812, Mexican War, Civil War, Spanish-American War, WWI. Books, newspapers, documents, images.

R. EWING
P.O. Box 993
Montpelier, VT 05602, USA
Military Award Certificates.
Send SASE for List

STEPHEN AUSLENDER
P.O. Box 122
Wilton, CT 06897, USA
(203)-762-3455
Old military models, books, picture postcards, photos.
S4; 20;
Catalogue available. As needed.
CONTACT: Stephen Auslender
SALES DEPT: Stephen Auslender
PURCHASING DEPT: Stephen Auslender
I sell old military: models (3 stamps for lists), out of print books and magazines ($1.00 for catalog), picture postcards (military, aviation, naval), photos, seals

POSTAL COVERS
P.O. Box 26
Brewster, NY 10509, USA
Postal items mail-order auction.
Catalogue $1.00.
CONTACT: Theo Van Dam

PAGES OF HISTORY
P.O. Box 656
Hudson, NY 12534, USA
Autographs & political items.
S4; 3;
CONTACT: Gerald J. Docteur
Sale of autographs, Presidents, Civil War, other paper items.

MR. 3 L
P.O. Box 35
2931 Lincoln Highway
Soudersburg, PA 17577, USA
717-687-6165
Military paper items
CONTACT: Leonard L. Lasko

W. GRAHAM ARADER,III
1000 Boxwood Court
King of Prussia, PA 19406, USA
215-825-6570
Paper items cleaned & restored.
Write for details & estimates.

MILITARY GRAPHICS
P.O. Box 228
Dunkirk, MD 20754, USA
Military paper items.
Catalogue available.

L. F. LIVINGSTON STAMPS,INC.
716 York Road
Towson, MD 21204, USA
301-823-1661
Postage stamps.
Catalogues available.
CONTACT: Vivien E. Pietro

F. DON NIDIFFER AUTOGRAPHS
P.O. Box 8184
Charlottesville, VA 22906, USA
(804)-296-2067
Military autograph & document appraisals.
S6,S12; 7;
Catalogue available.
CONTACT: Dr. F. Don Nidiffer
SALES DEPT: Don Nidiffer
PURCHASING DEPT: Don Nidiffer
I buy and sell autographs in all fields, but specialize in military and presidential material. Large collection or one item. Replies generally within 48 hours if xerox sent.

MILITARY PAPER ITEMS

AUTOS + AUTOS, AUTOGRAPHS
P.O. Box 280
Elizabeth City, NC 27909, USA
(919)-335-1117
Autographs & documents.
S4,S12,S13; 13;
Catalogue available. 3 times per year.
CONTACT: Dr. B. C. West,JR.,M.D.
SALES DEPT: B. C. West
PURCHASING DEPT: B. C. West
Selling & buying authentic autographs of all forms - Letters, documents, photos, etc. Specializing in Luftwaffe aces.

MARK ANGERT
2919 Commerical Blvd.
Fort Lauderdale, FL 33308, USA
305-944-9100
Bubble gum cards.

GORDON McHENRY
P.O. Box 117
Osprey, FL 33559, USA
(813)-966-5563
Military paper & postal items.
S6; 31;
SALES DEPT: Gordon McHenry
PURCHASING DEPT: Gordon McHenry
Dealer in military postal history (stamps & covers) and military paper items (letters, commissions, passes, autographs,etc.)

WILLIAM E. BARKER
3675 W. 130th Street
Cleveland, OH 44111, USA
Wartime postal covers.
List available.

IDENTITY CHECK PRINTERS
P.O. Box 149
Park Ridge, IL 60068, USA
Check forms with an aviation motif.

STEVEN ANDERSON
Rt. 1, Box 269
Sugar Grove, IL 60554, USA
312-466-4402
Military intelligence paper items.

JOHN FRY PRODUCTIONS
P.O. Box 9444
San Diego, CA 92109, USA
Military postcards

NIPPON PHILATELICS
Drawer 7300
Carnel, CA 93921, USA
(408)-625-2643 (405)-624-4617
Japanese propaganda leaflets, picture postcards, military mail & stamps.
S6; 9; 1200; $40.00
Catalogue available. As needed.
CONTACT: F. L. Allard,Jr.
SALES DEPT: F. L. Allard,Jr.
PURCHASING DEPT: F. L. Allard,Jr.
Japan specialized. All types of military paper material, buy and sell. Especially strong in WWII era military picture postcards and propaganda leaflets.

JIM LOGSDON
2495 Night Shade Lane
Fremont, CA 94539, USA

Military certificates

LEE POLESKE
P.O. Box 871
Seward, AK 99664, USA
(907)-224-5525
Comic WWII postcards
S4; 4;
Comic World War II postcards, sent on approval, no list.

MILITARY BOOK PUBLISHERS

CHRISTOPHER PUBLISHING COMPANY
P.O. Box 1014
West Hanover, MA 02339, USA
Military book publishers.
List available.

WEAPONS & WARFARE PRESS
218 Beech Street
Bennington, VT 05201, USA
(802)-447-0313
Military book publisher.
S6; 19; 4,000; $50.00
Catalogue available. Monthly.
CONTACT: Ray Merriam
Publish magazine, hundreds of books on every aspect of military history. Sell thousands of current titles of other publishers (retail only). See our advertisement.

DEVIN-ADAIR PUBLISHERS
P.O. Box A
Old Greenwich, CT 06870, USA
Military book publisher.

GREENWOOD PRESS
P.O. Box 5007
88 Post Road West
Westport, CT 06881, USA
Military book publisher.
Catalogue available.

WEBSTER'S UNIFIED,INC.
333 Post Road West
Westport, CT 06889, USA
Military book publisher.
Catalogue available.

VINTAGE CASTINGS,INC.
127 74th Street
North Bergen, NJ 07047, USA
201-861-2979
'Toy Soldier Review',publisher.
$12.00 annual subscription. Quarterly.
CONTACT: William Lango

ARCO PUBLISHING,INC.
215 Park Avenue South
New York, NY 10003, USA
Military book publisher.

FACTS ON FILE
460 Park Avenue South
New York, NY 10016, USA
800-322-8755
Military book publisher.

McKEE PUBLICATIONS
121 Eatons Neck Road
Northport, NY 11768, USA

(516)-757-8850
Old gun catalogs & gun parts, sporting & aviation books, target rifles & acces.
S4,S5,S12; 4;
Catalogue available. Quarterly.
CONTACT: Arthur McKee
SALES DEPT: Arthur McKee
PURCHASING DEPT: Arthur McKee

STACKPOLE BOOKS
P.O. Box 1831
Cameron & Kelker Streets
Harrisburg, PA 17105, USA
(717)-234-5041
Military & outdoors book publisher.
S6; 50;
Catalogue available. Bi-annually.
CONTACT: M. David Detweiler, President
SALES DEPT: Bruce D. Fleming
Publisher of military, woodworking, gun, hunting & fishing books. Retail discounts available.

ABOUT FACES PUBLISHING
913 Collins Drive
West Chester, PA 19380
(215) 692-9911
Publisher
Contact: Gil Thomas
Unique and unusual books -- not restricted to militaria

XENVIRONS
29 Poplar Street, Bldg. B-5
Hatfield, PA 19440
(215) 368-8693
Graphic Arts Service
S5; 17
Contact: Max Bendel
Complete design and copywriting service. Logos, brochures, direct mail packages, books, etc.

'THE PHOENIX EXCHANGE'
P.O. Box 66
163 Troutman Road
Arcola, PA 19420, USA
(215)-933-0909
Quarterly military collectors magazine.
; 2;
CONTACT: Terry Hannon,Publisher
SALES DEPT: W. R. Bendel
PURCHASING DEPT: Terry Hannon
AD MANAGER: William Bendel
'The Phoenix Exchange' functions as the military collector's & dealer's marketplace. Articles cover a wide spectrum of militaria, but attend to each subject in detail. Send $2.00 for a sample copy.

HOWARD UNIVERSITY PRESS
2900 Van Ness Street, N.W.
Washington, DC 20008, USA
Military book publisher

SMITHSONIAN INSTITUTION PRESS
P.O. Box 1579
Washington, DC 20013, USA
Military book publisher

ZENGER PUBLISHING COMPANY
P.O. Box 42026
Washington, DC 20015, USA
301-881-1471
Military book publisher

MILITARY BOOK PUBLISHERS

THE INTERNATIONAL RESEARCH
INSTITUTE FOR POLI. SCIENCE
P.O. Box 199
College Park, MD 20740, USA
Military book publisher

BIBLIOPHILE LEGION BOOKS, INC.
P.O. Box 612
Silver Spring, MD 20901, USA
Military book publisher

UNIVERSITY OF ILLINOIS PRESS
P.O. Box 1650
Hagerstown, MD 21741, USA
800-638-3030 301-824-7300
Military booksellers and publishers

ESSENTIAL PRESS C. L. Batson
P.O. Box 143
5512 Buggy Whip Drive
Centerville, VA 22020, USA
(703)-631-0884
Military book fair promoter. Booksellers & Publishers. American Civil War.
S6,S13; 7;
SALES DEPT: C. L. Batson
PURCHASING DEPT: C. L. Batson
Produce the annual "All Military Book Fair" and "Civil War Book Fair" in Fairfax, VA (DC); also annual "Gettysburg Civil War Book Fair".

QUEST PUBLISHING COMPANY
P.O. Box 2081
6801 Sue Paige Court
Springfield, VA 22152, USA
703-451-9113
Military reference publisher
CONTACT: Charles P. McDowell

GERHARD V. KELLNER
1608 Piccadilly Drive
Charlotte, NC 28211, USA
Military book publisher

DERBY PUBLISHING COMPANY
P.O. Box 221474
Charlotte, NC 28222, USA
Military book publisher

ARSENAL PRESS
P.O. Box 12244
575 Pharr Road
Atlanta, GA 30355, USA
Publisher of military collector books.

R. J. HUFF & ASSOCIATES
P.O. Box 40023
Sarasota, FL 34242, USA
Publisher of collecter reference books.

THE UNIVERSITY OF ALABAMA PRESS
P.O. Box 2877
Tuscaloosa, AL 35487, USA
(205)-348-5180
Publisher of scholarly and trade books.
S6; 45;
Catalogue available. Semi-annually.
CONTACT: The University of Alabama
SALES DEPT: Merrill G. Floyd
PURCHASING DEPT: Linda Sandford
A non-profit organization dedicated to the publication of scholarly books and books of local and regional interest.

THE UNIV. PRESS OF KENTUCKY
102 Lafferty Hall
Lexington, KY 40506-0024, USA
(606)-257-2951 (606)-257-8442
Military book publisher.
S6,S8,S9,S10,S11,S12;
Catalogue available.
CONTACT: Kenneth Cherry, Director

AGINCOURT PUBLISHERS
P.O. Box 4039
Somerset, KY 42501, USA
Collector reference book publisher.

HAAS PUBLICATIONS
P.O. Box 775
Worthington, OH 43085, USA
Military reference book publisher.
CONTACT: David L. Hartline

BOSTON PUBLISHING COMPANY
P.O. Box 16617
4343 Equity Drive
Columbus, OH 43216, USA
Military book publisher.

MOTORBOOKS INTERNATIONAL
P.O. Box 2-Rev.
729 Prospect Avenue
Osceola, WI 54020, USA
715-294-3345 800-826-6600
Military publisher & bookseller.
CONTACT: Rita M. Cederholm, publicity

KRAUSE PUBLICATIONS
700 East State Street
Iola, WI 54990, USA
800-258-0929
Collectors reference books & collector oriented magazines.

HAMILTON, CO.
P.O. Box 803
Lakeville, MN 55044, USA
Military booksellers & publishers.
List available.
CONTACT: Northstar Commemoratives

HANDGUN PRESS
5832 South Green Street
Chicago, IL 60621, USA
Firearms books publisher

CROWN/AGINCOURT
6860 West 105th Street
Overland, KS 66212, USA
913-341-6619
Book publishers

MA/AH PUBLISHING
Eisenhower Hall
Kansas State University
Manhattan, KS 66506, USA
913-532-6733
Publisher of 'Military Affairs' and 'Aerospace Historian'.
CONTACT: Ms. Honor B. Phillips

MCN PRESS
P.O. Box 702073
Tulsa, OK 74170, USA
(918)-743-7048
Military reference book publisher.

S6; 20;
Catalogue available. Annually.
CONTACT: Jack Britton
SALES DEPT: Jack Britton
PURCHASING DEPT: Jack Britton
Publisher of books of interest to collectors of military patches, medals, badges, insignia, etc. Free catalog.

MEDALS YEARBOOK (U.S.A.)
P.O. Box 1915
Arvada, CO 80001, USA
303-469-6227
Medal reference book publisher.

BOOMERANG PUBLISHERS
6164 West 83rd Way
Arvada, CO 80003, USA
Military manual publishers.

DER ANGRIFF PUBLICATIONS
310B Vista Street E Apt 3
Long Beach, CA 90803, USA
Military book publisher.
CONTACT: Richard Baumgartner

TAYLOR PUBLISHING COMPANY
2370 Riverside Drive
Santa Ana, CA 92706, USA
714-744-0487
Military book publisher
CONTACT: Jeffrey R. Millet

EDEN PRESS
P.O. Box 8410
11623 Slater 'C'
Fountain Valley, CA 92728, USA
Military book publisher

AMERICAN BIBLIOGRAPHICAL CENTER
Clio Press, Box 4397
2040 Alameda Padre Serra
Santa Barbara, CA 93103, USA
Publisher of historical abstracts.
CONTACT: Hope Smith, Librarian

HOOVER INSTITUTION PRESS
Stanford University, Room 21
Stanford, CA 94305, USA
415-497-3373
Military book publisher

PRESIDIO PRESS
P.O. Box 1764
Novato, CA 94948, USA
Military book publisher
CONTACT: Al Pirenian

R. JAMES BENDER PUBLISHING
P.O. Box 23456
San Jose, CA 95153, USA
Military reference book publisher.

MILITARY BOOK PUBLISHERS

JAN/DON ENTERPRISES
P.O. Box 16763
South Lake Tahoe, CA 95706, USA
(916)-544-8802
Ship plans book
S6; 4;
CONTACT: Donald R. Wesenberg
SALES DEPT: D. Wesenberg
PURCHASING DEPT: D. Wesenberg
Tin Cans - 1/300 scale ship plans. Destroyers - Destroyer Escorts 1917-1945 - 20 Drawings. Airplane Plans Book coming - 1/72 scale, 24 planes.

BEACON PUBLISHING COMPANY
P.O. Box
Redding, CA 96002, USA
Military book publisher.
Send for catalogue.

WORLD WIDE PUBLISHING CORP.
P.O. Box 105
Ashland, OR 97520, USA
(503)-482-3800
Publisher of books & tapes.
S3,S6,S9,S10,S11,S12; 10; 30,000; $55.00
Catalogue available. Quarterly.
CONTACT: Hans J. Schneider
SALES DEPT: Rose Schneider
PURCHASING DEPT: Rose Schneider
Publishers of survival, how-to, self-help and other useful non-fiction books. Deals with subjects such as: Life Under Hitler, The American Rut and Ensuing Chaos, etc.

ARMORY PUBLICATIONS
P.O. Box 44372
Tacoma, WA 98444, USA
Military reference book publisher.
Send for information.

INFOS A 1 INTERNATIONAL
P.O. Box 127
Paris, Cedex 12 75012, France
Directory publisher
S13; 19;
CONTACT: Pierre Birukoff, Editor
We publish specialized directories and released sources of any hard-to-find or very unusual items.

TOKEN PUBLISHING LIMITED
Crossways Road, Grayshott
Hindhead, Surrey GU26 6HF, Great Britain
42873 7242
Magazines & medal finding service.
S6,S11; 4;
CONTACT: John W. Mussell
SALES DEPT: John W. Mussell
AD MANAGER: Mary Woodrow
"Coin & Medal News" incorporating "Military Chest" is the world's leading independent medal magazine - now encompassing military history.

THE MEDALS YEARBOOK
63 Ermineside
Bush Hill Park
Enfield, Middlesex Great Britain
Military medal reference publisher.

JOURNAL-VERLAG SCHWEND GMBH
Postfach 100340
Schwabisch Hall, 7170, West Germany
Military book publisher.

RADIO/ELECTRONIC ITEMS

SCIENTIFIC SYSTEMS
P. O. Box 716
Amherst, NH 03031, USA
Electronic warfare systems.

VIET-AMERICAN
109 Fire Lane
N. Cape May, NJ 08204, USA
Military telephone.

STEEL SERVICES COMPANY
P.O. Box 1299
628 E. 9th Street
New York, NY 10009, USA
Antique scientific instruments.
Catalogue available.
CONTACT: Richard Strich

CCS COMMUNICATION CONTROL, INC.
633 Third Avenue
New York, NY 10017, USA
212-697-8140
Esoteric electronics for covert ops.

CHARLES J. HINKLE
Rt. 11, Box 3
Fredericksburg, VA 22405, USA
(703)-373-6546
Military radio equipment collector.
Collector buying WWI and earlier military radio and telegraph. Occasionally have items for sale or trade.

USI CORPORATION
P.O. Box 2052
Melbourne, FL 32901, USA
305-725-1000
Communications accessories.

MICROCOM TECH CORPORATION
P.O. Box 347341
Cleveland, OH 44134, USA
Surveillance equipment

BAYTRONICS
P.O. Box 591
Sandusky, OH 44870, USA
(419)-627-0460
Military personnel and vehicular radios and accessories.
S6; 12;
Catalogue available. 3 times per year.
CONTACT: Stuart Stephens
SALES DEPT: Stuart Stephens
Specializes in military personnel, field, and vehicular radios and accessories, all eras. Mail order sales. Interested in quanity purchases.

FAIR RADIO SALES COMPANY
P.O. Box 1105
1016 E. Eureka Street
Lima, OH 45802, USA
419 227-6573
Military vehicle radio gear.

P. SORENSON
878 E. Newark Drive
West Bend, WI 53095, USA
German WWII radio collector.

STEVE BARTKOWSKI
4932 W. 28th Street
Cicero, IL 60650, USA
312-863-3090 after 5
Military surplus electronics.

AVIONICS
P.O. Box 1397
Lewisville, TX 75067, USA
Aviation ground electronics

TNM ENTERPRISES
P.O. Box 2331
Anaheim, CA 92804, USA
Military electronic equipment.
Send for information.

MILITARY REPRODUCTIONS

CUSTOMCRAFT JEWELRY
21 Dartmouth Street
Hooksett, NH 03106, USA
(603)-627-1837
Custom reproductions in real cloisonne.
S6; 21;
Catalogue available. As needed.
CONTACT: Dick Marple
SALES DEPT: Dick Marple
We have no stock inventory. All are custom produced from Logo, design, sketch, or sample submitted with order. Minimum 50 pcs. Absolute.

NAVY ARMS COMPANY, INC.
689 Bergen Blvd.
Ridgefield, NJ 07657, USA
201-945-2500
Reproduction blackpowder weapons.
Catalogue $1.00.
CONTACT: Val Forgett

THE PATCH KING
P.O. Box 101
Madison Square Station
New York, NY 10010, USA
Reproduction insignia.

BILL COMBS
63 Towers Road
Lake Carmel, NY 10512, USA
Reproduction field gear.

FOREMOST INSIGNIA, INC.
P.O. Box 68
Red Hook, NY 12571, USA
Reproduction insignia manufacturer.
Discrete inquiries respected.

WORLD WAR I AEROPLANES
15 Crescent Road
Poughkeepsie, NY 12601, USA
914-473-3679 Aero 818-243-6820 Skyways
Provides service info on WWI aeroplanes.
S6,S10; 25; 2200; 100.00
CONTACT: Leonard E. Opdycke
SALES DEPT: L. E. Opdycke
AD MANAGER: Richard Alden
Provides service (information, names, projects, materials, books) to modellers, builders & restorers of aeroplanes (1900-1919, 1920-1940) through two journals, WW I Aero & Skyways.

MILITARY REPRODUCTIONS

CHAMPLAIN CANNON WORKS
121 Dixon Road
Glen Falls, NY 12801, USA
Artillery reproductions.

THE WILLOW FORGE
P.O. Box 523
Route 40 East
Brownsville, PA 15417, USA
(412)-785-6997
Edged weapon repair & restoration.
S6; 7;
Flyer available.
CONTACT: Daniel Tokar
SALES DEPT: Daniel Tokar
PURCHASING DEPT: Daniel Tokar
I will make or repair any style of edged weapons, armour or hardware. Forge my own Damascus steel. Can research period styles of weapons.

WOODKNOTS
314 Keystone Drive
Telford, PA 18969, USA
(215)-257-5634
Early American recreations of furniture and outdoors equipment.
CONTACT: Curtis Hillmantel

G. GEDNEY GODWIN, INC.
P.O. Box 100
Valley Forge, PA 19481, USA
(215)-783-0670
Militaria from French & Indian War to the Civil War.
S6,S13; 20;
Catalogue available.
CONTACT: G. Gedney Godwin, President
Suppliers to reenactors, National Park locations, museums, Fife & Drum corps. All items are fully functional.

CANNON, LTD.
1316 Lafayette Avenue
Baltimore, MD 21207, USA
Artillery carriages & cannons.
S1,S4; 5;
Catalogue available.
SALES DEPT: P. Miller
PURCHASING DEPT: P. Miller
We build quality and guaranteed black powder cannon and mortar reproductions. We sell cannon hardware, wooden parts and wheels in 1/2, 3/4, and full scale.

MAYHEW-REECE AND ASSOCIATES
P.O. Box 20081
Greensboro, NC 27420, USA
(919)-294-3226
Original & replica 3rd Reich militaria.
S6,S12,S13; 14;
Catalogue available. Bi-monthly.
SALES DEPT: A. D. Reece,Jr.
PURCHASING DEPT: S. D. Mayhew
We specialize in Third Reich collectibles from World War II including postcards, photocards, cancelled envelopes, stamps, banknotes, and coins. Collector leaflets, medals, membership pins, armbands, belt buckles, and other miscellaneous items are occasionally offered. Send $1.00 for list.

IRAC
P.O. Box 1273
Covington, KY 41012, USA
Replica machinegun accessories.

EAGLE LTD.
635 Jefferson Street
Fairborn, OH 45324, USA
513-879-5579
Repro militaria

PHYLLIS ROSE REPROS
7760 Somerville Drive
Dayton, OH 45424, USA
513-236-0472
Repro field gear & uniform Dealer.
CONTACT: Phyllis Rose

DAVE BERRY
1638 N. Longview
Dayton, OH 45432, USA
Repro field gear

HAMMER
P.O. Box 1393
Columbus, IN 47201, USA
German marches on cassettes.
S6,S11,S12; 6;
Catalogue available.
CONTACT: Karl Hammer
Manufacturer/Wholesaler: WW II German march cassettes, flags, pins, and posters. Custom capacity retailer: Videos and books. We buy pre-1945, 78 rpm German march records.

QUARTERMASTER SHOP
3115 Nokomis
Port Huron, MI 48060, USA
(313)-987-4127
Repro Civil War uniforms & clothing.
S4; 10;
Catalogue available. Annually.
CONTACT: Jeff O'Donnell
SALES DEPT: Jeff O'Donnell
PURCHASING DEPT: Jeff O'Donnell

BIG T PARTS COMPANY
P.O. Box 8129
West Bloomfield, MI 48304, USA
313 535-7855
Reproduction vehicle parts & weapons.

THE REICH ART
P.O. Box 285
122 1/2 East Main Street
Flushing, MI 48433, USA
(313)-659-8999
Military art prints.
S6; 6;
Illustrated catalogue available. Annually.
CONTACT: James G. Thompson
SALES DEPT: Sybille Palentyn
PURCHASING DEPT: Sybille Palentyn
Presently we are providing military, historical, or ethnic art prints, as well as commissioned art, books and replicas.

ROWDEN
P.O. Box 351
Owosso, MI 48867, USA
Military reproductions.

DER DIENST
P.O. Box 221
Lowell, MI 49331, USA
Replica military medals, badges and insignia.
S4,S6; 6; 5,000; $75.00
Catalogue available. Bi-annually.
CONTACT: Michael Eddy
SALES DEPT: Michael Eddy
PURCHASING DEPT: Michael Eddy
Full size exact replicas of military medals, badges and insignia. Our line includes 300 items from the Third Reich, Imperial Germany, and the British Empire. Catalog $1.00.

NANCY KRUGER
8779 Oakgreen Avenue South
Hastings, MN 55033, USA
Repro field gear & uniforms.

AARDVARK ENTERPRISES
P.O. Box 1046
Glendale Heights, IL 60139, USA
(312)-653-0133
Military vehicles & Jeep parts.
S6,S12; 6;
CONTACT: Dan Rhame
SALES DEPT: Dan Rhame
PURCHASING DEPT: Dan Rhame
Manufacturer of replica M60 MGs. Inf., A/B, patrolboat, & lightweight configurations; also limited quantities of early style models. Models are full-size, fit original GI mounts & incorporate aluminum and original GI parts, , but cannot fire and are completely legal. Prices from $500. - $1,000.

SAXONY HOUSE LTD.
P.O. Box 875
Mokena, IL 60448, USA
Reproduction medals.
Free list.

FRITZ PODDIG
113 Ridgemoor
Edwardsville, IL 62025, USA
618-288-5847
Repro field gear & uniform

MILITARY COLLECTIBLES
P.O. Box 971
Minden, LA 71055, USA
(318)-371-1229 6-10
General militaria.
S6,S11,S12; 14;
Catalogue available. Annually.
CONTACT: Lee C. Estabrook
SALES DEPT: Lee C. Estabrook
PURCHASING DEPT: Lee C. Estabrook
Museum quality reproduction flags & guidons. Colonial, American Revolution, Yankee and Confederate, Indian Wars thru Vietnam, State and foreign flags. Custom orders welcome.

DEVIL'S BRIGADE
P.O. Box 1625
El Dorado, AR 71730, USA
Reproduction documents

BRANT DeBLIECK
17328 E. Wagontrail Parkway
Aurora, CO 80015, USA
303-693-6385
Repro field gear & uniforms.

MILITARY REPRODUCTIONS

BJ's COLLECTORS REPLICAS
Box 591
1093 Broxton Avenue
Westwood Village, CA 90024, USA
Replica firearms.
Catalogue $1.00.

JOHN MEIX
Box 3221
Alhambra, CA 91803, USA
WWII & reproduction jump suits.
Send S.A.S.E. for list.

MILITARIA INTERNATIONAL
P.O. Box 5551
Concord, CA 94524, USA
Reproduction militaria

HISTORICAL MILITARY ART
P.O. Drawer 1806
Lafayette, CA 94549, USA
415-283-1771
Repro military insignia.
Free catalogue.

GERD SCHNEIDER
Sterenstrasse 40
Wurzburg, 8700, West Germany
(09 31) 8 18 15
Reproduction armour.

RESTORATION/REPAIRS

DIRECT SUPPORT
P.O. Box 317
Pittsfield, MA 01202, USA
(413)-443-4841
Military vehicle parts & accessories.
S4; 3;
Catalogue available. Annually.
CONTACT: Bob Kettler
SALES DEPT: Bob Kettler
PURCHASING DEPT: Bob Kettler
Specializing in New, NOS, and used parts for M-Series 2 1/2 ton, 5 ton, lot 6x6 military trucks. Also offer restorations on trucks and tracked armored vehicles.

HIGGINS ARMORY MUSEUM
100 Barber Avenue
Worcester, MA 01606, USA
(617)-853-6015
Medieval arms & armour museum.
S8; 55; 500; $60.00
Catalogue available.
SALES DEPT: The Director
Higgins Armory Museum exhibits mainly 16th and 17th century arms and armor and related artifacts. A metal conservator conducts restoration/repairs in newly equipped laboratory.

COLONIAL REPAIR
47 Navarre Street
Roslindale, MA 02131, USA
617-469-2991
Firearms repair.

NEVELLE ANTIQUES
HCR 68, Box 130-L
Cushing, ME 04563, USA
215-667-4740
Barometer repair.

Write for details & estimates.

ORUM SILVER COMPANY
P.O. Box 805
Meriden, CT 06450, USA
203-237-3037
Medal repair.
Write for details & estimates.

WALTER C. KAHN
76 N. Sylvan Road
Westport, CT 06880, USA
203-227-2195
Porcelain repairs.
Write for details & estimates.

BOB ARCH-ANTIQUE RESTORATION
36 East 12th Street
New York, NY 10003, USA
212-777-2967
General antique repair.
Write for details & estimates.

RON FOX
416 Throop Street
N. Babylon, NY 11704, USA
-669-7232
Beer stein restoration.
Write for details & estimates.

WORLD WAR I AEROPLANES
15 Crescent Road
Poughkeepsie, NY 12601, USA
914-473-3679 Aero 818-243-6820 Skyways
Provides service info on WWI aeroplanes.
S6,S10; 25; 2200; 100.00
CONTACT: Leonard E. Opdycke
SALES DEPT: L. E. Opdycke
AD MANAGER: Richard Alden
Provides service (information, names, projects, materials, books) to modellers, builders & restorers of aeroplanes (1900-1919, 1920-1940) through two journals, WW I Aero & Skyways.

CAMBRIDGE TEXTILES
Cambridge, NY 12816, USA
-677-2624
Fabric restoration.
Write for details & estimates.

CHARLES ERB
Route #1
Fredericktown, PA 15333, USA
-757-6811
Colt Revolver restoration.
Write for details & estimates.

THE WILLOW FORGE
P.O. Box 523
Route 40 East
Brownsville, PA 15417, USA
(412)-785-6997
Edged weapon repair & restoration.
S6; 7;
Flyer available.
CONTACT: Daniel Tokar
SALES DEPT: Daniel Tokar
PURCHASING DEPT: Daniel Tokar
I will make or repair any style of edged weapons, armour or hardware. Forge my own Damascus steel. Can research period styles of weapons.

PHILADELPHIA ORDNANCE, INC.

Oreland Industrial Park
Oreland, PA 19075, USA
215-576-0259
Firearms repair.
Write for details & estimates.

HARRY A. EBERHARDT & SON, INC.
2010 Walnut Street
Philadelphia, PA 19103, USA
Art restoration.
Write for details & estimates.

GERRY VERMEESCH
1630 Orlando Road
Pottstown, PA 19464, USA
215-326-4097
Firearms repair & restoration

SUNDANCE ENGRAVING
P.O. Box 912
Reading, PA 19603, USA
215-678-1832
Engraving restoration.
Write for details & estimates.

DONALD M. DAUGHERTY
4106 Dee Jay Drive
Ellicott City, MD 21043, USA
-465-6565
Beer stein restoration.
Write for details & estimates.

BELAIR ROAD GUNS SERVICE
5622 Arnhem Road
Baltimore, MD 21206, USA
Blueing of firearms & repairs to dealers only.
; 16;
CONTACT: Gene Ruby
SALES DEPT: Gene Ruby

RUSSEL PRATT
121 Bradley Circle
Durham, NC 27713, USA
-544-2603
Leather restoration.
Write for details & estimates.

ESTES SIMMONS
1168 Howell Mill Road N.W.
Atlanta, GA 30318, USA
-875-9581
Plating repair.
Write for details & estimates.

WALKERS ANTIQUE CLOCK REPAIR
351 Mears Street
Martinez, GA 30907, USA
-863-3938
General antique repair.
Write for details & estimates.

R.S. STEFFEN HISTORIC TEXTILES
P.O. Box 1302
Covington, KY 41012, USA
Uniform restoration.
Write for details & estimates.
CONTACT: R. S. Steffen

RESTORATION/REPAIRS

BECK & ORR, INC.
1640 Fairwood Avenue
Columbus, OH 43206, USA
-443-8481
Book rebinding.
Write for details & estimates.

WALTER WARD
Route #1, Box 10
West Alexandria, OH 45381, USA
-839-4147
Photograph restoration.
Write for details & estimates.

M.A.R.S. (MILITARY ACQUISITION, RESTORATION & SUPPLY)
Rt. 1, Box 182
Nashville, IN 47448, USA
(812)-988-2330
Completely remanufacture vintage Dodge trucks and military equipment.
S13;
CONTACT: Paul W. Caudell

ELBINGER LABORATORIES
220 Albert Street
East Lansing, MI 48823, USA
-332-1430
Photograph restoration.
Write for details & estimates.

AARDVARK ENTERPRISES
P.O. Box 1046
Glendale Heights, IL 60139, USA
(312)-653-0133
Military vehicles & Jeep parts.
S6,S12; 6;
CONTACT: Dan Rhame
SALES DEPT: Dan Rhame
PURCHASING DEPT: Dan Rhame
Manufacturer of replica M60 MGs. Inf., A/B, patrolboat, & lightweight configurations; also limited quantities of early style models. Models are full-size, fit original GI mounts & incorporate aluminum and original GI parts, , but cannot fire and are completely legal. Prices from $500. - $1,000.

INTERNATIONAL HISTORICAL FILMS
P.O. Box 29035
Chicago, IL 60629, USA
(312)-436-8051 (312)-436-0038
Military history video-cassettes.
S6,S12; 9;
Write or phone for free catalogue. 2 or 3 times annually. CONTACT: Peter P. Bernotas
SALES DEPT: Peter Bernotas
PURCHASING DEPT: Peter Bernotas
VIDEOCASSETTES FOR SALE - WWI through the Falklands Campaign. Specializing in WW II German Newsreels and feature films reproduced from the worlds largest privately owned collection of original 35MM source materials, including contemporary Soviet and British videocassettes.

BILL'S GUN REPAIR
1007 Burlington Street
Mendota, IL 61342, USA
(815)-539-5786
Firearms repair & gunsmithing.
S1,S6,S11,S13; 10;
Catalogue available. Annually.
CONTACT: William Neighbor
SALES DEPT: William Neighbor
PURCHASING DEPT: William Neighbor
All repair work, custom gunsmithing, antique restoration, rifles, pistols, shotguns. Mail order firearm sales. All work guaranteed. Bill's Gun Repair, 1007 Burlington St., Mendota, IL 61342.

FRANK JASEK
Route 1, Box 174
McGregor, TX 76657, USA
(817)-853-2561
Book restoration & fine binding.

JOHNSON WATCH REPAIR
2735 23rd Street
Greeley, CO 80631, USA
Watch repair & restoration.

MODERN GUN REPAIR SCHOOL
2538 N. 8th Street
Phoenix, AZ 85006, USA
Firearms repair & maintenance school.

ANTIQUE HARDWARE COMPANY
P.O. Box 1592
Torrance, CA 90505, USA
-378-5990
Furniture restoration.
Write for details & estimates.

SERVICES

GLOBAL SCHOOL OF INVESTIGATION
P.O. Box 191
Hanover, MA 02339, USA
Instruction in investigative techniques.

PARAPACK
RD 1, Box 214
Equinunk, PA 18417, USA
717-224-6207
Militaria & firearms services

WAR MUSEUMS SUPPLIES
4122 N. Richmond, 2nd Floor
Chicago, IL 60618-2614, USA
Museum support services.

TELE-OPTICS
5514 Lawrence Avenue
Chicago, IL 60630, USA
(312)-283-7757
Optical equipment, repairs & sales. Binoculars, telescopes, riflescopes.
S1,S4,S11; 37;
CONTACT: Herb Koehler
SALES DEPT: Herb Koehler
PURCHASING DEPT: Herb Koehler

ORANGE GUN CLUB
P.O. Box 400
Orangefield, TX 77639, USA
(409)-735-4428
Gun club
S1,S10,S11,S12; 30;
Orange Gun Club allows any type of firearm to be used safely on the club range. Machine guns permitted.

HIGHLINE MACHINE COMPANY
654 Lela Place
Grand Junction, CO 81504, USA
(303)-434-4971
Gunsmithing & other custom machine work.
S11; 5;
Catalogue available. Every two years.
CONTACT: Randall Thompson
SALES DEPT: Randall Thompson
PURCHASING DEPT: Randall Thompson
Gunsmith with full machine shop. Make, repair, rebuild metal parts. Also tool & die and investment mold work.

VIC COOPER FIREARMS DEALER
P.O. Box 1646
1601 1/2 - 19th Street
Eunice, NM 88231, USA
(505)-394-3237
Gunsmithing, firearms & accessories.
S1,S4,S11; 3;
CONTACT: Milton V. (Vic) Cooper, Jr.
SALES DEPT: M. V. Cooper (Vic)
PURCHASING DEPT: M. V. Cooper (Vic)
General gunsmithing services, retail firearms, ammo & related items. Special firearms transfer interstate. Special order of all hunting gear, clothing, guns, tents & camping equipment.

INTERNATIONAL BOOKFINDERS, INC.
P.O. Box 1
Pacific Palisades, CA 90272, USA
Bookfinding service.
S4; 36;
CONTACT: Richard Mohr
Free search for out-of-print military books. You name it - we find it! No obligation. International Bookfinders, Inc., P.O. Box 1-px, Pacific Palisades, CA 90272.

VETERANS
525 Mulberry Avenue
Patterson, CA 95363, USA
Military paperwork service.
Send SASE for information.
CONTACT: Jim Logsdon

BYGONE WARRIOR MILITARIA
P.O. Box 70211
Seattle, WA 98107, USA
(206)-283-5141 (206)-784-1700 Shop
General militaria.
S4; 6;
Catalogue available. 3 or 4 times annually.
CONTACT: Christopher C. Bruner
SALES DEPT: Christopher C. Bruner
Catalogue subscription $6.00 for four issues. Specializing in United States 1898-1945, German World War One and foreign. Uniforms, insignia, medals, field gear, ordnance, wings, paper.

MILITARY SURPLUS

P. HERZBERG
31 Pioneer Road
Hingham, MA 02043, USA
617-749-5336
U.S. military surplus.

MILITARY SURPLUS

MASS. ARMY & NAVY STORE
895 Boylston Street
Boston, MA 02115, USA
(617)-783-1250
Mail-order sales of Army-Navy, camping, surplus and clothing supplies.
S1,S4; ; 90,000; ; $65.00
Catalogue available. Quarterly.
CONTACT: David A. Glaser

VALLEY SURPLUS
P.O. Box 346
Tariffville, CT 06081, USA
203-658-6228
Military surplus.

NEWMAN'S
R.R. #1, Box 782
Augusta, NJ 07822, USA
201-875-3252
Military surplus.

RICK'S ARMY-NAVY SALES
P.O. Box 354
Marlton, NJ 08053, USA
Military surplus.

B. LARSSON
P.O. Box 782
Murray Hill, NY 10156, USA
Military surplus.

MILITARY SURPLUS DISTRIBUTORS
P.O. Box 190
Brooklyn, NY 11204, USA
(718)-376-6200
Military Surplus mail-order sales.
S4;
CONTACT: Bill DiVito

HAWKEYE SURPLUS INDUSTRIES
P.O. Box 321
Pawling, NY 12564, USA
(914)-855-5051
Surplus
S1; 10;
CONTACT: E. Walsh
SALES DEPT: J. Barsamian
PURCHASING DEPT: J. Barsamian

TRI STAR INTERNATIONAL
P.O. Box 96-LPO
Niagara Fallls, NY 14304, USA
(716)-731-5100
Military surplus
S3,S6; 15;
CONTACT: James A. Heider
SALES DEPT: James A. Heider
PURCHASING DEPT: James A. Heider
Buy & sell surplus

JOLLY ROGER SURPLUS COMPANY
P.O. Box 53
Roxbury, PA 17251, USA
Military surplus

DELK'S SURPLUS SALES,INC.
Rt #3, Box 114
Asheboro, NC 27203, USA
919 629-0991
Military surplus

WALTER BUDD
3109 Eubanks
Durham, NC 27707, USA
U.S. cavalry items.
Quarterly catalogue $5.00 year.

LOVE'S GUNS & MILITARY SURPLUS
Rt. 1, Box 643
Andrews, NC 28901, USA
704-321-3383 (WK) 704-321-5477 (HM)
Military surplus

BROCK'S ARMY SURPLUS,INC.
P.O. Box 33242
Decatur, GA 30033, USA
404-294-9500
Military surplus

THE SUPPLY SERGEANT
4300 Highway 20
Buford, GA 30518, USA
Military surplus. u
Catalogue $1.00.

S & S SUPPLY COMPANY
Box 1181
Waycross, GA 31502, USA
Military surplus.
List $.25.

SURPLUS
11350 N.E. 8th Court
Miami, FL 33161, USA
Surplus

J. A. S. ENTERPRISES
P.O. Box 38118
Germantown, TN 38183, USA
Military surplus

BACON CREEK GUN SHOP
P.O. Box 814
Corbin, KY 40701, USA
U.S. military surplus.

ANDY'S ARMY-NAVY SURPLUS
421 South Airport Road
Traverse City, MI 49684, USA
517-739-3340
Military surplus,
Catalogue $5.00.

JOHN KRUESEL
22 - 3rd Street, S.W.
Rochester, MN 55902, USA
507-289-8049
Military surplus

UNCLE DAN'S
2440 North Lincoln Avenue
Chicago, IL 60614, USA
312-477-1918
Military surplus & militaria.
Catalogue available.

RUVEL & COMPANY,INC
3037 North Clark Street
Chicago, IL 60657, USA
(312)-248-1922
Military surplus & sporting goods.
S4,S13; 21;
Catalogue available, $2.00. Annually.
CONTACT: Robert G. Ruvel
SALES DEPT: Robert G. Ruvel

PURCHASING DEPT: Robert G. Ruvel
Ruvel and Co., Inc. Army-Navy surplus, military & commercial & sporting goods. USN or USAAF/type leather jackets. Field jackets.

COMSEC INTERNATIONAL,INC.
826 Horan Drive
St. Louis, MO 63026-2478, USA
Military surplus & survival gear.

THE QUARTERMASTER SUPPLY DEPOT
RR. 1, Box 616
Washington, MO 63090, USA
314-239-7558
U.S. military surplus.
Send SASE with inquiries.
CONTACT: Gary L. Voelker

WORLD SURPLUS
10028 Manchester
Glendale, MO 63122, USA
314-821-4627
U.S. & European military surplus.
CONTACT: Pete Ranciglio

POOL SURPLUS
P.O. Box 370
Benton, AR 72015, USA
501-778-2260
Military surplus

C.J.W.
4240 South 36th Place
Phoenix, AZ 85040, USA
602-437-2724
Military surplus

KAUFMAN'S WEST
1660 Eubank N.E.
Albuquerque, NM 87112, USA
Military surplus & survival supplies.

SHERWOOD INTERNATIONAL CORP.
18714 Parthenia Street
Northridge, CA 91324, USA
818-886-1024
Military surplus
CONTACT: Mike Kokin, President

QUARTERMASTER STORES
3039 Adams Avenue
San Diego, CA 92116, USA
Military surplus

KEN NOLAN,INC.,P.M.DIV.
P.O. Box C-19555
16901 P.M. Milliken Avenue
Irvine, CA 92713, USA
(714)-863-1532
Surplus, insignia, medals & uniforms.
S4; 29; 100,000; $60.00
Catalogue $1.00. 3 or 4 times annually.
Serving individuals, military, police for 29 years. 90% of orders shipped within 24 hours. Military field clothing, boots, insignia, nameplates, etc. Send $1.00 for catalog.

MILITARY SURPLUS

INTERNATIONAL SURPLUS WHOLESALERS
P.O. Box 31
Oroville, CA 95965, USA
800-824-5115
Military surplus.
Catalogue available.
CONTACT: John Lane
SALES DEPT: S. Seidenglanz
AD MANAGER: Chuck Heindell

P. G. WING & CO.
Peggy's Walk
Littlebury
near Saffron Walden, Essex CB11 4TG, Great Britain 0799 21801 0799 21802
Over 400 lines government surplus clothing, equipment and collectibles.
S13; 15;
Catalogue available.
CONTACT: P. G. Wing
SALES DEPT: P. G. Wing
PURCHASING DEPT: P. G. Wing

CLUBS & SOCIETIES

AMERICAN ANTIQUARIAN SOCIETY
185 Salisbury Street
Worcester, MA 01609, USA
(617)-755-5221
Learned society.
Library, journal & newsletter.
CONTACT: Marcus A. McCorison, Director

VETERANS ASSOCIATION OF THE FIRST CORPS OF CADETS
227 Commonwealth Avenue
Boston, MA 02116, USA
617-267-1726
Regional military history association. Museum & library.
CONTACT: John F. McCauley

COLLECTOR ARMS DEALERS ASSOCIATION
P.O. Box 298
Falmouth, MA 02541, USA
617-548-0660
Firearms dealers association. Gun show promoters.
CONTACT: William 'Pete' Harvey

MAINE HISTORICAL SOCIETY
485 Congress Street
Portland, ME 04101, USA
207-774-1822
Regional historical society. Museum, library, quarterly journal & newsletter.

COMPANY OF MILITARY HISTORIANS
North Main Street
Westbrook, CT 06498, USA
Military history & militaria collectors association. Convention, journal.
CONTACT: William R. Reid

YE CONNECTICUT GUN GUILD, INC.
Brookfield Road
Harwinton, CT 06791, USA
203-485-1356
Firearms collectors association.
Gun show promoters.

CONTACT: Robert Dailey

NEW JERSEY HISTORICAL SOCIETY
230 Broadway
Newark, NJ 07104, USA
(201)-483-3939
Regional historical association. Museum, library, quarterly & newsletter.
CONTACT: Robert M. Lunny, Director

BROTHERHOOD OF XCCCERS
14 Orchard Street
Elmwood Park, NJ 07407, USA
Ex-Civilian Conservation Corps association. Newsletter.

MONMOUTH COUNTY HISTORICAL ASSOCIATION
70 Court Street
Freehold, NJ 07728, USA
(201)-462-1466
Regional historical association.
Museums, library, exhibits & newsletter.
CONTACT: Wilson E. O'Donnell

IMPERIAL GERMAN MILITARY COLLECTORS ASSOCIATION
82 Atlantic Street
Keyport, NJ 07735, USA
(201)-264-0802 (816)-455-3214
Imperial German militaria study group.
; 14;
CONTACT: Gen. R. H. Thompson
SALES DEPT: J. J. Daub
PURCHASING DEPT: J. J. Daub
AD MANAGER: J. J. Daub
IGMCA is a world-wide organization of collectors & museums devoted to the study of Imperial German history and accoutrements.

MORRISTOWN EDISON NATIONAL PARK SERVICE GROUP
P.O. Box 1136
Morristown, NJ 07960, USA
Park support association.

STEIN COLLECTORS INTERNATIONAL
P.O. Box 463
Kingston, NJ 08528, USA
(201)-329-2567
Beer stein collectors association. Quarterly magazine.
S7; 23;
CONTACT: Jack G. Lowenstein, Exec. Dir.
AD MANAGER: Jack Lowenstein
We are an organization of 1500+ serious collectors of antique beer steins (mostly German) and similar drinking vessels. We publish a quarterly magazine, "Prosit", 36 pp

JERSEY SHORE ANTIQUE ARMS COLLECTORS
P.O. Box 100
Bayville, NJ 08721, USA
201-269-3290
Firearms collectors association. Gun show promoters. Monthly meetings.
CONTACT: Joe Sisia

WAR COVER CLUB
P.O. Box 173
Jamesburg, NJ 08831, USA
Military postal history association.

NEW YORK HISTORICAL SOCIETY
170 Central Park West
New York, NY 10024, USA
Regional historical society

STATEN ISLAND HISTORICAL SOCIETY
441 Clarke Avenue
Staten Island, NY 10306, USA
718-351-1611
Regional historical association. Museum, quarterly, newsletter.
CONTACT: William McMillen

STRATFORD GUN COLLECTORS ASSOCIATION
P.O. Box 828
Harrison, NY 10528, USA
914-698-7228
Firearms collectors association. Gun show promoters.
CONTACT: Don Von Den Driesch

MILITARY BOOK CLUB
501 Franklin Avenue
Garden City, NY 11535, USA
Military book club.
Write for details.

NORTH EASTERN ARMS COLLECTORS
P.O. Box 185
Amityville, NY 11701, USA
Militaria & firearms collectors association. Gun show promoters.

LONG ISLAND ANTIQUE GUN COLLECTORS ASSOCIATION
52 Tompkins Street
E. Northport, NY 11731, USA
Firearms collectors association. Gun show promoters.
CONTACT: Ed Batcheller, Secretary

RANGER BOOK CLUB
600 Washington Court
Guiderland, NY 12084, USA
Military book club.

NORTH-EAST ARMS COLLECTORS ASSOCIATION, INC.
P.O. Box 385
Mechanicville, NY 12118, USA
(518)-664-9743 (518)-664-7610
Gun collector associates.
CONTACT: David Petronis
AD MANAGER: Cathy Petronis
NEACA, Inc. publishes "Arms Collectors Journal" monthly and administers "New East Coast Arms Collector Associates", a member orginization and sponsor of regional arms and militaria shows.

NEW YORK STATE MILITARIA COLLECTORS ASSOCIATION, INC.
1074 W. Genesee Street
Syracuse, NY 13204, USA
315-455-1716
Militaria & firearms collectors ass'n. 4 gun shows & 4 newsletters annually.
CONTACT: Edward Monarski

ORGANIZATIONS, CLUBS & SOCIETIES

NEW YORK STATE ARMS COLLECTORS ASSOCIATION,INC.
P.O. Box 312
Oxford, NY 13830, USA
607-843-6266
Antique firearms collectors association. Gun show promoters, newsletter
CONTACT: Ronald W. Bullock,Secretary

OLD FORT NIAGARA ASSOCIATION
P.O. Box 169
Youngstown, NY 14174, USA
716-745-7611
Regional historical association. Old Ft. Niagara, newsletter.

BUFFALO & ERIE COUNTY HISTORICAL SOCIETY
25 Nottingham Court
Buffalo, NY 14216, USA
Regional historical society

EMPIRE STATE ARMS COLLECTORS ASSOCIATION
P.O. Box 2328
Rochester, NY 14623, USA
716-334-5277
Firearms collectors association. Gun show promoters.

LAND OF THE SENECAS MUZZLE LOADERS
P.O. Box 476
Breesport, NY 14816, USA
Black powder shooters association.

PENNSYLVANIA GUN COLLECTORS ASSOCIATION
P.O. Box 3
Bethel Park, PA 15102, USA
412-854-0937
Militaria & firearms collectors ass'n. Gun show promoters, 5 per year.
CONTACT: Kenneth C. Smith,Secretary

INDEPENDENT MOUNTAIN MEN OF PA
18 Frazier Avenue
McKees Rocks, PA 15136, USA
Black powder shooters association.

TRI-STATE COLLECTORS ASSOCIATION
P.O. Box 6061
Pittsburgh, PA 15211, USA
412-488-8989
Firearms collectors association. Gun show promoters.
CONTACT: Ed Threfall,President

BLACK BOYS OF BLOODY RUN
RD 2, Box 615
Everett, PA 15537, USA
(814)-652-5816
Black powder shooters association.
CONTACT: Lois C. Stoner

FORT ERIE MUZZLELOADERS
P.O. Box 4046
Erie, PA 16512, USA
(814)-838-2228
Black powder shooters association.
CONTACT: Bob Harris

AMEIGH VALLEY IRREGULARS
Box 225 C, R.D. 2
Gillett, PA 16925, USA
(717)-537-6702 (717)-537-2887
Black powder shooters association.

U.S. ARMY MILITARY HISTORY INSTITUTE
Historical Reference Branch
Carlisle Barracks, PA 17013, USA
717-245-3611
U.S. Army historical research institute. Museum, library, publications.
CONTACT: John J. Slonaker,Chief

ANTIQUE AUTOMOBILE CLUB
501 W. Governor Road
Hershey, PA 17033, USA
717 534-1910
Antique vehicle collectors association.

AMERICAN SOCIETY OF MILITARY INSIGNIA COLLECTORS
1331 Bradley Avenue
Hummelstown, PA 17036, USA
(717)-566-3032
U.S. military insignia collectors ass'n. Journal, newsletter, convention.
CONTACT: James F. Greene,Jr.
AD MANAGER: James F. Greene,Jr.

EASTERN ARMS COLLECTORS
P.O. Box 445
Gettysburg, PA 17325, USA
717-334-4564
Firearms collectors association. Gun show promoter.

BLUE RIDGE RIFLES,INC.
Route 183, Wayne Township
Summitt Station, PA 17533, USA
Black powder shooters association.
CONTACT: John S. Bates, Pres.

INTERNATIONAL CARTRIDGE COLLECTORS ASSOCIATION
1211 Walnut Street
Williamsport, PA 17701, USA
Cartridge collector association.
CONTACT: Victor Engels

PENNSYLVANIA FEDERATION OF BLACK POWDER SHOOTERS, INC.
P.O. Box 150
Selinsgrove, PA 17870, USA
Black powder shooters association.
CONTACT: James Fulmer, Sec.

FORKS OF THE DELAWARE WEAPONS ASSOCIATION
R.D. 4, Box 4325
Bangor, PA 18013, USA
215-588-8305
Firearms & militaria collectors ass'n. Gun show promoters.
CONTACT: John F. Scheid

BOULDER VALLEY SPORTSMAN'S ASSOCIATION
Perkiomenville Road
Sumneytown, PA 18084, USA
(215)-234-4767
Black powder shooters association.

MID ATLANTIC ARMS COLLECTORS
P.O. Box 278
Carbondale, PA 18407, USA
717-282-2832
Firearms & militaria collectors ass'n. Gun show promoters.

COUNCIL CUP MUZZLE LOADERS, INC.
P.O. Box 65
Nescopeck, PA 18635, USA
Black powder shooters association.

WYOMING VALLEY MILITARY COLLECTORS ASSOCIATION
1 Holiday Drive
Duryea, PA 18642, USA
717-457-9473
General militaria collectors association. Military show promoters.
CONTACT: Ed Winn,Sr.

MINIATURE FIGURE COLLECTORS
102 St. Pauls Road
Ardmore, PA 19003, USA
Model soldiers collectors association.

DELAWARE COUNTY FIELD & STREAM ASSOCIATION
P.O. Box 1092
Brookhaven, PA 19015, USA
(215)-586-1799
Black powder shooters association.
CONTACT: William Cockerill

PENNSYLVANIA ANTIQUE GUN COLLECTORS ASSOCIATION
28 Fulmer Avenue
Havertown, PA 19083, USA
215-446-5973
Militaria & firearms collectors ass'n. Bi-annual gun shows. Monthly newsletter.
CONTACT: Kathleen Beyer,Secretary

MILITARY ORDER OF THE LOYAL LEGION
1805 Pine Street
Philadelphia, Pa 19103, USA
Association of descendants of Civil War U.S. Army officers. Museum & library.

LANCASTER MUZZLE LOADING RIFLE ASSOCIATION
Rt. 1
Cochranville, PA 19330, USA
(215)-593-6208
Black Powder firearms association. Gun show sponser.
CONTACT: Robert Rambo

BLUE MOUNTAIN MUZZLE LOADING RIFLE ASSN.,INC.
Shartlesville, PA 19554, USA
Black powder shooters association.

MORTON INSIGNIA GROUP
18 Bristol Knoll
Newark, DE 19711, USA
302-731-1307
Military insignia collectors ass'n. Insignia shows, newsletter.
CONTACT: Stanley Blake

ORGANIZATIONS, CLUBS & SOCIETIES

DELAWARE ANTIQUE ARMS COLLECTORS ASSOCIATION
231 Market Street
Wilmington, DE 19801, USA
(302)-652-0972 Days (302)-478-2341 Eves.
Antique firearms collectors association. Military relic & gun show promoters.
CONTACT: Jack Coonin

CATHOLIC WAR VETERANS OF U.S.A
2 Massachusetts Ave.,N.W.
Washington, DC 20001, USA
Military veterans association

NATIONAL GUARD ASSOCIATION
1 Massachusetts Avenue N.W.
Washington, DC 20001, USA
Military association.

THE SOCIETY OF THE CINCINNATI
2118 Massachusetts Ave., N.W.
Washington, DC 20008, USA
202-785-2040
Association of male descendants of Revolutionary War U.S. Army officers.
Museum & library.

NATIONAL RIFLE ASSOCIATION
1600 Rhode Island Avenue, N.W.
Washington, DC 20036, USA
Firearms owners association

NATIONAL GEOGRAPHIC SOCIETY
Library
16th & M Streets, N.W.
Washington, DC 20036, USA
Learned society, museum & library.

THE AMERICAN VETERANS COMMITTEE
1735 DeSales Street,N.W.
Washington, DC 20036, USA
(202)-639-8886
Veterans association
Veterans association whose principle concerns involve political issues, such as, women's rights, veteran's rights, minority veterans.

AIR FORCE HISTORICAL FOUNDATION
Newsletter, Building 819
Bolling Air Force Base
Washington, DC 20332, USA
Historical foundation.

COMMITTEE ON MUSEUMS IN AMERICA
Bldg. 58
Washing Navy Yard
Washington, DC 20374, USA
Museum association
CONTACT: C. A. Wood, Secretary

MARINE CORPS HISTORICAL FOUNDATION
Building 58, Navy Yard
Washington, DC 20374, USA
Items, books, graphics pertaining to Marine Corps history or memoribilia.
S1,S4; 10;
Catalogue available. Annually.
CONTACT: Col. W. E. Reynolds, USMC,Ret.
SALES DEPT: Col. W. E. Reynolds,USMC(Ret)
Supports Marine Corps Historical Program by operating museum store, offering awards for articles and art, giving research and thesis and disertation grants.

VETERANS OF WORLD WAR I
941 N. Capitol St. NE, #1201-C
Washington, DC 20421, USA
Veterans association.

AIR FORCE SERGEANTS ASSOCIATION
5211 Auth Road
Suitland, MD 20746, USA
Military active duty association.
CONTACT: Mr. Dave Givens, Mem. Sec.

MARYLAND ARMS COLLECTORS ASSOCIATION
P.O. Box 481
Normandy Branch
Elliott City, MD 21043, USA
301-877-2912
Firearms collectors association. Gun show promoters.

GREAT WAR ASSOCIATION
3608 Woodhome Drive
Jarrettsville, MD 21084, USA
301-357-8566
World War I collectors association.
CONTACT: Terry Daley

MARYLAND HISTORICAL SOCIETY
201 West Monument Street
Baltimore, MD 21201, USA
301-685-3750
Regional historical association. Museums, library, magazine & newsletter.
CONTACT: Mrs. S. Painter

BALTIMORE ANTIQUE ARMS ASSOCIATION
E 30 - 2600 Insulator Drive
Baltimore, MD 21230, USA
(301)-244-0225
Antique weapon collectors association. Gun show promoter.
CONTACT: Stanley I. Kellert

VIRGINIA GUN COLLECTORS ASSOCIATION
4205 Sleepy Hollow Road
Annandale, VA 22003, USA
703-256-0665
Firearms collectors association. Gun show promoters.
CONTACT: Marty Eakes

COUNCIL ON AMERICA'S MILITARY PAST
P.O. Box 1151
Fort myers, VA 22211, USA
(202)-479-2258 (703)-486-0444
U.S. military history preservationists. Newsletter.
; 20; ; 2600; ; 100.00
CONTACT: Herbert M. Hart, Secretary
SALES DEPT: Edward L. Boyer
AD MANAGER: Mark Magnussen
Only national military history/historic preservation organization interested in location and preservation of military sites no longer serving purpose for which originally established.

SHENANDOAH VALLEY GUN COLLECTORS ASSOCIATION
P.O. Box 288
Winchester, VA 22601, USA
Firearms collectors association. Gun show promoters.
CONTACT: Daniel E. Blye

VIRGINIA HISTORICAL SOCIETY
P.O. Box 7311
Richmond, VA 23221, USA
804-358-4901
Regional historical society. Museum, library, quarterly & newsletter.
CONTACT: Lauri J. Cohen,Develop. Ass't.

ARMY TRANSPORTATION MUSEUM FOUNDATION
P.O. Drawer D
Ft. Eustis, VA 23604, USA
Support museum building & development program.

CAROLINA GUN COLLECTORS ASSOCIATION
3231 7th St. Dr. N.E.
Hickory, NC 28601, USA
704-327-0055
Firearms collectors association. Gun show promoters.
CONTACT: Jerry M. Ledford,Secretary

SOUTH CAROLINA ARMS COLLECTORS ASSOCIATION
P.O. Box 6611, Station B
Greenville, SC 29606, USA
Firearms collectors association. Gun show promoters.
CONTACT: Bobby O'Shields

GEORGIA ARMS COLLECTORS ASSOCIATION
P.O. Box 277
Alpharetta, GA 30201, USA
404-992-6539
Firearms collectors association. Gun show promoters.
CONTACT: Michael S. Kindberg

SOCIETY OF VIETNAMESE RANGERS
P.O. Box 29965
Atlanta, GA 30359, USA
(404)-296-0800 (404)-998-2401
Elite forces military insignia.
S4,S11,S12;
CONTACT: McDonald Valentine
Our society has more than 500 of the most famous soldiers in elite forces, with most of them having been seen in elite forces books for years. We have aided many research scholars in their quest for information. Average 3 or 4 requests per month. Usual fee $350 - $500, satisfied.

CHEROKEE GUN CLUB
P.O. Box 941
Gainesville, GA 30503, USA
404-983-3869 (8-9)
Firearms collectors association. Gun show promoters.
CONTACT: Paul Watkins

HEART OF GEORGIA GUN CLUB
1398 Rocky Creek Road
Macon, GA 31206, USA
912-788-9291
Gun club & gun show sponsors.
CONTACT: E. W. Cater

ORGANIZATIONS, CLUBS & SOCIETIES

NORTH FLORIDA ARMS COLLECTORS ASSOCIATION
P.O. Box 52191
Jacksonville, FL 32201, USA
904-733-6881
Firearms collectors association. Gun show promoters. Newsletter.
CONTACT: Martha Pierce

MILITARIA COLLECTORS SOCIETY OF FLORIDA, INC.
P.O. Box 63/5036
Margate, FL 33063, USA
(305)-643-3764
Militaria collectors association.
CONTACT: Joe Hamilton, President
To promote the knowledge, study and preservation of military relics and encourage and support Florida militaria collectors in pursuit of their hobby.

8th AIR FORCE HISTORICAL SOCIETY
P.O. Box 3556
Hollywood, FL 33083, USA
Eighth Air Force historical society. Quarterly journal, reunions, library.

PALM BEACH GUN COLLECTORS ASSOCIATION
P.O. Box 6372
Lake Worth, FL 33461, USA
305-968-9563
Firearms collectors association. Gun show promoters.

NAPOLEONIC SOCIETY OF AMERICA
640 PM Poinsettia Road
Belleair, FL 33516, USA
(813)-586-1779 Days (813)-584-1255 Night
Napoleonic collectors asssociation. Quarterly bulletins published.
; 3;
CONTACT: Robert M. Snibbe, Mgr. Dir.
Members receive a membership card and pin, a Members Bulletin containing interesting articles, plus conducted study tours to France every spring and annual meetings.

FLORIDA GUN COLLECTORS ASSOCIATION
5700 Mariner Drive
Tampa, FL 33609, USA
813-879-3660
Firearms collectors association. Gun show promoters.
CONTACT: John Hammer

JAPANESE MILITARIA COLLECTORS ASSOCIATION
4001 Windermere Drive
Tuscaloosa, AL 35405, USA
(205)-556-0086
Collectors of Japanese militaria. Meetings, newsletter, library.
CONTACT: Kathleen Harper, Ed. & Pres.

ALABAMA GUN COLLECTORS ASSOCIATION
P.O. Box 5548
Tuscaloosa, AL 35405, USA
Firearms collectors association. Gun show promoters.

LITTLE BIGHORN ASSOCIATION
P.O. Box 663
Boaz, AL 35957, USA
George A. Custer collectors association.

CENTRAL ALABAMA GUN CLUB
P.O. Box 4961
Montgomery, AL 36101, USA
205-272-8445
Firearms collectors association. Gun show promoters.
CONTACT: Phil Mitchell

TENNESSEE GUN COLLECTORS ASSOCIATION
3556 Pleasant Valley Road
Nashville, TN 37204, USA
615-292-3020
Gun collectors association. Gun show promoterss, newsletter.

MOUNTAIN EMPIRE GUN COLLECTORS ASSOCIATION
P.O. Box 1471
Kingsport, TN 37662, USA
615-246-7788 (AM)
Gun collectors association. Gun show promoters.

TENNESSEE GUN & KNIFE ASSOCIATION
P.O. Box 787
Kingsport, TN 37662, USA
615-247-2406
Weapons collectors association. Gun show promoters

MEMPHIS MILITARY COLLECTORS ASSOCIATION
DINO'S, c/o Rudy
645 N. McNean
Memphis, Tn 38104, USA
Military collectors association. 12 meetings per year.
CONTACT: Dr. John Mitchel, Sec.

MEMPHIS ANTIQUE WEAPONS ASSOCIATION
P.O. Box 11137
Memphis, TN 38111, USA
Antique weapon collectors association. Gun show promoters.

MILITARY ORDER OF THE STARS & BARS
P.O. Box 5164, Southern Station
Hattiesburg, MS 39401, USA
Military society.

BLUEGRASS MILITARY COLLECTORS CLUB
P.O. Box 9
Ft. Knox, KY 40121, USA
502-351-1833
General militeria collectors association.
CONTACT: Ronald D. Koontz

KENTUCKY GUN COLLECTORS ASSOCIATION
P.O. Box 64
Owensboro, KY 42302, USA
502-729-4197
Firearms collectors association. Gun show promoters.

WWII NATIONAL HISTORICAL SOCIETY
3286 Winding Creek Drive
Columbus, OH 43223, USA
614-875-0111
WWII historical society.
CONTACT: Richard A. Pemberton

MAUMEE VALLEY GUN COLLECTORS ASSOCIATION
P.O. Box 492
Maumee, OH 43537, USA
(419)-893-5173
Firearms collectors association. Gun show promoters, newsletter.
; 29;
CONTACT: Richard Keith, President
350 table gun show, six times/year. (419)-893-5173.

ASSOCIATION OF AMERICAN MILITARY UNIFORM COLLECTORS
446 Berkshire Road
Elyria, OH 44035, USA
(216)-365-5321
Military uniform collectors association. Quarterly newsletter.
CONTACT: Gil Sanow, II, Editor
Organization seeks to further collecting, preservation & research of American military uniforms, primarily of the Twentieth Century. Publishes 14 page newsletter 'FOOTLOCKER' quarterly.

OHIO VALLEY MILITARY SOCIETY
P.O. Box 9246
Cincinnati, OH 45209, USA
513-922-6411
Militaria collectors association. Gun show promoters, newsletter.
CONTACT: Paul Peters

THE AIR FORCE MUSEUM FOUNDATION, INC.
P.O. Box 1903
Wright-Patterson, OH 45433, USA
513-258-1218
Friends of the U.S.A.F. Museum. A.F. Museum, newsletter, calender, book.

TIPPECANOE COUNTY HISTORICAL ASSOCIATION
10th & Southern Streets
Lafayette, IN 47901, USA
Regional historical society. Museums, library, exhibits, & newsletter.
CONTACT: MR. J. M. Harris

MICHIGAN ANTIQUE ARMS COLLECTORS, INC.
P.O. Box 1008
Rochester, MI 48063, USA
313-651-8407
Antique firearms collectors association.
; 39;
CONTACT: Ray & Winnie Russell

LOWER MICHIGAN GUN COLLECTORS
8639 Donna Drive
Westland, MI 48185, USA
313-427-8946
Firearms collectors association. Gun show promoters.

ORGANIZATIONS, CLUBS & SOCIETIES

ANTIQUE AIRPLANE ASSOCIATION
Route #2
Ottuma, IA 52501, USA
Vintage aircraft collectors association.

GREAT LAKES ARMS COLLECTORS ASSOCIATION, INC.
1439 North 50th Street
Milwaukee, WI 53208, USA
414-453-8231
Firearms collectors association. Gun show promoters.
CONTACT: Robert L. Selissen

BROOKFIELD COLLECTOR'S GUILD
P.O. Box 10200
1601 W. Greenfield Avenue
Milwaukee, WI 53210, USA
Collectors association.

WISCONSIN MILITARY COLLECTORS ASSOCIATION
1007 Tripoli Road
Janesville, WI 53545, USA
608-757-0219
Militaria collectors association. Gun show promoters.
CONTACT: Jeff W. Ashenfelter

THE STATE HISTORICAL SOCIETY OF WISCONSIN
816 State Street
Madison, WI 53706, USA
Regional historical association. Historic sites, quarterly, newsletter.
CONTACT: James Watson, Curator

WORLD WAR TWO HISTORICAL REENACTMENT SOCIETY, INC.
4725 Nora Lane
Madison, WI 53711, USA
(608)-222-6489
Organization of WW II reenactors.
; 10;
CONTACT: John Ong, Secretary-Treasurer
To honor: the servicemen & women of WWII. To preserve: the artifacts, equipment, memorabilia of WWII. To participate: in re-creating the atmosphere & events of WWII.

MINNESOTA HISTORICAL SOCIETY
1500 Mississippi Street
St. Paul, MN 55101, USA
612-296-0332
Regional historical association. Historic sites, quarterly, newsletter.

MINNESOTA WEAPONS COLLECTORS ASSOCIATION
P.O. Box 662
Hopkins, MN 55343, USA
Weapons collectors association. Gun show promoters.
CONTACT: Gail Foster

NORTH DAKOTA HERITAGE FOUNDATION, INC.
Box 1976
Bismarck, ND 58502, USA
Regional Historical Society.
CONTACT: Clifford B. Keller, COE

STATE HISTORICAL SOCIETY OF NORTH DAKOTA
North Dakota Heritage Center
Bismarck, ND 58505, USA
701-224-2666
Tax-funded state historical agency. Museums, journal, library, sites.
CONTACT: Larry Remele, Historian/Editor

ARNOLD AIR SOCIETY
P.O. Box 5134
Boseman, MT 59715, USA
406-587-1026
Firearms & militaria collectors association. Gun show promoters.
CONTACT: Ted Terrazas

WEAPONS COLLECTORS SOCIETY OF MONTANA
3100 Bancroft
Missoula, MT 59801, USA
Weapons collectors association. Gun show promoters. Newsletter.
CONTACT: R. G. Schipf, Executive Sec.

DOUGHBOY HISTORICAL SOCIETY
P.O. Box 3912
Missoula, MT 59806, USA
U.S. WWI collectors association.

FOX VALLEY ARMS FELLOWSHIP
110 Railroad
East Dundee, IL 60118, USA
312-426-3455
Firearms collectors association. Gun show promoters.
CONTACT: Fred Doederlein

G.A.R. MEMORIAL ASSOCIATION
P.O. Box 1043
Aurora, IL 60507, USA
G.A.R. collectors association.

PINE TREE PISTOL CLUB
332 E. Riverside Blvd.
Rockford, IL 61111, USA
815-229-1476
Firearms collectors association. Gun show promoters.
CONTACT: Greg Lyle

MIDWEST GUN COLLECTOR'S ASSOCIATION
1505 Sunset Drive
Spring Beach
East Peoria, IL 61611, USA
309-699-6760
Firearms collectors association. Gun show promoters, newsletter, range.
CONTACT: Gene Jordan

WW II HIS. REENACTMENT SOCIETY ST. LOUIS CHAPTER
815 St. William
Cahokia, IL 62206, USA
WWII Historical Reenactment Society chapter. WWII reenactment coordinator.
CONTACT: John Lampe

WW II HIS. REENACTMENT SOCIETY WEST CENTRAL ILLINOIS CHAPTER
2722 Monroe
Quincy, IL 62301, USA
WWII Historical Reenactment Society chapter.
WWII reenactment coordinator.
CONTACT: Joe Homberger

MISSOURI HISTORICAL SOCIETY
Jefferson Memorial Building
Forest Park
St. Louis, MO 63112, USA
314-361-1424
Regional historical society. Museums, library, quarterly, & newsletter.

MISSOURI ARMS COLLECTORS ASSOCIATION
122 N. Kirkwood Road
St. Louis, MO 63122, USA
314-821-8244
Firearms & militaria collectors association. 4 gun shows annually.
CONTACT: Ralph Persels

THE MISSOURI VALLEY ARMS COLLECTORS ASSOCIATION
P.O. Box 33033
Kansas City, MO 64114, USA
(913)-333-6509
Firearms collectors association Gunshow promoter, newsletter

MISSOURI SPORT SHOOTING ASSOCIATION
P.O. Box 6756
Jefferson City, MO 65102, USA
Firearms owners association. Civic involvement, newsletter.

WORLD ARCHAEOLOGICAL SOCIETY
Star Route, Box 445
Hollister, MO 65672, USA
Archaeological society.
CONTACT: Ron Miller, Director

CHISHOLM TRAIL ANTIQUE GUN ASSOCIATION
P.O. Box 12142
WICHITA, KS 67212, USA
Black powder & firearms collectors association. Matches & newsletter.
CONTACT: Bob Clevenger, Secretary

KANSAS STATE RIFLE ASSOCIATION
P.O. Box 503
Salina, KS 67402, USA
913-827-0027
Firearms owners association. Civic involvement, newsletter.

NEBRASKA STATE HISTORICAL SOCIETY
P.O. Box 82554
1500 R Street
Lincoln, NE 68501, USA
Regional historical society. Museums, library, quarterly, & newsletter.

NEBRASKA STATE HISTORICAL SOCIETY
Fort Robinson Museum
P.O. Box 304
Crawford, NE 69339, USA
Regional historical society.

ORGANIZATIONS, CLUBS & SOCIETIES

BAYOU GUN CLUB OF LOUISIANA
4449 W. Metairie Avenue
Metairie, LA 70001, USA
504-455-6571
Gun collectors association. Gun show promoters.

ATTAKAPAS GUN CLUB
P.O. Drawer Q
Lafayette, LA 70502, USA
Firearms collectors association. Gun show promoters.

ARK-LA-TEX GUN COLLECTORS ASSOCIATION
6150 Line Avenue
Shreveport, LA 71106, USA
318-865-9311 DAYS
Firearms collectors association. Gun show promoters.
CONTACT: Stella Chapman

ARKANSAS GUN AND CARTRIDGE COLLECTORS CLUB
Rt. 1, Box 1
Scott, AR 72142, USA
501-961-9284
Firearms collectors association. Gun show promoters.
CONTACT: Suzy Cotham

ORDER OF THE INDIAN WARS
P.O. Box 7401
Little Rock, AR 72217, USA
501-225-3996
Indian Wars historical association. Quarterly journal, annual meeting.
CONTACT: Jerry L. Russell, Chairman

CONFEDERATE HISTORICAL INSTITUTE
P.O. Box 7388
Little Rock, AR 72217, USA
Confederate historical association. Quarterly journal, newsletter.
CONTACT: Jerry L. Russell, Chairman

CIVIL WAR ROUND TABLE ASSOCIATES
P.O. Box 7388
Little Rock, AR 72217, USA
501-225-3996
U.S. Civil War historical association. Newsletter, annual meeting.
CONTACT: Jerry L. Russell, Chairman

VICTORIAN MILITARY HISTORY INSTITUTE
P.O. Box 7401
Little Rock, AR 72217, USA
501-225-3996
Victorian military historical ass'n. Quarterly journal & newsletter.
CONTACT: Jerry L. Russell, Chairman

FORT SMITH DEALERS & COLLECTORS ASSOCIATION
P.O. Box 941
Fort Smith, AR 72902, USA
Firearms collectors association. Gun show promoters.
CONTACT: John Ragains

OKLAHOMA GUN COLLECTORS ASSOCIATION
3001 The Paseo
Oklahoma City, OK 73103, USA
405-528-1222
Antique firearms collectors association. Gun show promoters, 3 meetings annually.
CONTACT: Claude A. Hall

45th INFANTRY DIVISION ASSOCIATION
2145 N.E. 36th Street
Oklahoma City, OK 73111, USA
Military history preservationists.
CONTACT: LTC Robert A. Wilson(Ret)

UNITED STATES FIELD ARTILLERY ASSOCIATION
P.O. Box 33027
Fort Sill, OK 73503, USA
(405)-355-4677
Professional association. Publisher of 'Field Artillery Journal'.
; 14;
CONTACT: Gen. (Ret.) W. T. Kerwin, Jr.
SALES DEPT: Mrs. Anna Lou Johnson
PURCHASING DEPT: Mrs. Anna Lou Johnson

FIELD ARTILLERY MUSEUM ASSOCIATION
Building 435
Fort Sill, OK 73503, USA
Museum support association.

THE DALLAS ARMS COLLECTORS ASSOCIATION, INC.
Route 1, Box 282B
Desoto, TX 75115, USA
Firearms collectors association. Gun show promoters, monthly meeting.
CONTACT: Joe W. Watson, Treasurer

DALLAS GUN COLLECTORS ASSOCIATION
P.O. Box 31094
Dallas, TX 75231, USA
214-956-9883
Firearms collectors association. Gun show promoters.

HOUSTON GUN COLLECTORS ASSOCIATION
P.O. Box 741429
Houston, TX 77274, USA
Firearms collectors association. Gun show promoters, newsletter.
CONTACT: James Suchma, President

SOUTHEAST TEXAS GUN & KNIFE ASSOCIATION
3420 9th Avenue
Port Arthur, TX 77642, USA
409-985-7701 (9-5)
Firearms collectors association. Gun show promoters.

TEXAS WEAPONS COLLECTORS ASSOCIATION
Star Route, Box 17-D
Devine, TX 78016, USA
512-663-5124
Firearms collectors association. Gun show promoters.
CONTACT: Mike Morris

82nd AIRBORNE DIVISION ASSOCIATION
Alamo Chapter
7403 Gallop Street
San Antonio, TX 78227, USA
Airborne militaria collectors association.

ALAMO ARMS COLLECTORS ASSOCIATION
28211 Bonn Mountain Drive
San Antonio, TX 78260, USA
512-438-2892
Gun collectors association. Gun show promoters.
CONTACT: Chip Swinney

BANZAI-JAPANESE MILITARY COLLECTORS
P.O. Box 393
San Diego, TX 78384, USA
W.W.II Japanese militaria collectors association.

SOUTH TEXAS GUN COLLECTORS, INC
P.O. Box 10815
Corpus Christi, TX 78410, USA
512-241-0282
Firearms collectors association. Gun show promoters.

CONFEDERATE AIR FORCE
P.O. Box CAF
Harlingen, TX 78551, USA
512-425-1057
Military aircraft preservationists. Air shows, 120 aircraft, museum & magazine.
CONTACT: Ralph Royce

FAYETTE COUNTY ARMS COLLECTORS ASSOCIATION
P.O. Box 1106
La Grange, TX 78945, USA
409-968-5056
Firearms collectors association. Gun show promoters.
CONTACT: James Johnson

THE INTERNATIONAL PLASTIC MODELERS SOCIETY
P.O. Box 480
Denver, CO 80201, USA
Plastic modelers association. Quarterly journal & six updates.

WW II HIS. REENACTMENT SOCIETY ROCKY MOUNTAIN CHAPTER
6840 Richthofen Parkway
Denver, CO 80220, USA
303-399-9264
WWII Historical Reenactment Society chapter. WWII reenactment coordinator.
CONTACT: Flint Whitlock

MILITARY VEHICLE COLLECTORS CLUB
P.O. Box 33697
Thornton, CO 80233, USA
(303)-450-9184
Military vehicle collectors association. Journal, newsletter, rallies.

FORT LARAMIE HISTORICAL ASSOCIATION
Fort Laramie National Historic Site
Fort Laramie, WY 82212, USA
National Park Service cooperating ass'n. Ft. Laramie support services.

ORGANIZATIONS, CLUBS & SOCIETIES

FORT FETTERMAN SPORTSMAN ASSOCIATION
P.O. Box 843
Douglas, WY 82633, USA
Outdoors & shooting club. Annual gun show, newsletter.
CONTACT: Keith A. Lengkeek, Secretary

PHOENIX SPORTSMANS CONGRESS
1201 E. Desert Cove
Phoenix, AZ 85020, USA
602-997-0798
Gun fanciers & shooters association. Gun show promoters.

S.S.M.S.
1725 Farmer Avenue, Dept. 26
Tempe, AZ 85281, USA
Organization honoring Gen. Sherman. Civil War Veterans service records.
CONTACT: Stan Schirmacher

ARIZONA HISTORICAL SOCIETY
949 E. 2nd Street
Tucson, AZ 85719, USA
602-623-8915
Regional historical association. Museums, library, quarterly, newsletter.

SHOTGUN NEWS TRADE & GUN SHOW
Box 397
Yerington, NV 89447, USA
(702)-463-3893
Trade & gun show promoter.
S7,S12,S13; 15; ; No
CONTACT: Jim Carpenter
SALES DEPT: Jim Carpenter
PURCHASING DEPT: Jim Carpenter
Shotgun News Trade and Gun Shows covers the gun world from the earliest antiques to the guns of tomorrow. Info - Jim Carpenter (702) 463-3893.

CIVIL WAR SKIRMISH ASSOCIATION 'POWDER & BALL'
2405 Fisk Lane, #A
Redondo Beach, CA 90278, USA
Civil War reenactment association. Journal, reenactment promoters.

SOCIETY OF WWI AERO HISTORIANS
10443 South Memphis Avenue
Whittier, CA 90604, USA
WWI Aviation collectors association.

THE TAILHOOK ASSOCIATION
P.O. Box 40
Bonita, CA 92002, USA
(619)-566-6019
Naval carrier aviation support group. Quarterly journal.
; 30;
CONTACT: Ron Thomas, Ex. Director
SALES DEPT: Ron Thomas
PURCHASING DEPT: Ron Thomas
AD MANAGER: Gary Beals
Dedicated to foster support for U.S. Navy carrier aviation.

THE TOKEN AND MEDAL SOCIETY
611 Oakwood Way
El Cajon, CA 92021, USA
Tokens, medals & exonumia collectors. Journal, library.

CONTACT: Dorothy C. Baber, Secretary

SOUTHERN CALIFORNIA ORDERS AND MEDALS SOCIETY
8831 Catalina Way
Yucca Valley, CA 92284, USA
(619)-365-4182
Orders & medals collectors association.

INTERNATIONAL MERC FORCES
2973 Harbor Road, Ste 300
Costa Mesa, AZ 92626, USA
Mercenary educational services.
S6,S11;
CONTACT: Bob Lewis
SALES DEPT: Bob Lewis
PURCHASING DEPT: Melvin Zurn
IMF offers education, service, products and prestige to the modern professional, as well as the "would if I could" thinking person. Associate membership $20.00.

AMERICAN AVIATION HISTORICAL SOCIETY
2333 Otis
Santa Ana, CA 92704, USA
U.S. aviation historical society. Newsletter, journal.

CALIFORNIA HISTORICAL ARMS ASSOCIATION
P.O. Box 5101
Vallejo, CA 94590, USA
707-642-5886
Firearms collectors association. Gun show promoters.
CONTACT: Norman Ferrando

WASHINGTON ARMS COLLECTORS, INC
P.O. Box 7335
Tacoma, WA 98407, USA
(206)-752-2268
Weapons collectors club.
Newsletter.
CONTACT: Dennis Cook

THE MILITARY HISTORICAL SOCIETY OF AUSTRALIA
P.O. Box 30
Garran, ACT 2605, Australia
Military collectors & historians association. Journal.
CONTACT: Federal Secretary

CANADIAN MILITARY HISTORICAL SOCIETY
R.R.#2
Rockwood, Ontario W0B 2KD, Canada
Regional military historical society. Publishes a journal.

MILITARY COLLECTORS CLUB OF CANADA
40 Greenwood Village
Sherwood Park, Alberta Canada
Military collectors association. Journal.

THE CANADIAN SOCIETY OF MILITARY MEDALS AND INSIGNIA
1 1/2 Kingsway Crescent
St. Catharines, Ontario L2N 1A5, Canada
Military medal & insignia collectors association. Journal.

CONTACT: The President

ORDENSHISTORISK SELSKAB
Pilegardsparken 65
Birkerod, 3460, Denmark
Military collectors & historians association.
CONTACT: Hans Levin Hansen, President

ESCADRON DE L'HISTOIRE
35 Avenue des Gobelins
Paris, 75013, France
(1) 331.47.62
Military vehicle preservationist association.
Escadron de l'Histoire is a club for people interested in the restoration of military vehicles, and in using them off-road.

ORDERS AND MEDALS RESEARCH SOCIETY
33, Berkeley Avenue
Greenford, Middlesex UB6 0NY, Great Britain
Military medal collectors & historians association. Journal.
CONTACT: N. I. Brooks, Membership Sec.

MILITARY HISTORY SOCIETY
44 Mansfield Road
Hampstead, London NW3 2HT, Great Britain
Military collectors and historians association. Journal.
CONTACT: C. B. Sanham, Secretary

THE PSYWAR SOCIETY
21 Metchley Lane
Harborne
Birmingham, B17 0HT, Great Britain
Aerial propaganda leaflets.
Quarterly magazine.
CONTACT: Hon. Sec. Dr. R. Oakland

NEW ZEALAND MILITARY HISTORICAL SOCIETY
P.O. Box 5123
Wellesley Street
Auckland, New Zealand
Military collectors & historians association. Journal.
CONTACT: The Secretary

THE MILITARY MEDAL SOCIETY OF SOUTH AFRICA
1 Jacqueline Avenue
Northcliff
Johannesburg, 2195, South Africa
Military collectors & historians association. Journal. Newsletter.
CONTACT: The Secretary

HISTORICAL FIREARMS SOCIETY OF SOUTH AFRICA
Box 145
Newlands, 7725, South Africa
Historical firearms, arms & armour collector's ass'n. Journal & newsletter.

BUND DEUTSCHER ORDENSAMMLER
Postfach 1260
Steinau, 6497, West Germany
0 66 63/67 16
German military collectors association. Journal.
CONTACT: Werner Sauer, President

TRAVEL/BATTLEFIELD TOURS

BATTLEFIELD TOURS

HAMILTON TRAVEL
3 E. 54th Street
New York, NY 10017, USA
Battlefield tours.
Send for information.

GREAT ADVENTURES UNLTD.,INC.
P.O. Box 11
Wisconsin Dells, WI 53965, USA
608 254-6080
Scenic DUKW tours.

NORTHSTAR TOURS
P.O. Box 810
Lakeville, MN 55044, USA
(612)-920-6060 (800)-344-9898
Military history tours of Europe.
CONTACT: Ray Cowdery

BOON-ERIK TOURS,INC.
4959 Olson Memorial Highwyay
Minneapolis, MN 55422, USA
612-340-1400 1-800-322-2378
WWII military campaign tours of Europe.

CORPORATE WEST INC.
2602 North 20th Avenue
Phoenix, AZ 85009, USA
Historical travel guide

NET TOURS,INC.
150 Powell St., Suite 307
San Francisco, CA 94102, USA
(800)-227-5464 (415)-433-1614
Special interest travel services.
S6,S13; 5;
CONTACT: Ms. Carroll Zensius
SALES DEPT: Ms. Carroll Zensius
NET TOURS,INC. specializes in arranging custom designed group and individual tours for veterans to re-visit Asia and the Pacific destinations.

VALOR TOURS,LTD.
P.O. Box 1617
Schoonmaker Bldg.
Sausalito, CA 94966, USA
415-332-7850
Military campaign tour promoter.

UNIFORMS & COMPONENTS

DELTA SUPPLY
P.O. Box 936
Boylston, MA 01505, USA
(617)-732-6881
U.S. & foreign elite unit insignia.
S6,S7,S11; 7;
Send $2.00 for List. 1 - 2 times annually, with suppliments. CONTACT: Hal S. Feldman
SALES DEPT: Hal S. Feldman
PURCHASING DEPT: Hal S. Feldman
Special Forces, airborne, "Elite" forces. All types of worldwide insignia and militaria. Wholesale and retail mail sales. Elite forces mail-auction starts Spring 1987.

O.K.W.
105 Granite St., 2nd Floor
Quincy, MA 02169, USA
(617)-471-2829 (617)-328-5992
German > English translations. Custom military artwork.
S4,S12; 2;
Catalogue available. Quarterly.
CONTACT: Gary Wilkins
Dealer in German/Russian militaria. Guaranteed pre-May 1945 goods. We carry a selection of uniforms, headgear, medals. Services: Translations from German; Custom military artwork.

RZM IMPORTS
P.O. Box 995
Southbury, CT 06488, USA
(203)-264-4036
German WWII militaria.
S4,S12; 1;
Catalogue available. Quarterly.
CONTACT: J. Tavolacci & R. Spezzano
SALES DEPT: Joseph Tavolacci
PURCHASING DEPT: Leslie Tavolacci
High quality German WWII uniforms, field equipment. Mostly Army (Herr) & Waffen SS. Rare and hard to get items.

EAVES & BROOKS COSTUMES,INC.
432 W. 55th Street
New York, NY 10019, USA
Costumer.
CONTACT: Attn: Harold D. Blumberg

WOODHAVEN MILITARY
75-17 Jamaica Avenue
Jamaica, NY 11421, USA
(718)-296-4982
U.S. WWII uniforms & accoutrements.
S6;
Catalogue available, send SASE or call. Annually.
CONTACT: Sam Naglieri
SALES DEPT: Sam
Woodhaven's U.S. military surplus. Genuine U.S. WWII, WWI uniforms, clothing & equipment. Everything for the U.S. WWII soldier. 800 items from the past are now available to you. A SASE or a call to 718-296-4982 will bring our catalogue to you.

PHOENIX MILITARIA CORPORATION
P.O. Box 66
163 Troutman Road
Arcola, PA 19420, USA
(215)-933-0909
General militaria & publisher of militaria collectors books & magazines.
S6; 19; 23,000; $60.00
Catalogue $3.00. Semi-annually.
CONTACT: Terry Hannon,President
SALES DEPT: Irene Karas
PURCHASING DEPT: Terry Hannon
Send $3.00 or 3 pieces of military insignia for our latest catalogue of U.S. & world-wide militaria. Thousands of items listed including: books, manuals, medals, insignia, equipment, uniform components, badges, etc.

'THE PHOENIX EXCHANGE'
P.O. Box 66
163 Troutman Road
Arcola, PA 19420, USA
(215)-933-0909
Quarterly military collectors magazine.
; 2;
CONTACT: Terry Hannon,Publisher
SALES DEPT: W. R. Bendel
PURCHASING DEPT: Terry Hannon
AD MANAGER: William Bendel
'The Phoenix Exchange' functions as the military collector's & dealer's marketplace. Articles cover a wide spectrum of militaria, but attend to each subject in detail. Send $2.00 for a sample copy.

CAMOUFLAGE
P.O. Box 427
Edinburg, VA 22824, USA
Military camouflage items.

WORLD CAMOUFLAGE UNIFORMS
20 Mass. Avenue
Pensacola, FL 32505, USA
904-438-9787
Camouflage uniforms.
List $2.00.

COMMAND POST
3025 Highway 31 South
Pelham, AL 35124, USA
(205)-663-5678
Army-Navy surplus & Rhodesian camo.
S1,S4; 3;
Catalogue available.
CONTACT: Nick Bondi
SALES DEPT: Nick Bondi
PURCHASING DEPT: Nick Bondi
Rhodesian camouflage - Absolutely genuine. Military insignia and prints of Rhodesian units, weapons, vehicles. $1.00 for price list.

ACT ONE COSTUME COMPANY
20 W. Fifth Street
Dayton, OH 45402, USA
Costumers.

AERO CLASSICS CORPORATION
P.O. Box 9222
Livonia, MI 48150, USA
Commemorative historical momentoes.

ROBERT C. MARSHALL
Rt. 2
Earlville, IL 60518, USA
WWI & WWII naval uniforms collector.

GEMINI INDUSTRIES
P.O. Box 20064
Oklahoma City, OK 73156, USA
(405)-521-8715
General military uniforms & equipment.
S6; 11;
Catalogue available. Annually.
CONTACT: Rita Scranton
SALES DEPT: Rita Scranton
PURCHASING DEPT: Rita Scranton
Commercial and military survival systems. We sell Nightstalker black combat uniforms and supplies, general military surplus, martial arts items, and equipment by Eagle Industries.

EAGLE ARMS
P.O. Box 270117, Weslayan Sta.
Houston, TX 77277, USA
Military uniforms.
Send $.22 for list.

UNIFORMS & COMPONENTS

AIRELICS
P.O. Box 2164
Huntington Beach, CA 92647, USA
(714)-848-4161
Appraisals, authentications, evaluations and identifications of Flying Badges.
S12; 20;
CONTACT: Art Grigg
SALES DEPT: Art Grigg
PURCHASING DEPT: Art Grigg
Wings, flight gear; Foreign, U.S. 1900-1945. Buy, trade helmets, goggles, float vests, parachutes, oxygen masks, flight suits.

KEN NOLAN,INC.,P.M.DIV.
P.O. Box C-19555
16901 P.M. Milliken Avenue
Irvine, CA 92713, USA
(714)-863-1532
Surplus, insignia, medals & uniforms.
S4; 29; 100,000; $60.00
Catalogue $1.00. 3 or 4 times annually.
Serving individuals, military, police for 29 years. 90% of orders shipped within 24 hours. Military field clothing, boots, insignia, nameplates, etc. Send $1.00 for catalog.

RICHARD COWELL
Box 538
El Dorado, CA 95623, USA
(916)-622-8333
Military fabrics

WHY NOT COLLECTABLES
701 North 6th Street
Kelso, WA 98626, USA
(206)-423-7931
U.S. womens militaria.
S7,S12;
CONTACT: Ken Kaighin
Wanted Womens uniforms & related items. Wave, WAC, Marine Spars, CAP, etc. Personal papers, books, medals, awards, 1914 to 1956.

MISCELLANEOUS

LAWRENCE AFFIAS
11 The Pines
Roslyn, NY 11576, USA
Printed military motif T-shirts.

COMMAND SHIRTS
8 Woodstock Court
Oyster Bay, NY 11771, USA
(516)-922-5481
Gun & para-military T-shirts.
S4; 5;
Catalogue available.
CONTACT: Virginia Commander
SALES DEPT: Mary Biggart
PURCHASING DEPT: Mary Biggart
Mail-order T-shirt business carrying para-military and pro-gun designs.

COOPERSTOWN TRADING POST,LTD.
Steiner Bridge Road
Valencia, PA 16059, USA
(412)-898-3146
Muzzleloader supplies & accessories.

NORTH EAST TRADE COMPANY
R.D. 2
Muncy, PA 17756, USA
(717)-546-2061
Muzzleloading supplies & accessories.

THE BLANKET BRIGADE
1204 Chidsey Street
Easton, PA 18042, USA
(215)-258-2615
Mountain man & black powder clothing & supplies.
CONTACT: Dale Harrison

THE PENNSYLVANIA RIFLE SHOP
RD #1, Box 166A
State Route 10 South
Cochranville, PA 19330, USA
(215)-932-8441 (215)-932-4994
Black powder supplies & accessories.

BIG HORN TRADE COMPANY
Box 88, R.D. #3
Honey Brook, PA 19344, USA
(215)-273-2337
Buckskin finished clothes and kits.

TECUMSEH'S FRONTIER TRADING POST
P.O. Box 369
Shartlesville, PA 19554, USA
(215)-488-6622
Handmade Indian and Mountainman buckskin garments and accessories.

SKIP HAMAKER
1309 N. 10th Street
Reading, PA 19604, USA
(215)-373-4419
Custom made scrimshawed powder horns.

WARREN RAYMOND
1808 Cullen Drive
Silver Springs, MD 20904, USA
Osteological preparations.

SWIFT HAWK TRADING COMPANY
59 Moyer Drive
Aberdeen, MD 21001, USA
(301)-272-6023
Muzzleloader, longhunter, and mountain man outfitters.
CONTACT: Trudy Stevenson

CREATIVE HORIZONS
P.O. Box 10952
Parkville, MD 21234, USA
Military motif jewelry.

J. P. GORDON COMPANY
781 Deer Lake Drive
Virginia Beach, VA 23462, USA
Military jewelry

CLOUDLAND ENTERPRISES
Star Route Box 78M
Cloudland, GA 30709, USA
T-shirt printer.

NEPTUNE FIREWORKS COMPANY
P.O. Box 398
Dania, FL 33004, USA
(305)-920-6770 (800)-835-5236
Fireworks (Class C).
; 3; ; 300,000; ; $40.00
Catalogue available. Annually.
SALES DEPT: Issac Dirkstein
PURCHASING DEPT: Mark Barnhart
Fireworks, Class C. Bottle rockets, firecrackers, Roman candles, M-60, pin wheels. One of the largest selection. Discount prices.

BAXENDALE'S CUSTOM WORK
4114 ML Sneed
Nashville, TN 37215, USA
615-297-1975
Custom bronze belt buckles.

TURTLE PRODUCTIONS
2112 Oklahoma Avenue
Davenport, IA 52804, USA
319-388-9621
Sculptor of military subjects.
CONTACT: Wayne R. Brouhard

HERITAGE BUCKLES
P.O. Box 50130
2049 Phoebe Drive
Billings, MT 59105, USA
406-252-4919
Commemorative belt buckle manufacturer.

HERITAGE BUCKLES
P. O. Box 50130
2049 Phoebe Drive
Billings, MT 59105-0130, USA
406-252-4919
Belt buckles

THE POST SUTLER
Route 1, Box 197
Waterman, IL 60556, USA
(815)-824-2829
Civil War related goods.
Catalogue available.
CONTACT: Bob & Joyce Trahan

M. J. SILBERT CO.,INC.
4032 N. Milwaukee Avenue
Chicago, IL 60641, USA
Dog tags

METAL ARTS,INC.
P.O. Box 19496
Oklahoma City, OK 73119, USA
Belt buckles

A. G. ENGLISH SALES COMPANY
708 S. 12th Street
Broken Arrow, OK 74012, USA
(918)-251-3399
Gun & fire safes.
S6,S12; 8;
Catalogue available. 2 or 3 times per year.
CONTACT: A. G. English,Jr.
SALES DEPT: Tony English
PURCHASING DEPT: Tony English
Browning Pro-steel, Ft. Knox, security products and Treadlok gun safes.

T-J JEWELRY COMPANY
P.O. Box Y
Apache Junction, AZ 85220, USA
Military rings & jewelry.

MISCELLANEOUS

ROYAL MILITARY JEWELRY
Apache Junction, AZ 85220, USA
602-982-2273
Military jewelry.
Catalogue available.

T R SUPPLY
P.O. Box 6600
Reno, NV 89513, USA
Belt buckles

THE DITTY BAG
P.O. Box 3156
Chula Vista, CA 92011, USA
(619)427-1747
Belt buckles.
S4; 3;
Catalogue available. Annually.
CONTACT: George B. Lusk
SALES DEPT: George B. Lusk
PURCHASING DEPT: George B. Lusk
Show your pride! Belt buckles for all branches of service $7.00 to $12.00. Send $1.00 for catalog.

SEA CREST ENTERPRISES
1951-A Yolanda Way
Tustin, CA 92680, USA
714-544-7241
Military commemorative belt-buckles.

HANKY PANKY ENTERPRISES
1126 West Ocean Avenue
Lompoc, CA 93436, USA
Space memorabilia.
Catalogue available.

EXECUTIVE PROTECTION PRODUCTS
1834 First Street
Napa, CA 94559, USA
(707)-253-7142
Personal protection & security products.
S4; 2;
Catalogue available. Annually.
CONTACT: G. Kelly
SALES DEPT: Mr. G. Kelly
State of the art personal protection & security products! Including; laser gunsights, bulletproof vests, countersurvaillance & privacy devices, tear gas, hidden vaults, nightvision, tactical equipment and much more. Send for catalog.

MILITARIA SHOW PROMOTERS

J. R. LaRUE
Box 4338
Portsmouth, NH 03801, USA
603-431-5334
Gun show promoter.

DONNA FRANSEN
491 Main Street
Bangor, ME 04401, USA
207-942-4861
Gun show promoter.

MEADOWLANDS MILITARIA
749 Seventh Street
Carlstadt, NJ 07072, USA
201-939-4342
Gun show promoter.
CONTACT: Robert Curran

MAX PROMOTIONS
P.O. Box 304
Clementon, NJ 08021, USA
609-784-6562
Gun & militaria show promoter.
CONTACT: Thomas T. Wittman, Co-ordinator

JOE SISIA
Box 100
Bayville, NJ 08721, USA
201-269-3290
Gun show promoter.

WESTCHESTER COLLECTORS, INC.
P.O. Box 162
Eastchester, NY 10709, USA
(914)-337-5998
Gun, knife, and militaria show promoter.
S12; 3;
Catalogue available. Quarterly.
CONTACT: Frank Fusco
SALES DEPT: Frank Fusco
Westchester Collectors, Inc. has periodic gun, knife and militaria shows in N.Y.; Yonkers Raceway, Yonkers, County Center, White Plains, and the Poughkeepsie Civic Center.

NORTH EASTERN ARMS COLL., INC.
Box 185
Amityville, NY 11701, USA
516-431-6628
Gun show promoter.

JOHN VANDER ZALM
437 Head of Neck Road
Bellport, NY 11713, USA
516-286-0816
Gun show promoter.

DR. ROBERT H. MacLEOD
7529 George Sickles Road
Saugerties, NY 12477, USA
914-246-5011
Gun show promoter.

STAN OLSON
Box 742
Newburgh, NY 12550, USA
914-561-1775
Gun show promoter.

JIM LaVALLEY
55 Park Street
Tupper Lake, NY 12986, USA
518-359-3328
Gun show promoter.

JACK ACKERMAN
24 S. Mountain Terrace
Binghampton, NY 13903, USA
607-723-5668
Gun show promoter

CHUCK MADSEN
343 Pearl Street
Rochester, NY 14607, USA
716-442-7086
Gun show promoter

VAUGHN CRISPIN
301 W. Monroe Street
Latrobe, PA 15650, USA
412-539-7050
Gun show promoter

GEORGE PARNAY
4348 Hollow Road
New Castle, PA 16101, USA
Gun show promoter

TERRY MARIACHER
Rt. 2, Box 2301
Fredonia, PA 16124, USA
412-475-2295
Gun show promoter

ROBERT ENGLISH
2516 Crawford Avenue
Altoona, PA 16602, USA
814-943-3800
Gun show promoter

CHRIS HEIMEL
Rt. 2, Box 165-B
Coudersport, PA 16915, USA
814-274-8643
Gun show promoter

RICHARD BUCKWALTER
1701 Bridge Road
Lancaster, PA 17062, USA
717-464-2581
Gun show promoter

JIM'S TRADING POST
5429 Philadelphia Avenue
Chambersburg, PA 17301, USA
717-267-2133
Gun show promoter

RODNEY STIGER
Rt. 2, Box 542
Muncy, PA 17756, USA
(717)-546-6211
Gun show promoter

H. A. HOFFMAN
RD #4, Box 4325
Bangor, PA 18013, USA
Gun show promoter

MID-ATLANTIC ARMS COLLECTORS
Rt. 2, Box 14 S & W
Uniondale, PA 18470, USA
(717)-679-2250 (717)-282-1561
Gun show promoter
CONTACT: Nicholas Jubinski

HENRY SIPPLE
117 Karlyn Drive
New Castle, DE 19720, USA
302-654-5750
Gun show promoter

MICHAEL DelGROSSO
806 N. Union Street
Wilmington, DE 19805, USA
302-571-1656
Gun show promoter

GREG VERSON
Box 305
Harrisonburg, VA 22801, USA
703-433-2488
Gun show promoter

GUN & MILITARIA SHOW PROMOTERS

COURTNEY SMITH,LTD.
Box 244
Highland Springs, VA 23075, USA
804-737-0484
Gun show promoter

DOUG McDOUGAL
Box 6601
Portsmouth, VA 23703, USA
804-481-2404
Gun show promoter

J. S. DAVIS
Box 1457
Galax, VA 24333, USA
703-236-7542
Gun show promoter

VAN THOMPSON
335 Ellison Avenue
Beckley, WV 25801, USA
304-252-7679
Gun show promoter

REPLICA PRODUCTS
P.O. Box 5232
Vienna, WV 26105, USA
304-295-7239
Display frames & gun show promoter.
CONTACT: Roger Crowley

CLARENCE GILLESPIE
16 Stonewall Street
Sutton, WV 26601, USA
Gun show promoter

K. W. FIELDS
Box 462
Garner, NC 27529, USA
919-772-5070
Gun show promoter

METROLINA GUN SHOWS
3452 Odell School Road
Concord, NC 28025, USA
(704)-786-8373
Gun show promoter
S12;

JERRY LEDFORD
3231 7th Street Dr. N.E.
Hickory, NC 28601, USA
704-327-0055
Gun show promoter

TRADE SHOWS,INC.
P.O. Box 769
Conover, NC 28613, USA
704-464-0655
Gun show promoter
CONTACT: Keith Eidson

H. W. BROWN
Rt. 3, Box 533-C
Morgantown, NC 28655, USA
704-437-3091
Gun show promoter

DON BROWN
20 Alpine Way
Asheville, NC 28805, USA
704-298-5457
Gun show promoter

PIEDMONT GUN SHOWS
1464 Boiling Springs Road
Spartanburg, SC 29303, USA
803-578-2939
Gun show promoter

MANN PROMOTIONS
Box 76
LaFrance, SC 29656, USA
803-646-3182
Gun show promoter

CECIL ANDERSON
Box 218
Conley, GA 30027, USA
404-366-5986
Gun show promoter

CHEROKEE GUN CLUB
Box 941
Gainesville, GA 30503, USA
404-534-7386
Gun show promoter.
CONTACT: William Hardman

E. W. CATER
1398 Rocky Creek Road
Macon, GA 31206, USA
912-788-9291
Gun show promoter.

MATT EASTMAN
Box 768
Fitzgerald, GA 31750, USA
912-423-4595
Gun show promoter.

JIMMY SHIERLING
P.O. Box 4418
Columbus, GA 31904, USA
404-323-6049
Gun show promoter.

ELMER KIRKLAND
4328 State Road 44
New Smyrna Beach, FL 32069, USA
904-427-6396
Gun show promoter.

DON SCOTT
P.O. Box 2226
Ormond Beach, FL 32074, USA
904-258-5663
Gun show promoter.

N. FLORIDA ARMS COLLECTORS
P.O. Box 52191
Jacksonville, FL 32201, USA
904-733-6881
Gun show promoter.
CONTACT: Martha Pierce

J. A. PATTERSON
Box 17035
Orlando, FL 32860, USA
305-298-3568
Gun show promoter.

TOM FISCHER
Box 361021
Melbourne, FL 32936, USA
305-777-0741
Gun show promoter.

NELS MICHELSON
1121 Country Club Drive
North Palm Beach, FL 33408, USA
305-968-9563
Gun show promoter.

CHARLIE BROWN
Box 13991
Tampa, FL 33611, USA
813-961-6297
Gun show promoter

JOHN TURVELL
2461 - 67th Ave., S.
St. Petersburg, FL 33712, USA
813-867-1019
Gun show promoter

E. L. LYON
Box 3486
Lakeland, FL 33802, USA
813-665-0092
Gun show promoter

J. A. PATTERSON
Box 17035
Orlando, FL 33860, USA
305-298-3568
Gun show promoter

GEORGE N. BOYD,JR.
1361 Federal Drive
Montgomery, AL 36106, USA
205-262-6936
Gun show promoter

SAM HUGGINS
Box 668
Andalusia, AL 36420, USA
205-222-7607
Gun show promoter

PAUL BARNETT
Box 6746
Mobile, AL 36660, USA
205-476-8828
Gun show promoter

M. H. PARKS
Box 40771
Nashville, TN 37204, USA
615-292-3020
Gun show promoter

WINCHESTER KNIFE COLL. CLUB
3550 Merritt Street
Memphis, TN 38128, USA
901-388-4465
Knife show promoter.
CONTACT: George Ashley

SAM SIMMONS
Box 40776
Memphis, TN 38174, USA
901-725-0633
Gun show promoter

D. A WALTERS & W. F. PIGG
Box 937
Columbia, TN 38401, USA
615-388-5581
Gun show promoter

GUN & MILITARIA SHOW PROMOTERS

CHARLIE McGEE
Box 173
Gulfport, MS 39501, USA
601-868-3242
Gun show promoter

TOM WEIR
Box 173
Gulfport, MS 39501, USA
601-863-3485
Gun show promoter

BILL GOODMAN
Box 5
Mt. Washington, KY 40047, USA
502-538-6900
Gun show promoter

KENTUCKIANA ARMS COLLECTORS
13305 Dixie Hwy.
Louisville, KY 40272, USA
(502)-937-1970
Gun show promoter
CONTACT: Nancy Sumner

McCANN SHOW COMPANY
Box 32
Symsonia, KY 42082, USA
618-337-7543
Gun show promoter

JOHN SNYDER
Box 492
Maumee, OH 43537, USA
419-893-5173
Gun show promoter.

VINCE GASPAR
3728 Perlawn Drive
Toledo, OH 43614, USA
419-385-3755
Gun show promoter

NORTHCOAST/SUNCOAST GUN
COLLECTORS
17800 Chillicothe Road
Chagrin Falls, OH 44022, USA
(216)-543-5500 (800)-443-GUNS
Gun show promoter
S13; 5;
CONTACT: Joseph Stegh
SALES DEPT: Joe Stegh,Jr.
Our shows are "Gun, Militaria and Wilderness Expos", and exhibits include: guns, militaria, military surplus, knives, hunting/camping gear and sports collectibles.

GARY NEDBALSKI
4797 State Road
Cleveland, OH 44109, USA
216-398-0446
Gun show promoter

SHELLY LICHTIG
16105 Fernway Road
Shaker Heights, OH 44120, USA
216-751-8980
Gun show promoter

OHIO CARTRIDGE COLLECTORS CLUB
8013 Norton Road
Garrettsville, OH 44231, USA
216-527-2254
Gun show promoter
CONTACT: Jack Salsgiver

JIM CONRAD
Box 76
Litchfield, OH 44253, USA
216-725-6083
Gun show promoter

TOM PAUMIER
3033 Leisure Road, N.W.
Minerva, OH 44657, USA
216-868-5165
Gun show promoter

DARRELL GASKEY
Box 97
Milan, OH 44846, USA
419-499-2948
Gun show promoter

RICHLAND COUNTY GUN SHOW
750 N. Home Road
Mansfield, OH 44906, USA
(419)-747-3717 (419)-884-1553 Res.
Gun show promoter
S12;
CONTACT: Roy E. Walter, Mgr.

STEVE HALL
200 W. Weller Street
Ansonia, OH 45303, USA
513-337-7022
Gun show promoter

DON INGLIS
Box 1201
Lima, OH 45801, USA
419-229-8407
Gun show promoter

TRI-STATE GUN COLLECTORS
Box 1201
Lima, OH 45802, USA
419-229-8407
Gun show promoter
CONTACT: Don Inglis

RON SHEPARD
Box 66
Arcadia, IN 46030, USA
317-984-3937
Gun show promoter

ART KESSLER
13203 Nottingham
Nobelsville, IN 46060, USA
317-842-4036
Gun show promoter

DAVE REECER
919 N. Main Street
Tipton, IN 46072, USA
317-675-6392
Gun show promoter

DICK BINGER
Rt. 2, Box 70
Morgantown, IN 46160, USA
317-878-5489
Gun show promoter

ED CRAFTON
1921 Winton Avenue
Speedway, IN 46224, USA
317-241-8037
Gun show promoter

BOB CARGILL
P.O. Box 2207
Hammond, IN 46323, USA
815-838-2969
Gun show promoter

NORBERT PAGELS
2521 Mishawaka Avenue
South Bend, IN 46615, USA
219-289-8121
Gun show promoter

NORTH INDIANA GUN SHOWS,INC.
P.O. Box 3890
South Bend, IN 46628, USA
(219)-272-9397
Gun & knife show promoter
S13; 12;
CONTACT: Ron
SALES DEPT: Ron
PURCHASING DEPT: Ron

GENE NEHER
Rt. 2
Churubusco, IN 46723, USA
219-693-2769
Gun show promoter

WILLIAM CRONE
521 N. 3rd Street
Decatur, IN 46733, USA
219-724-3123
Gun show promoter

CLASS PRODUCTIONS
4207 N. Clinton Street
Ft. Wayne, IN 46805, USA
219-483-6144
Gun show promoter
CONTACT: Bob Parker

BERTHA E. DAVIS
2909 S. Chippewa Lane
Muncie, IN 47302, USA
317-289-3057
Gun show promoter

W. J. KING
339 W. Beech
Sullivan, IN 47882, USA
812-268-5759
Gun show promoter

MID AMERICAN ANTIQUE ARMS SOC.
Box 148
Dearborn, MI 48121, USA
313-425-0578
Gun show promoter
CONTACT: Bill Johnson

LOWER MICHIGAN GUN COLLECTORS
8639 Donna Drive
Westland, MI 48185, USA
313-427-8946
Gun show promoter
CONTACT: Joseph Rhoney

GUN & MILITARIA SHOW PROMOTERS

LEN IRELAND
Box 6442
Flint, MI 48508, USA
313-232-8066
Gun show promoter

WARREN GOEMAERE
300 Lumm Street
Gladwin, MI 48624, USA
517-426-9885
Gun show promoter

LARRY COIN
Box 453
Portage, MI 49081, USA
616-327-4557
Gun show promoter

RAY CHARLIER
Box 9
Hermansville, MI 49807, USA
906-498-2279
Gun show promoter

JIM REINART
Rt. 2, Box 46
Earlham, IA 50072, USA
515-758-2436
Gun show promoter

STEVE PENLAND
1910 Payton
Des Moines, IA 50315, USA
515-287-2516
Gun show promoter

DENNIS GREISCHAR
Box 841
Mason City, IA 50401, USA
515-424-2826
Gun show promoter

LOIS SCHWADE
633 - 3rd St., S.E.
Mason City, IA 50401, USA
515-424-9234
Gun show promoter

RANDALL HIRSCH
1421 - 4th Avenue, N.
Esterville, IA 51334, USA
Gun show promoter

STEVE PARKER
3306 8th Avenue
Council Bluffs, IA 51511, USA
712-328-8307
Gun show promoter

JIM WYMORE
Box 97
New Market, IA 51646, USA
712-585-3519
Gun show promoter

DARYL KLEIN
Rt. 1, Box 198
Durango, IA 52039, USA
319-552-1716
Gun show promoter

DON JOHNSON
1297 - 30th St., N.E.
Cedar Rapids, IA 52402, USA
319-362-1988
Gun show promoter

HAWKEYE GUN SHOP
602 S. 9th Street
Burlington, IA 52601, USA
319-754-8961
Gun show promoter

DAN GLEASON
Box 1033
Brookfield, WI 53005, USA
414-781-1328
Gun show promoter

ROBERT SELISSEN
1439 N. 50th Street
Milwaukee, WI 53208, USA
414-453-8231
Gun show promoter

ROBERT SCHRAP
7024 W. Wells Street
Wauwatosa, WI 53213, USA
414-771-6472
Gun show promoter

JEFF ASHENFELTER
1007 Tripoli Road
Janesville, WI 53545, USA
608-757-0219
Gun show promoter

ARVID SCHULZ
10133 Hwy. 151
Manitowoc, WI 54220, USA
414-758-2822
Gun show promoter

JOHN DAUMGART
2475 Hemlock Drive
Green Bay, WI 54301, USA
(414)-468-4267
Gun show promoter.

FRED GOLDBERG
4221 - 82nd St. S.
Wisconsin Rapids, WI 54494, USA
715-424-2516
Gun show promoter

DUKE FREZEL
1320 Oak Street
Wisconsin Rapids, WI 54494, USA
715-423-7523
Gun show promoter

DAVE IPPOLITE
4777 Menard Drive
Eau Claire, WI 54701, USA
715-874-6311
Gun show promoter

WAYNE ZANDTKE
1100 - 4th Street
Pine City, MN 55063, USA
612-629-6764
Gun show promoter

MINNESOTA WEAPONS COLLECTORS
P.O. Box 662
Hopkins, MN 55343, USA
612-721-8976
Gun show promoter
CONTACT: Gail Foster

DALE HARBARTH
Hwy. 7
Hutchinson, MN 55350, USA
612-587-8989
Gun show promoter

JIM MURRAY
4619 Munger Shaw Road
Saginaw, MN 55779, USA
218-729-8064
Gun show promoter

JOHN BRIGEMAN
2002 7th Ave., S.E.
Austin, MN 55912, USA
507-433-2509
Gun show promoter

JAMES THULIEN
122 - 8th Ave., N.
St. James, MN 56081, USA
507-375-3984
Gun show promoter

DON ANDERSON
810 - 2nd St., N.W.
Wascea, MN 56093, USA
507-835-5667
Gun show promoter

KEN BAHR
Rt. 9, Box 141
Bemidji, MN 56601, USA
218-759-1822
Gun show promoter

DAKOTA TERRITORY GUN COLL.
Rt. 2, Box 169
Castlewood, SD 57223, USA
605-793-2347
Gun show promoter
CONTACT: Curt Carter

CURT CATER
Rt. 2, Box 169
Castlewood, SD 57223, USA
605-793-2347
Gun show promoter

WEAPONS COLL. SOC. OF MONTANA
P.O. Box 814
Billings, MT 59103, USA
406-259-3374
Gun show promoter
CONTACT: Arnold Bejot

DALE BENDER
Rt. 1, West No. 550
Glasgow, MT 59230, USA
406-228-8610
Gun show promoter

DALE HOLIDAY
P.O. Box 814
Bozeman, MT 59715, USA
406-586-6179
Gun show promoter

GUN & MILITARIA SHOW PROMOTERS

DENNIS GORDON
Box 3912
Missoula, MT 59806, USA
406-549-6280
Gun show promoter

PAUL WILLIS
P.O. Box 653
Kalispell, MT 59903, USA
406-755-9169
Gun show promoter

NORBERT ERTEL
Box 2580
Des Plaines, IL 60017, USA
603-835-2331
Gun show promoter.

JOHN CARTER
Box 102
Island Park, IL 60042, USA
815-653-9351
Gun show promoter

DENNIS KATALLO
900 Rohlwing Road
Addison, IL 60101, USA
312-620-8557
Gun show promoter

SCOTT GOLDEN
Box 763
Dekalb, IL 60115, USA
815-895-6809
Gun show promoter

FRED DOEDERLAIN
110 Railroad
East Dundee, IL 60118, USA
312-426-3455
Gun show promoter

PETE TROVATO
Box 131
Bedford Park, IL 60499, USA
312-599-6014
Gun show promoter

GEORGE HARVESTINE
Box 1098
North Riverside, IL 60546, USA
312-357-8444
Gun show promoter

CHICAGO MILITARIA COLLECTORS
SOCIETY
2341 W. Belmont
Chicago, IL 60618, USA
312-477-7096
Gun show promoter
CONTACT: Don Anderson

RICHARD FORD
Rt. 2
Oregon, IL 61061, USA
815-946-3769
Gun show promoter

PINETREE PISTOL CLUB
332 E. Riverside Blvd.
Loves Park, IL 61111, USA
815-229-1476
Gun show promoter

CONTACT: Gregory Lyle

LAWRENCE MAYNARD
Rt. 2, Box 101
Aledo, IL 61231, USA
309-582-7889
Gun show promoter

GORDY MATSON
Box 1057
Milan, IL 61264, USA
309-787-7020
Gun show promoter

FRED HORN
Box 828
Galesburg, IL 61402, USA
309-343-8585
Gun show promoter

WILLIAM FRITZ
Rt. 1, Box 116-B
Edelstein, IL 61526, USA
309-274-6096
Gun show promoter

GENE JORDAN
1505 Sunset Drive
East Peoria, IL 61611, USA
309-699-6760
Gun show promoter

EBERHARD GERBSCH
416 South Street
Danville, IL 61832, USA
217-442-0757
Gun show promoter

HOWARD'S GUN SHOP
805 N. State
Freeburg, IL 62243, USA
618-539-3374
Gun show promoter

ROBINSON GUNSHOWS
312 E. Main Street
Robinson, IL 62454, USA
618-546-1133
Gun show promoter

BOB LAPPIN
Box 1006
Decatur, IL 62523, USA
217-428-2973
Gun show promoter

RUSS GARDNER
P.O. Box 875
Jacksonville, IL 62651, USA
217-245-7479
Gun show promoter

HAROLD BLY
Rt. 1
New Berlin, IL 62670, USA
217-483-3968
Gun show promoter

LARRY HANCOCK
Rt. 8
Mt. Vernon, IL 62864, USA
618-242-4514
Knife show promoter

A. W. STEPHENSMEIER
1055 Warson Woods Drive
St. Louis County, MO 63122, USA
314-966-8975
Gun show promoter

D. RADCLIFFE
122 N. Kirkwood
Kirkwood, MO 63122, USA
314-821-4545
Gun show promoter

BOB MORTON
Box 11058
Ferguson, MO 63135, USA
314-521-9398
Gun show promoter

MIDWEST REGIONAL GUN & KNIFE
SHOWS
2608 N.W. 86th
Kansas City, MO 64154, USA
816-436-7662 816-436-0515
Gun show promoter

JOE & PEGGY PENROD
Box 23
Gravios Mills, MO 65037, USA
314-372-5531
Gun show promoter

CHARLES BROWN
Box 75
Kaiser, MO 65047, USA
314-348-5610
Gun show promoter

DEAN SMITH
3406 Bonny Linn
Columbia, MO 65202, USA
314-474-8686
Gun show promoter

JIM DAVIDSON
Rt. 2, Box 90-B
Paris, MO 65275, USA
816-327-5871
Gun show promoter

P. K. GRAGG
Box 8255
Topeka, KS 66617, USA
913-286-1221
Gun show promoter

JERRY BARNETT
1262 Thompson
Emporia, KS 66801, USA
316-342-7257
Gun show promoter

ELMDALE TRADING POST
Rt. 1
Elmdale, KS 66850, USA
316-273-6350 316-273-6837
Gun show promoter
CONTACT: Kip Wells

GUN & MILITARIA SHOW PROMOTERS

PETE BUREMAN
Box 98
1402 'I' Street
Bellville, KS 66935, USA
913-527-2873
Gun show promoter

WICHITA MILITARIA COLL. CLUB
2805 S. 147th
E. Wichita, KS 67232, USA
316-733-1915
Gun show promoter
CONTACT: Bill Quint

KANSAS CARTRIDGE COLLECTORS ASSOCIATION
204 - 5th Street, S.E.
Plainville, KS 67663, USA
913-434-2012
Gun show promoter
CONTACT: Don Meistrell

JACK BRUNA
310 W. 2nd Street
Wayne, NE 68787, USA
402-375-3712
Gun show promoter

JOE PAURATORE
Box 19394
New Orleans, LA 70179, USA
504-643-0153
Gun show promoter

RICHARD HOENER
440 Ridgewood Drive
Mandeville, LA 70448, USA
504-845-3252
Gun show promoter

WILLIAM STADLER
106 W. Pinewood Drive
Slidell, LA 70458, USA
504-538-6900
Gun show promoter

ALEX CHAUFFE
P.O. Box 265
Maringouin, LA 70757, USA
504-625-2126
Gun show promoter

DR. H. CHAPMAN
6150 Line Avenue
Shreveport, LA 71106, USA
318-868-2853
Gun show promoter

MIKE ROBERTS
P.O. Box 4176
Monroe, LA 71203, USA
318-397-2476
Gun show promoter

JACKIE GREEN
Box 972
West Monroe, LA 71291, USA
318-396-2722
Gun show promoter

L. W. WALLMAN
223 - 3rd Street
Hot Springs, AR 71913, USA
501-321-1231
Gun show promoter

JOHN PRINCE
Smokie Hollow Drive
Norman, AR 71960, USA
501-326-4617
Gun show promoter

JACK PIKE
Rt. 7, Box 242
Conway, AR 72032, USA
501-327-7849
Gun show promoter

SUZY COTHAM
Rt. 1, Box 1
Scott, AR 72142, USA
501-961-9284
Gun show promoter

GREATER OKLAHOMA TERRITORIES GUN SHOWS
Box 1131
Oklahoma City, OK 73101, USA
405-376-3833 405-528-6449
Gun show promoter
CONTACT: Randy Thomas

CLAUDE HALL
3001 The Pasco
Oklahoma City, OK 73103, USA
405-528-1222
Military unit year books. 101st A/B 1968,1969. 82nd A/B yearbook 1969.
; 4;
SALES DEPT: Claude Hall
PURCHASING DEPT: Claude Hall

WORLD CLASS GUN SHOWS
P.O. Box 14194
Oklahoma City, OK 73113, USA
405-340-1333
Gun show promoter

RON VETTER
Box 693
Lawton, OK 73502, USA
405-492-4664
Gun show promoter

DAVID WHITETURKEY
Box 3203
Bartlesville, OK 74006, USA
918-333-6762
Gun show promoter

DON WELLS
139 E. Main
Pawhuska, OK 74056, USA
918-287-4606
Gun show promoter

BLACK HAT PRODUCTIONS
Box 1157
Tulsa, OK 74101, USA
918-336-1444
Gun show promoter

MIDWAY-USA GUNSHOWS
4721 S. Madison Avenue
Tulsa, OK 74105, USA
Gun show promoter.
CONTACT: Rocky L. Baker

JOE WANEMACHER
Box 4491
Tulsa, OK 74105, USA
918-745-9141
Gun show promoter

TULSA MILITARY & GUN CLUB
7240 E. 46th Street
Tulsa, OK 74145, USA
918-663-9790
Gun show promoter
CONTACT: Bob Machlan

MIKE WATKINS
Box 1477
Tahlequah, OK 74465, USA
918-456-1472
Gun show promoter

JACK HELTON
1308 Charles Drive
Shawnee, OK 74801, USA
405-275-5164
Gun show promoter

SOONER GUN SHOWS
13815 S. Luther Road
Newalla, OK 74857, USA
405-391-7303
Gun show promoter
CONTACT: R. D. Diener

RICHARD SHEA
Rt. 1, Box 282-B
DeSoto, TX 75115, USA
214-223-3066
Gun show promoter

DAVID COOK
Box 31094
Dallas, TX 75231, USA
214-341-2895
Gun show promoter

EMBERLIN PROMOTIONS
Box 7681
Tyler, TX 75711, USA
214-581-6674
Gun show promoter

JAMES HOLSOMBACK
321 S. First Street
Lufkin, TX 75901, USA
409-634-4977
Gun show promoter

TROY HIGHNIGHT
13 Linderhof Circle
Bedford, TX 76022, USA
214-255-9111
Gun show promoter

FRED CARTER
2303 Dorothy
Wichita Falls, TX 76306, USA
817-855-4027
Gun show promoter

GUN & MILITARIA SHOW PROMOTERS

BILL STELMA
12496 Bellaire Blvd, Suite 157
Houston, TX 77072, USA
713-232-6700
Gun show promoter

HOUSTON GUN & KNIFE SHOW
P.O. Box 231183
Houston, TX 77223, USA
(713)-641-3008
Gun show promoter.

HOUSTON GUN COLLECTORS ASSOCIATION
Box 741429
Houston, TX 77274, USA
409-948-2609
Gun show promoter

BILL GROVER
Box 703
Richmond, TX 77469, USA
713-341-0775
Gun show promoter

J. P. VOIGHT
3420 - 9th Avenue
Port Arthur, TX 77642, USA
409-985-7701
Gun show promoter

TEXAS KNIFE COLLECTORS ASSOCIATION
707 Inwood
Bryan, TX 77802, USA
409-846-8325
Knife show promoter
CONTACT: M. R. Herron

MIKE MORRIS
Star Route, Box 17-D
Devine, TX 78016, USA
512-663-5124
Gun show promoter

TEXAS GUN & KNIFE ASSOCIATION
126 Cedar Knoll
Kerrville, TX 78028, USA
512-257-5844
Gun show promoter
CONTACT: Don & Deressa Hill

SOUTH TEXAS GUN COLLECTORS
Box 10815
Corpus Christi, TX 78410, USA
512-241-0282
Gun show promoter

SOUTHERN STATES ARMS COLL.
7500 West Hwy. 71
Austin, TX 78643, USA
512-288-4780
Gun show promoter
CONTACT: Bill Blessing

TOMMY NOLLEY
Box 4712
Austin, TX 78765, USA
512-837-2590
Gun show promoter

BO FRANKS
Box 9326
Austin, TX 78766, USA
512-258-7663
Gun show promoter

FAYETTE COUNTY ARMS COLLECTORS
Box 1106
LaGrange, TX 78945, USA
409-242-5023
Gun show promoter

L & L ENTERPRISES
Box 2718
Pampa, TX 79066, USA
806-665-6127
Gun show promoter

TOM PACE
5142 LeLand
Amarillo, TX 79110, USA
806-353-4239
Gun show promoter

JERRY BRANDON
Rt. 2
Quanah, TX 79252, USA
817-674-2426
Gun show promoter

JOHN POKORNY
2229 Hanley
Odessa, TX 79762, USA
915-362-7453
Gun show promoter

TOM DAVIS
6812 Cielo Vista Drive
El Paso, TX 79925, USA
915-772-5223
Gun show promoter

J. D. TANNER GUN SHOWS
Denver Merchandise Mart
451 E. 58th Ave., Suite 1105
Denver, CO 80216, USA
303-292-8416
Gun show promoter

LEADVILLE ROD & GUN CLUB
1420 Mt. Elbert Drive
Leadville, CO 80461, USA
303-486-0755
Gun show promoter
CONTACT: Orville Liend

RALPH AAB
Box 5161
Greely, CO 80631, USA
303-352-2324
Gun show promoter

LINDA HARWELL
204 W. 9th Street
Julesburg, CO 80737, USA
303-474-2449
Gun show promoter

JIM KEENEY
2305 Busch Avenue
Colorado Springs, CO 80904, USA
303-471-4067
Gun show promoter

TERRY ELLISON
Box 1644
Montrose, CO 81402, USA
303-249-9336
Gun show promoter

RONALD BROCK
125 N.W. Park Square
Fruita, CO 81521, USA
303-858-3631
Gun show promoter

DIETER STURM
2057 N. 16th Street
Laramie, WY 82070, USA
307-745-4372
Gun show promoter

WILLIAM HAAS
Box 91
Torrington, WY 82240, USA
307-532-2773
Gun show promoter

BOB FUNK
Box 1805
Riverton, WY 82501, USA
307-856-0061
Gun show promoter

W. TOM OUSLEY
P.O. Box 324
Rock Springs, WY 82901, USA
307-382-4133
Gun show promoter

LEWIS-CLARK TRADER
P.O. Box 219
Lewiston, ID 83501, USA
208-746-5555
Gun show promoter
CONTACT: Paul L. C. Snider

CROSSROADS OF THE WEST
332 Mountain Road
Kaysville, UT 84037, USA
801-544-9125
Gun show promoter
CONTACT: Bob Templeton

TER-MARK,INC.
Box 5106
Mesa, AZ 85201, USA
602-962-6372
Gun show promoter

PHOENIX SPORTSMANS CONGRESS
822-A South Mill Ave.,Ste. 113
Tempe, AZ 85281, USA
602-997-0798
Gun show promoter

FRED ROHRS
Box 652
Glendale, AZ 85311, USA
602-931-9927
Gun show promoter

C. R. HANSON
Box 5700
Yuma, AZ 85364, USA
602-782-1638
Gun show promoter

GUN & MILITARIA SHOW PROMOTERS

ARIZONA ARMS ASSOCIATION
Box 17061
Tucson, AZ 85731, USA
602-886-6876
Gun show promoter
CONTACT: Clay Fobes

DON BULLOCK
8291 Carburton Street
Long Beach, CA 90808, USA
213-430-5112
Gun show promoter

SHOWMASTERS ASSOCIATION
Box 7878
Van Nuys, CA 91409, USA
818-716-9114
Gun show promoter

WEST COAST PRODUCTIONS
Box 1294
Alta Loma, CA 91701, USA
714-989-1587
Gun show promoter

WALLACE BEINFELD
2012 Sangor Gonro Road
Palm Springs, CA 92262, USA
Gun show promoter.

DAN DELAVAN
9322 Comstock Drive
Huntington Beach, CA 92646, USA
714-964-0816
Gun show promoter

SANTA BARBARA HISTORICAL ARMS ASSOCIATION
Box 6291
Santa Barbara, CA 93111, USA
805-968-6386
Gun show promoter

V. H. WACKER
2309 Cipriani Blvd.
Belmont, CA 94002, USA
415-591-0839
Gun show promoter

TOM LINDER
Box 2249
Dublin, CA 94568, USA
415-829-0359
Gun show promoter

CALIFORNIA HISTORICAL ARMS ASSOCIATION
Box 5101
Vallejo, CA 94591, USA
707-642-5886
Gun show promoter
CONTACT: Norm Ferrando

BILL PASSALACQUA
Box 25
Bodega Bay, CA 94923, USA
707-875-3468
Gun show promoter

CENTRAL VALLEY PROMOTERS
Box 6067
Stockton, CA 95206, USA
209-478-5039
Gun show promoter
CONTACT: Ray Wood

49er WESTERN FRONTIER GUNSHOWS
Box 214503
Sacramento, CA 95821, USA
(916)-929-7766 (916)-934-4159
Gun show promoter.

GREAT WESTERN FAIR
Box 65
North Hollywood, CA 96103, USA
818-766-1665
Gun show promoter

DALLES RIFLE & PISTOL CLUB
418 W. 2nd Place
The Dalles, OR 97058, USA
503-296-2664
Gun show promoter
CONTACT: Boen DeLay

ROSE CITY GUN COLLECTORS
P.O. Box 16754
Portland, OR 97216, USA
503-254-5782
Gun show promoter
CONTACT: Ken Glass

OREGON ARMS COLLECTORS
Box 25103
Portland, OR 97225, USA
503-292-6145
Gun show promoter
CONTACT: Ted Dowd

MURRY BROOKS
Box 5191
Eugene, OR 97405, USA
503-345-4227
Gun show promoter

JIM KIRK
Box 246
Cheshire, OR 97419, USA
503-998-6227
Knife show promoter

LARRY DeSPAIN
3114 Boardman
Klamath Falls, OR 97603, USA
503-884-6544
Gun show promoter

AL ABELLERA
Box 25532
Seattle, WA 98125, USA
206-483-5321
Gun show promoter

VERN BOYLE
6514 River Road, East
Puyallup, WA 98371, USA
206-922-1717
Gun show promoter

NORTH WEST KNIFE COLLECTORS
Box 7216
Tacoma, WA 98407, USA
206-759-8264
Knife show promoter
S10;

BUD O'TOOLE
P.O. Box 64385
Tacoma, WA 98464, USA
Camouflage fabrics.
Illustrated list for 15 $.22 stamps.

WESTERN COLLECTORS
Box 1083
Ephrata, WA 98823, USA
509-754-4023
Gun show promoter
CONTACT: Jack Park

FRED McPHERSON
Box 1592
Walla Walla, WA 99362, USA
509-529-8329
Gun show promoter

WILLARD JOHNSON
Box 699
Palmer, AK 99645, USA
907-745-3446
Gun show promoter

BRITANNIA ENTERPRISES
33, Victoria Road
South Woodford, London E18 1LJ, Great Britain
01-530 6229
Medal fair promoter
S12;
CONTACT: Mrs. Jeannie Gordon

ARMS FAIRS LTD.
121 Fawnbrake Avenue
London, SE24, Great Britain
Arms fair promoters.

MILITARY MODELS/TOY SOLDIERS

AMERICA IN MINIATURE
Box 151-A, HC 75
Mattawamkeag, ME 04459, USA
Toy soldiers.
Write for information.

FRANK SALVIA
P.O. Box 271
Orange, NJ 07050, USA
Warship model builder.
Send for details.

TASK FORCE II
P.O. Box 1148
Orange, NJ 07050, USA
Models.

DUNBAR MINIATURES
3 Bywood Lane
Trenton, NJ 08628, USA
Modeling supplies.

COASTAL ENTERPRISES
P.O. Box 1053
Bricktown, NJ 08723, USA
(201)-477-7948
Miniature figures & supplies.
S4; 6;
Catalogue available.
CONTACT: Howard Wehner

MILITARY MODELS/TOY SOLDIERS

TED PRZYCHODA
Route #2, Box 58-H
Waretown, NJ 08758, USA
Military toy collector.

MICRO-MARK
P.O. Box 5112-33
24 East Main Street
Clinton, NJ 08809, USA
800-225-1066
Modeling supplies.

MODEL RECTIFIER CORPORATION
2500 Woodbridge Avenue
Edison, NJ 08817, USA
Modeling supplies.

THE TOY SOLDIER COMPANY
100 Riverside Drive
New York, NY 10024, USA
Toy soldiers.
Send SASE for List.

DI MARCO
51 Bay Street
Bronx, NY 10464, USA
Model kits.
Catalogue $1.00.

1ST ARMORED MODEL SUPPLY CO.
34/20 32nd Street, Apt. 3G
Astoria, NY 11106, USA
Modeling supplies.
Catalogue $3.00.

ENOLA GAMES
P.O. Box 1900
Brooklyn, NY 11201, USA
Modeling supplies.

HOWARD EINBINDER
303 Clinton Street
Brooklyn, NY 11231, USA
Model Kits.
Send SASE for List.

I/R MINIATURES
P.O. Box 89
Burnt Hills, NY 12027, USA
Toy soldiers.
$3.00 for catalogue.

WORLD WAR I AEROPLANES
15 Crescent Road
Poughkeepsie, NY 12601, USA
914-473-3679 Aero 818-243-6820 Skyways
Provides service info on WWI aeroplanes.
S6,S10; 25; 2200; 100.00
CONTACT: Leonard E. Opdycke
SALES DEPT: L. E. Opdycke
AD MANAGER: Richard Alden
Provides service (information, names, projects, materials, books) to modellers, builders & restorers of aeroplanes (1900-1919, 1920-1940) through two journals, WW I Aero & Skyways.

MODELER'S DISCOUNT HOBBY SHOP
P.O. Box 105
E. Rochester, NY 14445, USA
Modeling supplies

JOANNE E. RUDDELL
278 Walton Street
Demoyne, PA 17043, USA
215 966-3234
Toy soldiers & models.

SANTOS MINIATURES
P.O. Box 4062
Harrisburg, PA 17111, USA
Modeling supplies

LONDON BRIDGE COLLECTOR'S TOYS
P.O. Box 426
1344 Route 100 South
Trexlertown, PA 18087, USA
(215)-395-2500
Toy soldiers & miniatures.
S1,S4; 7;
Catalogue available. Annually with supplements.
CONTACT: Ronald P. Ruddell
SALES DEPT: Ronald P. Ruddell
PURCHASING DEPT: J. E. Ruddell or A. Soos

ROCKETSHIPS AND ACCESSORIES
623-625 S. 4th Street
Philadelphia, PA 19147, USA
Space toys, ray guns, & robots.
Catalogue available.
CONTACT: Joel Spivack

APC HOBBIES
11A Fence Road
Earlysville, VA 22936, USA
804-973-2705
Modeling supplies

NEW ERA METAL MODELS
1612 N.W. 55th Place
Gainesville, FL 32606, USA
904-372-7815
Toy soldiers.
Send $2.00 for catalogue.

HOLT'S HOBBIES
19800 SW 180th Ave., Box 40
Miami, FL 33187, USA
305-253-4251
Toy soldiers.
Send for current listing.

MODELERS MART 3
2071 Range Road
Clearwater, FL 33575, USA
813-443-3822
Models, miniatures & accessories.
72 page catalogue $2.00.
CONTACT: Al Younghaus

CENTENNIAL GUNS
P.O. Box 14314
Huntsville, AL 35815, USA
Artillery models.

A. AND J. MINIATURES
24 Woodhaven Street
Vicksburg, MS 39180, USA
601-636-0037
Toy soldiers.
Write for brochure.

STONE CASTLE IMPORTS
804 North Third Street
Bardstown, KY 40004, USA
(502)-897-0207 (502)-348-4879
Toy soldiers, miniatures, models & books.

S4;
Catalogue available.

K. W. MITCHELL
5148 Longbranch, Apt C
Columbus, OH 43213, USA
Toy soldiers.
Send SASE for catalogue.

W. MITCHELL
5880 Yorkland Court
Columbus, 43232, USA
Toy soldiers.
Catalogue $5.

A. J. FRICKO COMPANY
P.O. Box 43276
Cincinnati, OH 45243-0276, USA
Modeling supplies

BRIAN K. MITCHELL
P.O. Box 52414
Livonia, MI 48152, USA
Modeling supplies.

BATTLE HOBBIES
P.O. Box 184
St. Paul Park, MN 55071, USA
Modeling supplies

TRADITION USA
12924 Viking Drive
Burnsville, MN 55337, USA
(612)-890-1634
Military miniatures.
S6; 6;
Catalogue available.
CONTACT: Ed & Miriam Studer

MIDWEST MINIATURES
4645 Lilac Avenue
Glenview, IL 60025, USA
312-296-5465
Toy soldiers & miniatures.
Send for current listing.
CONTACT: Gus Hansen

H-R PRODUCTS
P.O. Box 67
McHenry, IL 60050, USA
Modeling supplies & accessories.

OLD TOY SOLDIER NEWSLETTER
209 North Lombard
Oak Park, IL 60302, USA
Toy soldiers
CONTACT: Steve Sommers

MILITARY COLLECTORS GUILD, LTD
12747 Olive Street Road, #190
St. Louis, MO 63141, USA
Military minatures & model soldiers.
Catalogue available.

MILITARY MODELS/TOY SOLDIERS

ARMCHAIR GENERAL LTD.
906 S. Main Street
St. Charles, MO 63301, USA
(314)-946-5997 (314)-469-0666
Toy soldiers & accessories.
S1,S4,S7; 9; 6500; 100.00
Catalogue $2.00. Quarterly.
CONTACT: Mr. William P. McMahon
SALES DEPT: Martha McMahon
Premiere toy soldier shop offers collectibles of military miniatures, antiques, books, posters, etc. Retail shop near St. Louis Airport. Catalog available - $2.00.

INTERNATIONAL PLASTIC MODEL SOCIETY
1615 Calvert
Lincoln, NE 68502, USA
Modeling supplies.

SQUADRON MAIL ORDER
1115 Crowley Drive
Carrollton, TX 75011, USA
214-242-8663
Models & modeling supplies.
Catalogue available.

THE DUNKEN COMPANY,INC.
P.O. Box 95
509 Texas
Calvert, TX 77837, USA
(409)-364-2020
Toy soldiers & accessories.
S4,S8; 12;
Catalogue available.
CONTACT: Bert Dunken
Lead soldier molds. Scarce collector molds.

D.M.F. HOBBIES
7128 La Tijera Blvd.
Los Angeles, CA 90045, USA
Military models & books.
CONTACT: Dwight French

THE MODEL SHOPPE
323 1/2 Richmond Street
El Segundo, CA 90245, USA
Modeling supplies & accessories.

THE COMMAND POST
P.O. Box 4403
Carson, CA 90749, USA
213-549-0006
Military models.
Price list $1.00.

CAMEO
P.O. Box 3035
Glendale, CA 91201, USA
Modeling supplies & accessories.

SERIES 77 MINIATURES
7861 Alabama Avenue, #14
Canoga Park, CA 91304, USA
818-340-5117
Modeling supplies & accessories.

VALLEY PLAZA HOBBIES
12160 Hamlin Street
N. Hollywood, CA 91606, USA
818-762-1927
Military miniatures & models.

BROOKHURST HOBBIES
12741 Brookhurst Way
Garden Grove, CA 92641, USA
714-636-3580
Modeling supplies & accessories.

COMBAT SERIES
P.O. Box 475
Pinedale, CA 93650, USA
Modeling supplies & accessories.

THE COMMAND POST
P.O. Box 4403
Carson, CA 94749, USA
Military models & accessories.

THE ARMORED TRAIN
P.O. box 478
Astoria, OR 97103, USA
Books, models, supplies.

GENNART LTD.
4/F Al Aqmar House
30 Hollywood Road
Hong Kong, Hong Kong
Model soldiers.
Free brochure.

AUCTION HOUSES

JAMES C. TILLINGHAST
P. O. Box 405
Hancock, NH 03449, USA
Militaria mail-auction.
4 auctions annually, catalogue $6.00.
CONTACT: James C. Tillinghast

WINTER ASSOCIATES
P.O. Box 823
21 Cooke Street
Plainville, CT 06062, USA
203-793-0288 203-525-6116
Auctioneers & appraisers.

UNCLE HANK'S
92 Swain Avenue
Meriden, CT 06450, USA
203-237-5356
Auction and information service.
Send SASE for information.
CONTACT: Hank Anton

HENRY CHRISTENSEN,INC.
P.O. Box 1732
Madison, NJ 07940, USA
201-822-2242
Coin mail-order auction.
CONTACT: William B. Christensen,Pres.

CHRISTIE'S EAST
Arms & Armour Department
219 East 67th Street
New York, NY 10022, USA
212-606-0540
Antique auction house.
Catalogues available.
CONTACT: Nicholas McCullough

RINSLAND AMERICANA MAIL AUCTION
P.O. Box 265
Zionsville, PA 18092, USA
215-966-3939 DAY 215-966-5544 NIGHT
Americana mail-auction house.
Catalogues available.

AAG INTERNATIONAL
20 Grandview Avenue
Wilkes-Barre, PA 18702, USA
(717)-829-1500
Military mail-auction house.
$18.00 for 4 auction catalogues.
CONTACT: Steve Flood

A. F. GLEIM
915A South Rolfe Street
Arlington, Va 22204, USA
U.S. medal auction.

PRESIDENTIAL COIN AND ANTIQUE
6204 Little River Turnpike
Alexandria, VA 22312, USA
Coin mail-order auction
CONTACT: COMPANY,INC

FRANKS ANTIQUES
P.O. Box 516
Hilliard, FL 32046, USA
904-845-2870
Mail-order militaria auction.

KURT R. KRUEGER
160 N. Washington Street
Iola, WI 54945, USA
715-445-3845
Exonumia mail-order auction.
Catalogues available.

MANION'S INTERNATIONAL AUCTION HOUSE
P.O. Box 12214
Kansas City, KS 66112, USA
Militaria mail-order auction house.
Catalogue available.

RIK LOOSVELT
Ringlaan 12
Kuurne, 8720, Belgium
Ammo & ordnance auction sales.
Catalogue $6.00.

INVICTA INTERNATIONAL
740 Gladstone Avenue
Ottawa, Ontario K1R 6X5, Canada
1-613-232-2263
General militaria mail auction.
Catalogue $3.00, Want lists solicited.

ACORN PHILATELIC AUCTIONS
66 Linden Grove
Great Moor
Stockport, Cheshire Great Britain
Philatelic mail-order auction.
Catalogue available.

KENT ARMS SALES
Kent House
4 New Road
South Darenth, Kent DA4 9AB, Great Britain
0322-864919
Military auction house.
Catalogue $1.50.

AUCTION HOUSES

SOTHEBY'S
Bloomfield Place
New Bond Street
London, W1A 2AA, Great Britain
Medals & Militaria auctions.
Catalogues available.

WALLIS & WALLIS
West Street Auction Galleries
Lewes, Sussex BN7 2NJ, Great Britain
0273-473137
Military auction house.
Catalogue available.

WELLER & DUFFY, LTD.
141 Bromsgrove Street
Birmingham, W. Midlands B5 6RQ, Great Britain
Military auction house.
Catalogues available.

CITY COINS
Ryk Tulbagh Square
Foreshore
Cape Town, South Africa
021-25-2639 021-25-3939
Military medals & coins auction house.
Catalogue available.
CONTACT: Mr. M. G. Hibbard

AUKTIONSHAUS PETER INEICHEN
CH-8027 Zurich, Postfach
C. F. Meyer-Strasse 14
Zurich, 8002, Switzerland
01-201-30-17
Arms & militaria auction house.
Catalogues available.

JAN K. KUBE-MILITARIA
Oskar-von-Miller-Ring 33
Munchen 2, 8000, West Germany
089 / 28 38 91
Militaria auction house.
4 auction catalogues, $26.00 annually.
CONTACT: Jan K. Kube

HANSEATISCHES AUKTIONSHAUS
Neuer Wall 75
Hamburg 36, 2000, West Germany
040/36 31 37-38
Military auction house.
Catalogues available.
CONTACT: Husken/Schafer OHG

WARGAMERS

THE COMPLEAT STRATEGIST
201 Massachusetts Avenue
Boston, MA 02115, USA
617-267-2451 800-225-4344
War games & accessories.
Free catalogue.

THE COMPLEAT STRATEGIST
215 Glenridge Avenue
Montclair, NJ 07042, USA
201-744-6622 800-225-4344
War games & accessories.
Free catalogue.

THE COMPLEAT STRATEGIST
11 East 33rd Street
New York, NY 10016, USA
212-685-3880 800-225-4344
War games & accessories.
Free Catalogue.

THE COMPLEAT STRATEGIST
320 West 57th Street
New York, NY 10019, USA
212-582-1272 800-225-4344
War games & accessories.
Free Catalogue.

THE COMPLEAT STRATEGIST
2011 Walnut Street
Philadelphia, PA 19103, USA
215-563-2960 800-225-4344
War games & accessories.
Free catalogue.

THE COMPLEAT STRATEGIST
254 W. DeKalb Pike
King of Prussia, PA 19406, USA
215-265-8562 800-225-4344
War games & accessories.
Free catalogue.

THE COMPLEAT STRATEGIST
231 Park Avenue
Baltimore, MD 21201, USA
301-752-1493 800-225-4344
War games & accessories.
Free catalogue.

THE COMPLEAT STRATEGIST
103 East Broad Street
Falls Church, VA 22046, USA
703-532-2477 800-225-4344
War games & accessories.
Free catalogue.

THE ULTIMATE GAME
P.O. Box 1856
Ormond Beach, FL 32075, USA
904-677-4358
War/combat games promoter.
Send for information.

ZOCCHI DISTRIBUTORS
1512 30th Ave
Gulfport, MS 39501, USA
601-896-8600
War board games.
CONTACT: Lou Zocchi

REENACTMENTS/LIVING HISTORY

WW II HIS. REENACTMENT SOCIETY NEW ENGLAND CHAPTER
153 Maynard Road
Framingham, MA 01701, USA
WWII Historical Reenactment Society Chapter.
WWII battle reenactment coordinator.
CONTACT: Jon Gawne

916th GRENADIER REGIMENT
401 Ferry Street, #3
Everett, MA 02149, USA
WWII German reenactment group.
WWII battle reenactment participant.
CONTACT: Frank J. Sorrento

COMPANY 'A', 643 TD BN.
299 Maplewood Avenue
Bridgeport, CT 06605, USA
203-334-3041
WWII reenactment group.
WWII battle reenactments, newsletter.
CONTACT: Dennis Ambrusco

44th INFANTRY DIVISION HISTORICAL REENACTMENT SOCIETY
P.O. Box 71
Manahawkin, NJ 08050, USA
(609)-597-6497
World War II reenactors. 4 battles per year, newsletter.
CONTACT: Arthur Smith

23rd REGIMENT OF FOOT, ROYAL WELSH FUSILIERS IN AMERICA
7 Manor Place
Huntington Station, NY 11746, USA
516-271-8083
Revolutionary War reenactment group.
CONTACT: Frederick J. Denton

THE BLACK WATCH ROYAL HIGHLAND REGIMENT OF CANADA
204 Martin Drive
Syosset, NY 11791, USA
516-681-8980 (HM)
WWII British reenactment group. WWII battle reenactment participant.
; 4;
CONTACT: Lee Benjamin
We are affiliated with the W.W.II H.R.S., W.W.II H.R.F, M.V.C.C. We are authenticity oriented & have direct liason with the actual regiment in Montreal, Canada.

COMPANY 'A', 2nd BN., 329th INF., 83rd DIV.
18 Schryver Court
Kingston, NY 12401, USA
914-338-0648
WWII American reenactment group. WWII battle reenactments, newsletter.
CONTACT: Mark H. Feldbin

GERMAN MOUNTAIN TROOPS
P.O. Box 51
Cherry Valley, NY 13320, USA
WWII reenactment group.
CONTACT: Spence Waldron

11th PENNSYLVANIA VOLUNTEERS, COMPANY I
818 Jane Avenue
Charleroi, PA 15022, USA
(412)-483-0108 (412)-838-6697
Civil War reenactment group. Reenactments, Living history.
CONTACT: Edgar Thomas Sheperd, Jr.
Recruiting for active, authentic style Union reenactment unit. Unit area includes all Southwestern Pennsylvania. Will assist volunteers to obtain uniform, rifle, and other gear.

501st PIR, 101st ABN. DIV.
220 Fifth Street
Oakmont, PA 15139, USA
WWII reenactment group. WWII battle reenactment participant.
CONTACT: Richard Ayars

REENACTMENT & LIVING HISTORY GROUPS

WW II HIS. REENACTMENT SOCIETY GREATER PITTSBURGH AREA CHAP.
6 Oak Drive
Pittsburgh, PA 15214, USA
412-931-5952
WWII Historical Reenactment Society Regional Chapter.
CONTACT: Herb Linde

51st HIGHLAND DIVISION
1105 Bingay Drive
Pittsburgh, PA 15237, USA
412-321-9317
WWII British reenactment group. Battle reenactments, newsletter.
CONTACT: Richard J. Baier
WWII Historical Reenactment Society portrays infantrymen of WWII using period weapons and uniforms. Battles held monthly across the country. Enthusiasts welcome.

FIRST PENNSYLVANIA REGIMENT OF THE CONTINENTAL LINE
2280 Old Forty Foot Road
Harleysville, PA 19438, USA
(215)-584-4394
Revolutionary War reenactment group. Competitive Black Powder shoots.
CONTACT: Theodore Heske,Jr.

NATIONAL WWII HISTORICAL FEDERATION
1823 Arabian Way
Falliston, MD 21047, USA
301-877-7070
WWII reenactment association. WWII battle participants, journal.
CONTACT: Tony Lambros

U.S. 29th INFANTRY DIVISION
1823 Arabian Way
Falliston, MD 21047, USA
301-877-7070
WWII American reenactment group. WWII battle reenactments, journal.
CONTACT: Tony Lambros

KAMFGRUPPE SKORZENY, H.G. DIV. 8th KAVALLERIE DIVISION
7325 Woodley Place
Falls Church, VA 22046, USA
WWII German reenactment group. WWII battle reenactment participant.
CONTACT: William Miller

6th ARMD. DIV. WELSH GUARDS
2812 Jermantown Road
Oakton, VA 22124, USA
WWII British reenactment group. WWII battle reenactment participant.
CONTACT: Edward Franzosa

GERMAN 38th DIVISION
339 Main Street
Smithfield, VA 23430, USA
WWII reenactment group.
CONTACT: Scott Shean

WW II HIS. REENACTMENT SOCIETY GEORGIA CHAPTER
339 Stafford Street
Marietta, GA 30064, USA
WWII Historical Reenactment Society Chapter. Reenactment coordinator.
CONTACT: Jerry May

509th PIR, 82nd ABN. DIV.
Rt. 10, Box 18
Cumming, GA 30130, USA
WWII reenactment group. WWII battle reenactment participant.
CONTACT: Terry Dougherty

SS PANZER GREN. RGT. 3 'DEUTSCHLAND'
161 Wilbur Lane
Lawrenceville, GA 30245, USA
WWII German reenactment group. WWII battle reenactment participant.
CONTACT: Paul Kallio

FALSCHRIM. ABTEIL. 500 SS
104-B Rio Delmar
St. Augustine, FL 32084, USA
WWII German reenactment group. WWII battle reenactment participant.
CONTACT: Clark Wullenweber

17th AIRBORNE DIVISION
P.O. Box 981
Douglasville, GA 33133, USA
WWII reenactment group. WWII battle reenactment participant.
CONTACT: Fred Schaller

30th INFANTRY DIVISION
3680 Richbriar Court
Nashville, TN 37211, USA
WWII reenactment group. WWII battle reenactment participant.
CONTACT: Richard Cornwell

1st GUARDS SHOCK ARMY
3680 Richbriar Court
Nashville, TN 37211, USA
WWII Russian reenactment group. WWII battle reenactment participant.
CONTACT: Richard Cornwell

WW II HIS. REENACTMENT SOCIETY BLUEGRASS CHAPTER
P.O. Box 9
Ft. Knox, KY 40121, USA
WWII Historical Reenactment Society chapter. WWII reenactment coordinator.
CONTACT: Bluegrass Military Col. Club

GROSSDEUTSCHLAND
259 Univ. Dr. #1
Radcliff, KY 40160, USA
WWII German reenactment group. WWII battle reenactment participant.
CONTACT: Barry D. Willard

83rd RECON BN., 3rd ARMD. DIV.
338 Redmar Blvd.
Radcliff, KY 40160, USA
WWII reenactment group. WWII battle reenactment participant.
CONTACT: Don Crum

1st INFANTRY DIVISION
866 Kyle Avenue
Columbus, OH 43207, USA
WWII American reenactment group. WWII battle reenactment participant.
CONTACT: Gene Stacy

3rd INFANTRY DIVISION
1283 Bryson Road
Columbus, OH 43224, USA
WWII American reenactment group. WWII battle reenactment participant.
CONTACT: Rick Kanoski

1st SPECIAL AIR SERVICE
3675 Clague Road, Suite 515
North Olmstead, OH 44070, USA
WWII British reenactment group. WWII battle reenactment participant.
CONTACT: Jim Russell

167th VOLKSGRENADIER
3607 Oak Road
Stow, OH 44224, USA
216-688-2135
WWII German reenactment group. WWII battle reenactment participant.
CONTACT: Phil Narzisi

9th SS PANZER DIVISION
5106 W. Caven Street
Indianapolis, IN 46241, USA
WWII German reenactment group. WWII battle reenactment participant.
CONTACT: Jeff Niese

22nd I.P.C. 6th ABN. DIV.
1100 W. Wells St. #1102
Milwaukee, WI 53233, USA
WWII British reenactment group. WWII battle reenactment participant.
CONTACT: Michael A. Ferenz

FIRST MINNESOTA VOLUNTEER INFANTRY,INC.
c/o Fort Snelling History Center
St. Paul, MN 55111, USA
Reenactment group. Historical society concerned with Minnesotans in Civil War.

GREN. RGT. 48, 12th. DIV. 2nd KP.
1201 Court C
Hanover Park, IL 60103, USA
WWII German reenactment group. WWII battle reenactment participant.
CONTACT: Gary Arden

110th PANZER GREN. RGT.
9864 Garden Court
Schiller Park, IL 60176, USA
312-671-0224
WWII German reenactment group. WWII battle reenactment participant.
CONTACT: George Krupa

3rd SS PANZER DIVISION
1221 Kings Court, #33
West Chicago, IL 60185, USA
WWII German reenactment group. WWII battle reenactment participant.
CONTACT: Ronald D. Kewin

504th PIR, 82nd ABN. DIV
431 South Cass Avenue
Westmont, IL 60559, USA
312-968-1378
WWII American reenactment group. WWII battle reenactment participant.
CONTACT: Ken Thompson

REENACTMENT & LIVING HISTORY GROUPS

34th INFANTRY DIVISION
238 N. Wilmette Street
Westmont, IL 60559, USA
WWII reenactment group. WWII battle reenactment participant.
CONTACT: Dave Serikaku

DEVILS BRIGADE
624 Bohm Avenue
Rockford, IL 61107, USA
WWII reenactment group. WWII battle reenactment participant.
CONTACT: Randy Nielsen

RGT. BRAND. ZBV 800
624 Bohm Avenue
Rockford, IL 61107, USA
WWII German reenactment group. WWII battle reenactment participant.
CONTACT: Randy Nielson

1st SS L.A.H.
113 Ridgemoor Drive
Edwardsville, IL 62025, USA
WWII German reenactment group. WWII battle reenactments, newsletter.
CONTACT: Fred Poddig

HQ. TRP. 6th CAV. 3rd ARMY
2722 Monroe
Quincy, IL 62301, USA
WWII reenactment group. WWII battle reenactment participant.
CONTACT: Joe Homberger

NO. 6 COMMANDO
124 North 12th Street
Quincy, IL 62301, USA
217-224-3990
WWII British reenactment group. WWII battle reenactment participant.
CONTACT: Robert C. Scholz, Jr.

2nd SPECIAL AIR SERVICE
P.O. Box 487
Du Quoin, IL 62832, USA
WWII British reenactment group. WWII battle reenactment participant.
CONTACT: Bob Chapman

3rd GEBIRGS. DIVISION
6136 Alaska
St. Louis, MO 63111, USA
WWII German reenactment group. WWII battle reenactment participant.
CONTACT: James Ghiglione

2nd U.S. RANGER BN.
924 Scott Avenue
KIRKWOOD, MO 63122, USA
WWII reenactment group. WWII battle reenactment participants.
CONTACT: John Strohm

2nd ARMORED DIV., HQ. CO.
5908 Brownleigh
St. Louis, MO 63134, USA
WWII reenactment group. WWII battle reenactment participant.
CONTACT: Carol Venable

2nd GUARDS TANK BRIGADE
10184 Winkler Drive
St. Louis, MO 63136, USA
WWII Russian reenactment group. WWII battle reenactment participant.
CONTACT: Jim Otto

11th FIELD HOSPITAL
3448 Tedmar Avenue
St. Louis, MO 63139, USA
WWII reenactment group. WWII battle reenactment participant.
CONTACT: Agnes Kalal

38th JAGER REGIMENT
5254 Fairview Avenue
St. Louis, MO 63139, USA
314-351-6991
WWII German reenactment group. WWII battle reenactment participant.
CONTACT: Art Obermeyer

163rd SIG. PHOTO CO.
4026 San Fernando
ST. Charles, MO 63303, USA
WWII American reenactment group. WWII battle reenactment participant.
CONTACT: Richard Horrell

28th SS 'WALLONIE' DIVISION
P.O. Box 945
Cape Girardeau, MO 63701, USA
618-661-1611
WWII German reenactment group. WWII battle reenactments, newsletter.
CONTACT: Jay Sproat

128th RGT. 48th INF. DIV.
4840 Cleveland Ave. #10
Lincoln, NE 68504, USA
402-464-2596
WWII German reenactment group. WWII battle reenactment participant.
CONTACT: Randy Butts

GREN. RGT. 48, 12th. DIV. 1st KP.
4000 N. 63rd Street
Lincoln, NE 68507, USA
WWII German reenactment group. WWII battle reenactment participant.
CONTACT: Neil Mitchell

45th INFANTRY DIVISION
4813 Newport Drive
Del City, OK 73115, USA
WWII reenactment group. WWII battle reenactment participant.
CONTACT: Charles Primeaux

ROYAL ARTILLERY (WWII)
7500 Maplewood #250
Ft. Worth, TX 76118, USA
WWII British reenactment group. WWII battle reenactment participant.
; 5;
CONTACT: Jerry McGhee
A recreation of a W.W.II British artillery battery which participates in mock battles with other British, American and German units.

6th FALLSCHRIM.
1539 Castle Court #2
Houston, TX 77006, USA
WWII German reenactment group. WWII battle reenactment participant.
CONTACT: J. R. Van DeGrift

2nd SS 'DAS REICH' DIVISION
9718 Pine Lake Road
Houston, TX 77055, USA
WWII German reenactment group. WWII battle reenactment participant.
CONTACT: Ken Austin

10th MOUNTAIN DIVISION
6840 Richthofen Parkway
Denver, CO 80220, USA
303-399-9264
WWII American reenactment group. WWII battle reenactment participant.
CONTACT: Flint Whitlock

5th SS 'WIKING' DIVISION
P.O. Box 827
La Porte, CO 80535, USA
WWII German reenactment group. WWII battle reenactment participant.
CONTACT: Harry Connors

FALLSCHIRMJAGER - REGT 1 1.
FALLSCHIRM DIVISION
P.O. Box 13382
Ft. Carson, CO 80913, USA
303-392-1300
WWII German reenactment group. WWII battle reenactment participant.
CONTACT: Dean Byrne

WW II HIS. REENACTMENT SOCIETY
ARIZONA STATE CHAPTER
1225 N. Oregon Street
Chandler, AZ 85224, USA
WWII Historical Reenactment Society chapter. WWII reenactment coordinator.
CONTACT: Don Cluff, Jr.

60th PANZER GREN. REGT.
4 S. 132nd Street
Chandler, AZ 85224, USA
WWII German reenactment group. WWII battle reenactment participant.
CONTACT: Chris Lares

WW II HIS. REENACTMENT SOCIETY
CALIFORNIA CHAPTER
11041 Dutton Drive
La Mesa, CA 92041, USA
WWII Historical Reenactment Society chapter. WWII reenactment coordinator.
CONTACT: Monte Mattson

2nd PANZER GREN. RGT. 4
P.O. Box 3238
Fremont, CA 94539, USA
WWII German reenactment group. WWII battle reenactment participant.
CONTACT: Tony Smith

Phoenix Militaria's
American Militaria Sourcebook & Directory
BUSINESS CARD DIRECTORY

Cook's Arsenal Works

Lawrence E. Cook
642 Johnson Ave.
Meriden, Ct. 06450
203-237-7997

Careers in Secret Operations

Finding employment with CIA, FBI, DEA, the Secret Service, National Security Agency, and all services involved in clandestine operations. A "must" book, says ex-CIA Director Richard Helms.

By David Atlee Phillips, a 25-year CIA veteran and founder of the Association of Former Intelligence Officers. Job descriptions and application tips and more. Phillips knows his stuff and tells you how to become a secret operative. Chapters on undercover operations, hazards of the trade, agent handling, opportunities for those without college degrees, and the agent's jargon. In spy lingo, this book shows you how to "put in the plumbing" for an exciting and rewarding intelligence career.

$9.95 prepaid:

Phoenix Militaria
Postal Box 66
Arcola, Pennsylvania 19420

SPECIALIZING IN 6 X 6
2½T · 5T · 10T · 'M'-Series

Direct Support
Military Vehicle Restorations
Parts - Service - Rebuilding
Bought & Sold

BOX 317
PITTSFIELD, MASS. 01202

BOB KETTLER
413-443-4841

The Phoenix Exchange

'The Phoenix Exchange' functions as the military collector's & dealer's marketplace. Articles cover a wide spectrum of militaria, but attend to each subject in detail. Send $2.00 for a sample copy.

Postal Box 66/Arcola/Pennsylvania 19420

F.F.L. DEALER   LICENSED GUNSMITH

HARRY G. CAKOUNES
D/B/A
NOAH'S MOTORS
622 BROADWAY (U.S. RT. 1N)
SAUGUS, MASS. 01906
617-233-1616
U.S. SERVICE RIFLES MATCH CONDITIONED

M1 GARANDS   M14/1A   SPRINGFIELDS

MILITARY SURPLUS SUPPLY

U.S. MILITARY SURPLUS
WWII THRU PRESENT

Send For Listing Today!
$1.00 (refundable)

5594 AIRWAYS ROAD
MEMPHIS, TN. 38116

901-346-0090

PHOENIX MILITARIA

MILITARIA & WAR RELICS

Send $3.00 or 3 pieces of military insignia for our latest catalogue of U.S. & world-wide militaria. Thousands of items listed including: books, manuals, medals, insignia, equipment, uniform components, badges, etc.

Postal Box 66/Arcola/Pennsylvania 19420

U.A.C.C.   The Manuscript Society

AMERICAN HISTORICAL DOCUMENTS

DR. F. DON NIDIFFER
P. O. Box 8184
Charlottesville, VA 22906
(804) 296-2067

Presidential • WWII • Civil War
Revolutionary War • Miscellaneous

NARIAKI ITOH
AVIATION COLLECTOR

0574-25-8343
26-7644

256-17, TORIKUMI, SAKAHOGI-CHOH, KAMO-GUN, GIFU-KEN, 505, JAPAN.

FLYING LEATHER GIFU HEAD QUARTER

PRESIDENT NARIAKI ITOH

**WANTED !! AVIATION ITEMS from pre-W.W.1 to post-W.W.2 all kinds of all nations.
A-2. B-3 jackets, helmet, goggles, DOOLITTLES TOKYO RAIDERS, AVG. THE FLYING TIGERS. etc······**

Large (approx. six square feet) flags. $25.00@ retail. Wholesale $75.00 per half dozen!

Including: Imperial Prussian Battle, Third Reich Battle, NSDAP High Party Member's, Japanese Battle, Fuehrer, Police, Luftwaffe, Allgemeine SS, SS HQ, Swastika.

Also: Mini-flags, armbands, pins, posters, and cassettes at retail and wholesale. Dealer inquiries invited!

Send SASE for Free List: Hammer, Box 1393-PMD, Columbus, IN 47201

MILITARY VEHICLE COLLECTORS CLUB

**The club with the most for MILITARY VEHICLE enthusiasts worldwide.
Over 4000 members in 26 countries.**

Army Motors: Our technical magazine is published quarterly.
Supply Line: Our advertising journal comes out six times a year.

U.S. Membership is $20.
Membership outside the U.S. is $35.

**MILITARY VEHICLE COLLECTORS CLUB
Box 33697AF
Thornton, Colorado 80233
U.S.A.**

WORLD WAR II HISTORY

★★★★★★★★★★★★★★★

We carry many unusual and hard-to-get books... also all issues of *After the Battle Magazine* and *Wheels and Tracks*. PLUS a large selection of books and manuals on military and civilian jeeps, military trucks and armored vehicles. We also stock many titles on World War Two and Post-War military subjects... some only available through us! Send $2 today for our latest 36-page catalog, sent First Class.

★★★★★★★★★★★★

PORTRAYAL PRESS

P.O. Box 1913-28
Bloomfield, N.J. 07003
(201) 743-1851

Member: American Booksellers Association

VISIT MAS PALEGRY

The most bucolic **Aviation Museum** in the **South of France**
Be also welcome to visit the adjacent wine cellars

Charles NOETINGER a former Armée de l'Air Reconnaissance Pilot before becoming a viticulturist in the Vines of **MAS PALEGRY** has re-created for you his world of Air Force Pilot in a very personal **Private Museum** with his planes and a lot of items.

Ch NOETINGER - MAS PALEGRY - (4 km Sud PERPIGNAN)
Route d'ELNE - 66000 PERPIGNAN - **Tél.: 68.54.08.79.**

MS 733 ALCYON — CM170 FOUGA MAGISTER — DE HAVILLAND VAMPIRE F5 — REPUBLIC RF84 F — COCKPIT OF F84G — FULLSCALE MOCK UP OF THE MIRAGE F1 — ASSORTED MISSILES — GLIDER CAUDRON C800 — FUSELAGE OF CAUDRON LUCIOLE — MS 603 — HUNDREDS OF PICTURES — STICKERS AND INSIGNIA AND AN EXTENSIVE COLLECTION OF HUNDREDS BALSA WOOD MODELS IN SCALE 1/200 TH HAND MADE BY **CHARLES NOETINGER**.

Careers in Secret Operations:
HOW TO BE A FEDERAL INTELLIGENCE OFFICER
By DAVID ATLEE PHILLIPS

A reliable guide for men and women who contemplate intelligence careers and, for those who have made a decision, advice on how to find employment with the CIA, the FBI, the Defense Intelligence Agency, the National Security Agency, the Secret Service, and other federal departments which engage in clandestine operations.

"A 'must' for those Americans interested in joining any U.S. intelligence organization". **RICHARD HELMS**
FORMER CIA DIRECTOR

"A wonderful new book. Phillips knows his stuff and tells you how to become a federal intelligence officer in a variety of government agencies. This guide is more than a summary of job descriptions and application procedures, Phillips treats the ethics and morality of secret operations and the dilemmas they raise for a free and open society. He discusses affirmative action programs, clandestine careers, opportunities for those without college degrees and the jargon of the intelligence agent. In spy lingo, this book shows you how 'to put in the plumbing' for an intelligence career."
JOYCE LAIN KENNEDY
SYNDICATED CAREER COUNSELOR

"Answers questions sure to be of interest to potential applicants (are lie detectors used? will drug use or homosexuality disqualify an applicant?) while maintaining a firm grip on reality. Sincere practical information on intelligence jobs is scarce; Phillips' book answers a definite need. His clear, readable style, and the excellent annotated bibliography put this a notch above the average career book. Recommended." **LIBRARY JOURNAL**

Questions about *Careers in Secret Operations* are answered by David Atlee Phillips, 25-year CIA veteran and founder of the Association of Former Intelligence Officers, a national group of ex-intelligence people from all services. He is the author of five books, including *The Night Watch*, a CIA memoir.

Phoenix Militaria
Postal Box 66
163 Troutman Road/Arcola/Pennsylvania 19420, U.S.A.

ISBN 0-932123-00-7

Send check/money order payable to:

Rush _____ softback copies of *Careers in Secret Operations* $9.95 ea   Total: $ _____

SEND TO _____
STREET & NUMBER _____ APT _____
CITY _____ STATE _____ ZIP _____

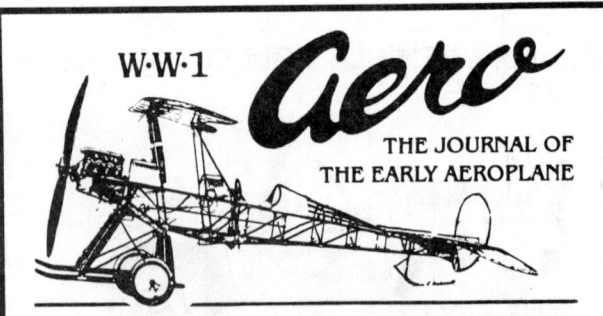

SOLD WITH A 30-DAY MONEY-BACK GUARANTEE

W·W·1 Aero
THE JOURNAL OF THE EARLY AEROPLANE

Join collectors, restorers, replica builders, historians, and modellers all over the world in the only authoritative organization devoted to Those Magnificent Flying Machines.

Receive five issues a year of WWI AERO featuring all types of aircraft 1900-1919, each issue containing histories, photographs, construction drawings, engines, performance data, models, and lore written by experts. Learn about projects all over the world. Discover the most complete source for information, parts, models. Published regularly since 1961.

NAME _____
STREET _____
CITY _____ STATE _____ ZIP _____

SAMPLE ISSUE $4.00

WORLD WAR I *Aeroplanes*, **INC.**
15 Crescent Road, Poughkeepsie, NY 12601, USA
Dept. P

Hey You!
Join us Today

Arms Collectors Journal
Serving An Area of Over 50 Million People

- NEWS & VIEWS OF THE GREAT NORTHEAST -
- Up to Date Reporting For Sportsmen & Women -

LISTING OF ALL MAJOR EVENTS & SHOWS
... For Who's Where & What's Happening

COLLECTOR ARTICLES • POLITICAL NEWS • OUTDOOR TIPS
• INTERESTING COMMENTARY & BOOK REVIEWS •
Plus
FEATURE STORIES & 'ANTIQUE FIREARMS INFORMATION'
'THE TACKLE BOX' 'FANG'S ADVICE' 'CARTRIDGE CORNER'
& Views From 'BOTH SIDES OF THE TABLE'

BECOME A PART OF THE ONLY NEWS & COLLECTOR JOURNAL DEVOTED TO THE **NORTHEAST** - A MARKET OF OVER ONE-THIRD OF THE POPULATION OF AMERICA & CANADA - AN AREA WITH CONSTANT RESIDENCY TURNOVER - A VACATIONING PUBLIC WITH THE HIGHEST PER CAPITA INCOME ...ASSOCIATES ON THE MOVE - **JOIN US TODAY!**
CALL 518-664-9743 TO FIND OUT WHAT YOU'VE BEEN MISSING.

Published Monthly - Year's Subscription $12.00 U.S. ($18.00 Canada)
Arms Collectors Journal; NEACA, Inc., PO Box 385, Mechanicville, NY 12118

Military History At A Discount!

Over 1,000 books, publications, reprints, and video and audio cassettes are listed in the W&WP Master Catalog. Every subject and period covering the entire history of 'weapons and warfare' is covered - especially the 20th century, and particularly World War Two.

You can receive the Master Catalog - plus a year's worth of Catalog Supplements - for only $5.00 (foreign, $6.00). And you'll also receive a **FREE gift**.

You will also be able to save up to 25% off the retail prices of practically everything listed!

And, we'll include for your inspection a **FREE issue** of **War And Resources** - the bi-weekly publication that provides you with the latest information on hundreds of sources of all types of reference materials and hobby products.

You won't find a better source for new military history products than W&WP. A large selection - fast and efficient service - quality products - low prices. A hard-to-beat combination!

Won't you join our 'family'?

WEAPONS and WARFARE PRESS

218-X35 Beech Street
Bennington VT 05201

THE COURIER

IF YOU ARE A CIVIL WAR BUFF,
COLLECTOR, DEALER, HISTORIAN,
OR GENEALOGICAL RESEARCHER,
OR
IF YOU BELONG TO A C.W.R.T.,
A "SONS/DAUGHTERS"
ORGANIZATION,
RE-ENACTMENT GROUP
OR THE N.S.S.A.,
THE COURIER WILL HELP YOU IN
PURSUING YOUR INTERESTS!

SEND FOR YOUR FREE SAMPLE COPY
TODAY:
THE COURIER
PO BOX 1863
WILLIAMSVILLE,
NEW YORK, 14221

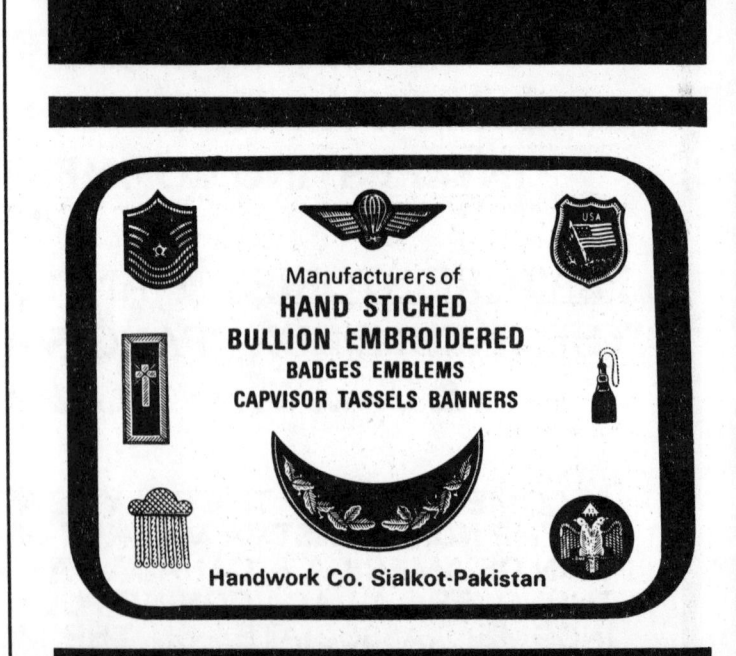

G. GEDNEY GODWIN
SUTLER

Whose Establishment is against Bass on Welsh Road, at the Valley Forge, in Pennsylvania Province,

WISHES to announce that he is importing a complete Line of Items useful to the Military Man; such as Muskets, Bayonets, Leather Goods, Uniforms, Cocked Hats, divers sorts of brazen Sundries, &c., all done in the neatest manner. Those Gentlemen who will favor him with their custom, may depend on their work being dispatched and their favors gratefully acknowledged, by their very humble Servant.

G. Gedney Godwin.

We are specialists in historically accurate, fully functional reproductions from the French & Indian War through the Civil War. An 85-pg. illustrated catalog costs $3.85 - refundable.

G. Gedney Godwin Inc. • P.O. Box 100 • Valley Forge, PA 19481

Manufacturers of
**HAND STICHED
BULLION EMBROIDERED**
BADGES EMBLEMS
CAPVISOR TASSELS BANNERS

Handwork Co. Sialkot-Pakistan

BACK ISSUES NOW AVAILABLE!!!

MILITARY COLLECTOR'S JOURNAL

The Phoenix Exchange has acquired the assets of the late Military Collector's Journal. This acquisition included a large inventory of back issues which are now available to you. Collectors and historians will find this publication both good reading and a valuable reference. For your convenience we have listed the content of each available issue........... $2.00 each

(1) Volume I, No. 2
   Guderian's Pennant
   U.S. Army Overseas Cap
   OSS Weapons, Part II
   Queen's Royal Irish Hussars
   1896 Mauser

(2) Volume I, No. 3
   The Blue Max
   U.S. Colt .25 Automatics
   Pearl Harbour
   Scharfs schutzen Ab Zeichen
   The Free Slovak-State Army Labor
   Presentation Luger
   Kit Carson Scouts

(3) Volume I, No. 4
   Albanian Irregulars in WWI
   Pearl Harbour
   Nazi Diplom~~ ~~
   Gune ~~SOLD OUT~~
   ~~ ~~ge Uniforms WWII
   ~~ ~~gnia
   ~~Im~~perial German Inf. Uniforms WWI
   P-38, Collector's Review

(4) Volume II, No. 1
   Imperial German Badge
   Long Range Desert Patrol
   U.S. ARMY Chevrons
   Vietnamese Communist Militaria
   1892 Krag
   SS Honor Dagger
   Japanese Samuri

(5) Volume II, No. 3
   Japanese Aircraft, 1941-1942
   WWI, American Infantryman
   Japanese GXP Submachine Gun
   Deutschlandlug Plaque
   WWI Re-enactment
   Equip the Band
   Polish Second Corps, 1942-1945
   I Thought It Was Time To Get Out

(6) Volume II, No. 4
   U.S. Army Staff
   Dress Dagger Kaiser William II
   Soldier Folk-Art
   The 1911 AI
   Junkers Ju 52
   Chinese Victory Medal

(7) Volume II, No. 5
   Dien Bien Phu
   WWI German Uniform
   Viet-Nam, The River War
   King's Royal Rifle Corps
   M3 Greasegun
   8/16/45
   The Meaning of Alsos
   Adolph Hitler's Letter Opener

(8) Volume II, No. 6
   Bureau of Insular Affairs
   The War in the Falklands
   World of Semi-autos
   Dien Bien Phu
   Horse Marines
   Dairy of Leonard Stiegerwalt
   Kaiserlichen Marine
   Hamilton International Air Show

(9) Volume III, No. 1
   Inchon
   World of Semi-autos
   Verdun
   Those Short-legged Soldier Boys
   A Medal for a Spy
   The War in the Falklands
   Edged Weaponry of the Nazi Hunt

(10) Volume III, No. 2
   Victory-Falklands
   Eastern Front Re-enactment
   Vietnamese Croix de Guerre
   Confederate Air Force Flies Again
   SS Tropical Tunic is Rare Find
   1st Calvary Patch
   A Kindness Repaid for Eddie Butler

(11) Volume III, No. 3
   (Note: Misprinted Vol. III, No. 2)
   Early Turkish Air Force Pilot's Badge
   The Polish Virtuti Militari
   Aircraft Insignia
   Skyraider
   German Machine Gun Units of WWI
   Carlisle Barracks

(12) Volume III, No. 4
   "Bulge" Re-enactment Cold and Foggy
   The Chained SA Dagger
   The USARV Training Group
   Airborne Jump Wings
   E.R. Reedstrom, Collector
   P-38, Lightning-Forktailed Devil

(13) Volume III, No. 5
   Hac Bao: The Demise of a Once-Proud Unit
   The German 'Coal Scuttle' Steel Helmet
   Soldiers at Play
   The Royal Corps of Transport
   96th Aero Squadron, Aug. 10, 1918
   A Medal for Lance Corporal Kuklinski
   D-Day Plus Forty
   Patch Collector—Take Notice

(14) Volume III, No. 6
   Weapons of the Falkland Conflict
   1915—Uniform of the French Army
   U.S. Coast Guard and It's Insignia
   The Disappearance of 'Peking' Man...
   The Skyraider—It's Operational Use
   The Heirl-Roche Labor Corps Dagger

How to Order

Fill out the form below. Circle the order number of each issue you wish to receive. Enclose $2.00 for each issue ordered.

Add $1.00 Postage & Handling for (1) issue & 50¢ for each additional issue.

The Phoenix Exchange
Postal Box 66/163 Troutman Road/Arcola/Pennsylvania 19420

Circle Order Number
(1) (2) (3) (4)
(5) (6) (7) (8)
(9) (10) (11)
(12) (13) (14)

Name _____
Address _____
City _____ State ___ Zip ___
Country _____

Total Issues _____ Amount Enclosed _____

Overseas orders must include an additional $1.25 Air Mail Post for each issue ordered. Payment must be in United States Dollars Bank Draft or International Money Order.

NEW!

Phoenix Militaria's
AMERICAN MILITARIA
SOURCEBOOK & DIRECTORY

"The Who's Who In Militaria"

Over *4,500 Listings

8½" X 11"

For Military Collectors & Businesses...

This easy-to-use reference lists over ~~3,500~~ Militaria Dealers, Federally Licensed Firearms Dealers, Manufacturers, Service Companies, Museums, Libraries, Auction Houses, Book Publishers, Military Collector Publications, Clubs, Associations, Societies, Film/Video Sources, and other militaria organizations throughout the United States.

Each listing includes all the information we could gather—Business Names, Addresses, Telephone Numbers, Products & Services, Key People, and Years in Business. Never before has there been a comprehensive easy-to-use militaria sourcebook & directory with over ~~3,500~~ listings, and valuable collector/business information. *

Whether you're a militaria collector or in business, you'll find this valuable book an important addition to your reference library. And we guarantee it! If for any reason you are not completely satisfied return it within 15 days for a full refund or credit. To order, simply fill out the form below and mail today!

★

☐ Please rush _____ copy(s) of Phoenix Militaria's AMERICAN MILITARIA SOURCEBOOK & DIRECTORY at $19.95 each postage-paid.

PLEASE PRINT

Name _____
Address _____
City _____ State ___ Zip ___
Country _____

Overseas orders must include an additional $3.00 for postage. Total payment must be in U.S. funds, U.S. Dollars Bank Draft or International Money Order.

Phoenix Militaria
163 Troutman Road
Postal Box 66
Arcola, Pennsylvania 19420, U.S.A.

(215) 933-0909

The Military Collector's Marketplace

A First-Class Money-Saving Introductory Offer for Militaria Collectors and Historians...

Right now you can guarantee every issue of The Phoenix Exchange reaches your mailbox in style. Subscribe today and we'll mail your one year subscription by first-class mail (with protective envelope) for only $6. You save $3.00 off the regular first-class subscription rate.

And if you're already a subscriber you can still take advantage of this special offer. But don't waste any time, we can't do this forever—subscribe or renew today!

Mail To:
The Phoenix Exchange
Postal Box 66/Arcola/Pennsylvania 19420, U.S.A.
(Sorry, this offer is not available to new or current overseas customers.)

PRICE SUBJECT TO CHANGE WITHOUT NOTICE

The PHOENIX EXCHANGE

☐ Yes, I want to guarantee my one year subscription to The Phoenix Exchange arrives in style—and save $3.00 off the regular first-class subscription rate.
  ☐ New Subscription   ☐ Renewal

Total Enclosed _____$6_____
(Regular First-Class Subscription Rate $9.00)

Name _____
Address _____
City _____ State _____ Zip _____

Polish Orders, Medals, Badges and Insignia

by
Prof. Dr. Zdzislaw P. Wesolowski

A comprehensive one volume encyclopedic research work on Polish military and civilian decorations. The book covers a period from 1705 to 1985 and includes 1,631 photographs with 404 pages. The decorations are identified and described with historical notes on each Polish order, medal, regimental and commemorative badge and national insignia. It is the most complete book ever published in English on the subject. There are 19 color plates.

Library of Congress Card Catalog No. 86-90004   ISBN NO. 0-937527-00-9

PRICE & SHIPPING: $29.50 for orders in the United States which include shipping charges. Canada, please add $2 for postage and handling; outside U.S. and Canada, add $6 for postage and handling. Discounts are available as follows: 5 books, 10%; 6 or more, 15%; 11 or more, 20%; 21 or more, 25%. Payment must be made in United States dollars and accompany book orders. Please send your check or money order to Printing Services, Inc.

Printing Services, Inc.
3249 N.W. 38th Street
Miami, Florida 33142, U.S.A.

POLISH ORDERS, MEDALS, BADGES AND INSIGNIA

Number of Books Desired _____

Name _____
Address _____
City and Zip Code _____

OPPORTUNITY BOOK BONANZA

IMPORT-EXPORT

HONG KONG TRADE DIRECTORY. Now you can buy direct more than 10,000 products at cheap Hong Kong prices. Full, complete 96 pages, with photos on every page. Find a new mail order item, or a source to have a product made to order for you. Get exclusive import deals. Locate nearly 300 different manufacturers of watches, radios, fashion accessories, jewelry, toys, calculators, car parts, lamps, etc.
S261 $10.00

TAIWAN TRADE DIRECTORY. Now you'll be able to deal directly with Taiwan manufacturers who produce thousands of products for the U.S. market. You can order samples and even have products designed especially for you at very low cost. The names and addresses in this photo-illustrated directory are guaranteed 97% accurate. Take advantage of the rich rewards of the Orient now.
S1728 $10.00

IMPORTER'S GUIDE TO SINGAPORE PRODUCTS. Shows you how to purchase products from Singapore at low factory prices. Solid information on products and services offered for export. Indexed alphabetically. Includes a list of trade associations, organizations and Chambers of Commerce.
P1501 $15.00

PASSPORT TO PROFITABLE IMPORTING. Share in the amazing profits earned through importing by learning what products to import, how to locate foreign suppliers, and obtain samples, where to sell your imports, how to import with a minimum investment, plus novel ideas that can start you on the road to even bigger success.
G15 $5.00

FOREIGN TRADE CONTACTS. At your fingertips, you'll find hundreds of world-wide agencies providing important export-import services such as: trade leads, prospect lists, publicity for your offers or needs, magazines and publications, statistics, data and any information and assistance you may need. Amazingly, most of these services are free when you know who to contact.
G26 $5.00

FORTUNES FROM FOREIGN INVENTIONS. An exciting, easy-to-follow program that shows you how to become a royalty-partner to a foreign inventor. Discover how and where to locate foreign inventions, how to obtain exclusive rights to an invention, how to evaluate its potential, and much more. Includes a Directory of Foreign Patent Offices, Foreign Inventors Associations, and Foreign Patent Attorneys and Agents.
G25 $12.00

THE UNABASHED SELF-PROMOTER'S GUIDE. What every man, woman, child and organization in America needs to know about getting ahead by exploiting the media. You'll learn about producing and arranging articles for yourself, creating and maintaining a Self-Promotion Network and much more. Packed with valuable source materials and samples of successful promotions.
JLA20 $30.00

MAIL ORDER

FORMAIDES FOR DIRECT RESPONSE MARKETING. Loaded with practical, easy-to-use information that can make the difference between success and failure in your mail order business. Includes forms, sample letters, worksheets, charts, formulas plus instructions to help you design and develop a successful direct marketing program.
ADL-92 $12.00

SELLING TO CATALOG HOUSES. Now you can sell thousands of books or other products to the large mail order houses. Learn how to price your products and what discounts large mail order firms expect. Find out about invoicing, credit terms, shipping and more. Includes an address directory of major catalog houses.
RD105 $15.00

500 MONEYMAKING MAIL ORDER IDEAS. Discover over 500 ways to make money in your very own mail order business. Start small and expand. Work right out of your home. Handle a wide variety of products, books or services. It's your choice. Over half of the ideas are entirely new. The other half are seasoned, successful plans that have been working year after year.
S201 $10.00

EXCLUSIVE TECHNIQUES THAT CREATE MAIL ORDER FORTUNES. Now the most important element of mail order success is fully revealed. With this extraordinary guide, you will learn how to become exclusive. Hundreds of clever techniques, methods, procedures and ideas are shown so that you can hit a mail order bonanza. Includes secrets of how ideas were developed into fortune-making one man businesses.
G-21 $15.00

SIMPLIFIED MAIL ORDER BOOKKEEPING SYSTEM. Designed especially for the small mail dealer. Provides accurate journals and records to record all your transactions in one handy book.
E202 $4.00

Start your own business. Send for your Free work-at-home business idea catalog. Over 60 business sources. Hundreds of ideas. Valuable start-up information. No obligation - write today!

XENVIRONS
29 Poplar Street, Building B-5, Hatfield, Pennsylvania 19440

☐ **FREE CATALOG**

Please send: Minimum Order: $10.00

Cat. Number	Book Title	Cost

Name _____   Total Books _____
                                          Postage $3.00
Address _____   PA Residents
                                          Add 6% Sales Tax _____
City _____ State ____ Zip ____   Total Order _____

If you're not satisfied with any book, return it within 30 days for a full refund.
CANADIAN CUSTOMERS: Money Orders Only made payable in U.S. funds.

The PHOENIX EXCHANGE

The Phoenix Exchange is published four times each year in the Spring, Summer, Fall and Winter seasons and mailed directly to over 12,000 active military collectors throughout the United States.

Classified — Display Advertising

(215) 933-0909

Postal Box 66/163 Troutman Road/Arcola/Pennsylvania 19420, U.S.A.

Display & Business Card Advertising

Phoenix Militaria's

American Militaria Sourcebook & Directory
RATE CARD
PRICES SUBJECT TO CHANGE
DISPLAY ADVERTISING

All display advertising rates are based on your ad(s) being camera ready. We'll be happy to typeset and layout your ad on request. Ad production is based on our cost.

¼ Page   (3½" Wide x 4⅞" Deep) ..... $ 65
½ Page   (7½" Wide x 4⅞" Deep) ..... $ 95
Full Page (7½" Wide x 10" Deep) $145

BONUS: When you select any one of the above ad sizes you'll receive a FREE copy of the 1988-89 edition of Phoenix Militaria's American Militaria Sourcebook & Directory...a $19.95 value!

BUSINESS CARD DIRECTORY

Super low cost advertising for your products and services. For quality reproduction be sure your card or similar artwork is currently printed in black or a dark colored ink. (Example: Black, Red, Dark Blue or Green, etc.)

Card Size (3½" Wide x 1⅞" Deep) $35

For an additional $10 typesetting fee we'll set-up your card from scratch or make any necessary changes.

IMPORTANT: Limited Display & Business Card advertising space is available. All space orders are on a First Received... First Reserved basis!

PHOENIX MILITARIA
MILITARIA & WAR RELICS

Postal Box 66
163 Troutman Road/Arcola/Pennsylvania 19420

DATA SHEET ★ ★ ★ DATA SHEET

Phoenix Militaria's
AMERICAN MILITARIA SOURCEBOOK & DIRECTORY

YOUR LISTING IS ABSOLUTELY FREE! Include all important information. If you wish to exclude information put N.A. (Not Applicable) on the appropriate line. Complete this Data Sheet as soon as possible and return it in the enclosed postage-paid envelope.

PRESIDENT/OWNER _____

BUSINESS NAME _____

ADDRESS (Post Office Box#) _____

STREET ADDRESS _____
(Street Address is required for our files. If you want only your Post Office Box # in your Free listing check here ☐)

CITY _____ STATE _____ Zipcode _____

TELEPHONE #1: (   ) _____ TELEPHONE #2: (   ) _____

RETAIL & WHOLESALE SALES CATEGORIES

Sales Categories show how you sell your products and services. Check all appropriate categories... read carefully before checking.

- ☐ S1 Retail Sales (over-the-counter only)
- ☐ S2 Wholesale Sales (over-the-counter only)
- ☐ S3 Retail & Wholesale Sales (over-the-counter only)
- ☐ S4 Retail Mail Order Sales
- ☐ S5 Wholesale Mail Order Sales
- ☐ S6 Retail & Wholesale Mail Order Sales
- ☐ S7 Auction Sales
- ☐ S8 Museum
- ☐ S9 Library
- ☐ S10 Organization/Club/Society, etc.
- ☐ S11 Special Order Services
- ☐ S12 Militaria Trade Shows (Gun & Militaria, etc.)
- ☐ S13 Other _____

MILITARIA & SOURCEBOOK LISTING CATEGORIES

Choose only (1) category for your FREE listing. Additional category listings are $10 each. For (5) or more listings take a 20% Discount (include your free listing in total count). Please carefully read all categories before making your choice(s)!

Total number of additional category listings: _____

- ☐ C1 Accoutrements
- ☐ C2 Aircraft/Aviation
- ☐ C3 Ammunition
- ☐ C4 Books/Manuals
- ☐ C5 Civil War Items
- ☐ C6 Edged Weapons
- ☐ C7 Firearms/Ordnance
- ☐ C8 Film/Video
- ☐ C9 Flags/Banners (All Types)
- ☐ C10 Fraternal Items (Military)
- ☐ C11 General Militaria
- ☐ C12 Headgear (All Types)
- ☐ C13 Insignia (All Types)
- ☐ C14 Library
- ☐ C15 Magazine/Newspaper/Newsletter Publisher
- ☐ C16 Manufacturer (Militaria & Related Items)
- ☐ C17 Medals/Orders/Decorations
- ☐ C18 Military Art/Posters/Photographs
- ☐ C19 Military Currency
- ☐ C20 Military Equipment
- ☐ C21 Military Vechicles/Parts
- ☐ C22 Museum
- ☐ C23 Optical Equipment
- ☐ C24 Paper Items
- ☐ C25 Publisher (Books, Calendars, etc.)
- ☐ C26 Radio/Electronic Items
- ☐ C27 Reproductions
- ☐ C28 Restoration/Repairs
- ☐ C29 Services (All Types)
- ☐ C30 Surplus
- ☐ C31 Organization/Club/Society, etc.
- ☐ C32 Travel (Battlefield Tours, etc.)
- ☐ C33 Uniforms/Components
- ☐ Other _____

— GENERAL BUSINESS INFORMATION —

Person to contact in Sales Department? _____

Purchasing Department contact name? _____

Hours Open: _____

Products or Services you offer? _____

Years in Business: _____

Do you publish a Catalogue?   ☐ YES   ☐ NO

How Often? _____

Do you rent your Mailing List?   ☐ YES   ☐ NO

Cost per Thousand? _____

Mailing List Size? _____

Free _____
25 WORD BUSINESS DESCRIPTION (Includes all types — Militaria & Firearms Dealers, Publishers, Libraries, Museums, Manufacturers, Service Companies, Clubs, etc.) You can add an additional (25) words to this description for only $10:

_____

_____

_____

_____

_____

_____

_____

_____

ORDER FORM

ORDER FORM—(Use this order form for your additional Category Listings and for your additional 25 Word Business Description)

CHECK APPROPRIATE BOX(S):

- ☐ Additional Category Listings - Total _____ X $10 .... _____
  (Do not include your Free listing in this total)
- ☐ (5) or more Category Listings DEDUCT 20% .... (−) _____
  (Include your Free listing in the total count)

Sub-Total _____

- ☐ Additional 25 Word Business Description $10.00

Total Enclosed _____

PHOENIX MILITARIA

Postal Box 66/163 Troutman Road
Arcola/Pennsylvania 19420, U.S.A.
Telephone: (215) 933-0909

Reach Thousands of Militaria Collectors & Businesses with Your Products & Services...